TECHNOLOGY COMPUTER AIDED DESIGN

Simulation for VLSI MOSFET

T0225450

TECHNOLOGY COMPUTER AIDED DESIGN

Simulation for VLSI MOSFET

Edited by

Chandan Kumar Sarkar

CRC Press
Taylor & Francis Group
Boca Raton London New York

CRC Press is an imprint of the
Taylor & Francis Group, an **informa** business

CRC Press
Taylor & Francis Group
6000 Broken Sound Parkway NW, Suite 300
Boca Raton, FL 33487-2742

First issued in paperback 2017

© 2013 by Taylor & Francis Group, LLC
CRC Press is an imprint of Taylor & Francis Group, an Informa business

No claim to original U.S. Government works

ISBN-13: 978-1-4665-1265-8 (hbk)
ISBN-13: 978-1-138-07575-7 (pbk)

Library of Congress Cataloging-in-Publication Data

Technology computer aided design : simulation for VLSI MOSFET / editor, Chandan Kumar Sarkar.
    pages cm
    Summary: "MOSFET and related high-speed semiconductor devices are spearheading the drive toward smaller, faster, and lower-power electronics. This work concentrates on technology computer aided design (TCAD) and its integration into the IC fabrication process flow. It presents modeling techniques and concepts involved with the TCAD simulation of MOSFET devices. The book describes basic concepts and background related to popular commercial TCAD software as well as recent technologies to improve device performance such as multiple gate MOSFET, FINFET, SOI devices, and high-k gate material devices"-- Provided by publisher.
    Includes bibliographical references and index.
    ISBN 978-1-4665-1265-8 (hardback)
    1. Integrated circuits--Very large scale integration--Computer-aided design. 2. Metal oxide semiconductor field-effect transistors--Computer-aided design. I. Sarkar, Chandan Kumar, editor of compilation.

TK7874.75.T43 2013
621.39'50285--dc23
                            2013008022

Visit the Taylor & Francis Web site at
http://www.taylorandfrancis.com

and the CRC Press Web site at
http://www.crcpress.com

To our family members

Contents

Preface

In the era of System-on-Chip (SoC) fabricated with ultra-deep sub-micron complementary metal-oxide-semiconductor (CMOS) technology, the semiconductor industry has to keep pace with the increased performance-capacity demands from consumers. The current sub-90 nm CMOS technology poses several critical challenges to very large scale integrated (VLSI) circuit and device designers regarding the characteristic behavior and performance of MOS transistors. In addition, there is huge pressure on the designers for reduction of the design time for economic reasons. To cope with issues like design productivity gap and yield drop and also to ensure correct behavior of each device, there is a need to involve computer aided design (CAD) methodologies and design automation tools. Technology computer aided design (TCAD) is an electronic design automation (EDA) tool that models the semiconductor device operation and fabrication based on fundamental physics through the computer aided simulations for design and optimization of semiconductor processing technologies and devices. Research and development (R&D) cost and time for electronic product development can be reduced by taking advantage of TCAD tools, thus making TCAD indispensable for modern VLSI devices and process technologies. TCAD simulation is gaining importance because it is prohibitively expensive for academia as well as for the several fab-less design industries to use silicon prototype for design verification.

The motivation for editing this volume on TCAD simulation of VLSI MOSFET came about when I was looking for student-level reference materials on TCAD simulation tools while teaching courses for postgraduate and senior undergraduate students. However, I could not find an appropriate book and was thus compelled to teach the course by consulting various user guides and manuals supplied with TCAD software and some semiconductor device textbooks. However, none of these caters to the intended purpose of introducing the physics of TCAD simulations to the students. Therefore, I found that there is a need for a book to provide the most up-to-date and comprehensive source of TCAD simulation of VLSI MOSFETs with an emphasis on the fundamental theory and the underlying physics.

This book is intended for senior undergraduate and postgraduate students of electrical and electronics engineering disciplines and for researchers and professionals working in the area of electronic devices. The purpose of the present edited volume is to rapidly disseminate to them the fundamental concepts and underlying physics of TCAD simulation of MOSFETs. The objectives of this book are to introduce the advantages of TCAD simulations for device and process technology characterization, to introduce the fundamental physics and mathematics involved with TCAD tools, to expose readers to two of the most popular commercial TCAD simulation tools, Silvaco and Sentaurus, to characterize performances of VLSI MOSFETs through TCAD tools, and to familiarize readers with compact modeling for VLSI circuit simulation. This volume provides the reader with comprehensive information and a systematic approach to the design, characterization, fabrication, and computation of VLSI MOS transistors through TCAD tools.

The chapters contain different levels of difficulty. Several example programs are supplied for illustration of the software tools and related physics. The book therefore provides a desirable balance between the basic concepts, equations, physics, and recent technologies of MOS transistors through TCAD simulation. The book is organized into eight chapters that encompass the field of TCAD simulation for VLSI MOSFET.

Chapter 1 provides an overview of the role, need, and advantages of TCAD tools in design, characterization, and fabrication of VLSI MOS transistors. The evolution of modern TCAD and its challenges are also provided. Chapter 2 reviews and analyzes the basic concepts and physics involved with nanoscale MOS transistors. The physical approach to the tools is described by basics of band theory, Poisson's equation, continuity equation, drift diffusion (DD), and hydrodynamic models. The physics of the scattering mechanisms and different mobility models used in TCAD simulations are discussed. Chapter 3 describes the basics and importance of numerical solution techniques applicable to TCAD. The numerical solution of the DD equations coupled with Poisson's equation and its application to semiconductor device modeling is described. Chapter 4 provides a detailed overview of the two-dimensional/three-dimensional (2D/3D) device simulator Synopsys Sentaurus TCAD. The various tools involved are introduced in a comprehensive manner. The different software-related aspects of this tool for device simulations are described. Complete design examples explaining step by step the construction, simulation, and performance extraction of MOS transistors are provided. Chapter 5 attempts to present the MOSFET simulation using Silvaco TCAD tools. From basic syntax to choice of a complex model is presented in this chapter with emphasis on the usage of the SILVACO simulation software. An overview of the software developed by Silvaco to meet simulation needs of researching conventional and advanced MOSFET structures is also presented. The discussion is presented with examples to perform simulations of different types of MOSFETs. Chapter 6 discusses in detail the physics of nanoscale MOS transistors through TCAD simulations. The various short channel effects involved with nanoscale MOS transistors are demonstrated through actual TCAD simulation results. Different technology aspects and engineering techniques for future MOSFETs are also introduced. Chapter 7 presents a comprehensive overview of compact modeling of MOS transistors for use in VLSI circuit simulation. The mathematical models for characterizing various ultra-deep sub-micron effects of sub-65 nm MOS transistors are introduced. The effects are demonstrated through actual simulation program with integrated circuit emphasis (SPICE) simulation results using industry standard compact models. The various circuit performances of MOS transistors are discussed. Chapter 8 addresses process simulation of MOSFET using TSUPREM-4. The chapter is devoted to bringing the key fabrication issues and their implementation in TCAD process simulation tool TSUPREM-4. It also considers how the output of process simulator TSUPREM-4 can be linked to device simulator MEDICI in order to analyze the performance of the fabricated device. An extensive list of references is provided at the end of each chapter for more elaborate discussion of the issues and to motivate readers to engage in further research.

I wish to congratulate all contributors and their peers. Their convictions and efforts were key to the success of this enterprise. The compilation of this book would not have been possible without the dedication and efforts of all the contributing authors. Special thanks go to Gagandeep Singh and Laurie Schlags and staff members of CRC Press for their responsiveness and immense patience demonstrated throughout the publishing process of this book.

C.K. Sarkar

The Editor

Chandan Kumar Sarkar is Professor of Electronics and Telecommunication at Jadavpur University, Calcutta, India. He received B.Sc. (Hons.) and M.Sc. degrees in Physics from Aligarh Muslim University, Aligarh, India, in 1975, was awarded a Ph.D. degree in Radio Physics from Calcutta University in 1979, and was awarded the D.Phil. degree from Oxford University, Oxford, United Kingdom, in 1984. He has been teaching for over 22 years.

In 1980, Sarkar received the British Royal Commission Fellowship to work at Oxford University. He worked as a visiting scientist at the Max Planck Laboratory in Stuttgart, Germany, as well as at Linko Pink University in Sweden. Sarkar also taught in the Department of Physics at Oxford University, and was a distinguished lecturer of the IEEE EDS.

Sarkar has served as a senior member of IEEE and was chair of the IEEE EDS chapter, Calcutta Section, India. He served as Fellow of IETE, Fellow of IE, Fellow of WBAST, and member of the Institute of Physics, United Kingdom.

His research interests include semiconductor materials and devices, VLSI devices, and nanotechnology. Sarkar has published and presented more than 300 research papers in international journals and conferences and also mentored 21 Ph.D.s.

Contributors

Swapnadip De
Electronics and Communication
  Engineering Department
Meghnad Saha Institute of
  Technology
Kolkata, India

Kalyan Koley
Electronics and Telecommunication
  Engineering Department
Jadavpur University
Kolkata, India

Atanu Kundu
Electronics and Communication
  Engineering Department
Heritage Institute of Technology
Kolkata, India

N. Mohankumar
Electronics and Communication
  Engineering Department
SKP Engineering College
Tiruvannamalai, Tamil Nadu,
  India

Soumya Pandit
Institute of Radio Physics and
  Electronics
University College of Science and
  Technology
University of Calcutta
Kolkata, India

Srabanti Pandit
Electronics and Telecommunication
  Engineering Department
Jadavpur University
Kolkata, India

Samar K. Saha
Compact Modeling
SuVolta, Inc.
Los Gatos, California
and
Electrical Engineering Department
Santa Clara University
Santa Clara, California

Angsuman Sarkar
Electronics and Communication
  Engineering Department
Kalyani Government Engineering
  College
Kalyani, India

1

Introduction to Technology Computer Aided Design

Samar K. Saha

CONTENTS

1.1 Technology Computer Aided Design (TCAD)

Technology computer aided design (CAD), commonly known as *technology CAD* or *TCAD,* is the electronic design automation that models integrated circuit (IC) fabrication and device operation. TCAD is the art of abstracting IC electrical behavior by critical analysis and detailed understanding

1

of process, device, and circuit simulation data. In general, TCAD includes *lithography modeling* to simulate the imaging of the mask by the lithography equipment, photoresist characteristics, and processing; *front end process modeling* for simulating the physical effects of manufacturing steps used to build transistors up to metallization; *device modeling* using hierarchy of physically based models for the operational description of active devices; *compact modeling* for active, passive, and parasitic circuit components; *interconnect modeling* to analyze the operational response of back-end architectures; *reliability modeling* for simulating the reliability and related effects on process, device, and circuit levels; *equipment modeling* for simulating the local influence of the equipment on each point of the wafer, especially in deposition, etching, and chemical-mechanical polishing (CMP) processes; *package simulation* for electrical, mechanical, and thermal modeling of chip packages; *materials modeling* to predict the physical and electrical properties of materials; *modeling for design robustness, manufacturing, and yield* to simulate the impact of process variability and dopant fluctuations on IC performance and determine design specifications for manufacturability and yield of ICs; and *numerical techniques* including grid generators, surface-advancement techniques, solvers for systems of partial differential equations (PDEs), and optimization routines [1,2].

TCAD offers capabilities to analyze how structural factors such as the geometry and processes conditions influence the electrical behavior of devices and circuits. Simulation data help in quantifying the details of the behavioral models for ICs at the transistor and circuit levels and show physical limitations at the processing and manufacturing levels [3,4]. By reverse modeling, the extended TCAD tools can be used to develop IC fabrication technology from product concept (i.e., from the product specification to IC fabrication technology as shown in Figure 1.1) [1]. The extended TCAD tools can also be used to assess the manufacturability of IC fabrication processes as shown in Figure 1.2 [1,5].

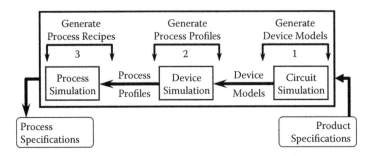

FIGURE 1.1
Extended TCAD use to generate product-specific IC process recipe by reverse modeling; in this approach, the sequential steps 1, 2, and 3 represent use of circuit, device, and process CAD, respectively. (From S.K. Saha, Managing technology CAD for competitive advantage: An efficient approach for integrated circuit fabrication technology development, *IEEE Trans. Eng. Manage.*, vol. 46, no. 2, pp. 221–229, May 1999. With permission.)

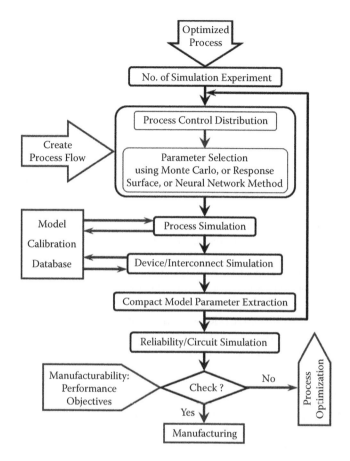

FIGURE 1.2
Flowchart showing the use of extended TCAD to evaluate the manufacturability of IC fabrication technology with respect to the target specifications. (From S.K. Saha, Managing technology CAD for competitive advantage: An efficient approach for integrated circuit fabrication technology development, *IEEE Trans. Eng. Manage.*, vol. 46, no. 2, pp. 221–229, May 1999. With permission.)

In the microelectronics industry, TCAD is widely referred to as *front-end process modeling* or *process CAD* and *device modeling* or *device CAD*. Therefore, unless otherwise specified, hereafter in this chapter TCAD refers to IC front-end process CAD and device CAD only.

1.1.1 Process CAD

Process CAD refers to front-end process modeling that includes the numerical simulation of the physical effects of IC processing steps used to fabricate transistors up to metallization. The process CAD is used to simulate the semiconductor processing steps such as oxidation and diffusion, deposition

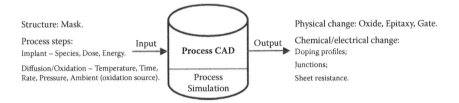

FIGURE 1.3
IC fabrication process simulation using process CAD to generate input file for device simulation; the "physical change" refers to the structural change of the device such as the oxide growth, whereas the "chemical change" refers to impurity diffusion; process CAD includes physical process models to perform numerical process simulation.

and etching, ion implantation, and annealing, and generate input data files for the device simulator as realistically as possible based on the microscopic information as shown in Figure 1.3 [1–5]. In the area of process simulation, physical models for process technology were rather limited in the 1960s and 1970s, and the sophisticated process simulators with multidimensional models were not necessary for the large device geometry used during that time. In recent years, the physical understanding of IC processes has advanced significantly. Moreover, the current evolution of IC devices into the nanoscale regime necessitates accurate multidimensional process models.

1.1.2 Device CAD

Device CAD refers to numerical simulation of IC device operation as shown in Figure 1.4. In general, the device CAD includes a suite of physical models describing carrier transport in materials. Device models range from the simple drift diffusion, which solves Poisson and continuity equations, to more complex and computationally challenging models such as the energy balance, which solves some higher moment simplification of the Boltzmann transport equation (BTE). In addition, the complex physics of today's nanoscale devices mandates the use of Monte Carlo (MC) codes, which stochastically solves BTE, and the use of Schrödinger solvers that account for quantum

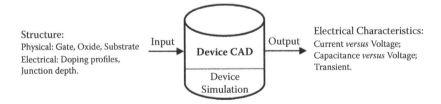

FIGURE 1.4
IC device simulation using device CAD to generate electrical characteristics of IC devices for circuit analysis; device CAD includes physical device models to perform numerical device simulation.

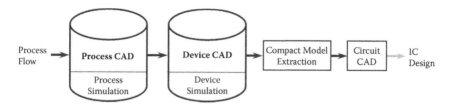

FIGURE 1.5
Process and device CAD are synergistically linked to predict the influence of various IC processing steps on device and circuit performance; here circuit CAD represents a modeling tool to analyze circuit performance.

mechanical (QM) effects in metal-oxide-semiconductor field-effect transistor (MOSFET) devices. The choice of the appropriate model depends on the problem and the level of detail required. Despite the significant advances of both numerics and physics, continuous development is required to meet the increasingly challenging industry needs for device exploration, scaling, and optimization. Therefore, the ability of device CAD to accurately model today's device performance and predict tomorrow's device limitations is of utmost importance [3,4].

Though the process CAD and device CAD refer to numerical simulation of different areas of computational electronics, they are synergistically linked. An understanding of the various IC processing steps is crucial to predicting device and circuit performance as shown in Figure 1.5.

1.2 A Brief History of TCAD

During the last four decades, technology CAD has evolved from one-dimensional (1D) Direct current (DC) or steady-state simulation only during the 1970s to full two-dimensional (2D) and three-dimensional (3D) simulation of today's complex very large scale integrated (VLSI) circuit fabrication processes and devices. This evolution of process and device CAD has been possible due to advancements in the computer technology and mathematical models, thus providing cost-effective and efficient multidimensional numerical analysis of IC fabrication processes and devices. A large number of researchers have contributed to the evolution of TCAD. In this section, we present only a brief history of the major development in the device and process CAD leading to commercial TCAD tools.

1.2.1 History of Device CAD

The seminal work of Gummel in 1964 led to the foundation of device CAD. For the first time, Gummel reported 1D simulation results of bipolar junction transistors (BJTs) by sequentially solving the three PDEs in the drift-diffusion system using an iterative method [6]. Gummel's numerical approach was further developed and applied to simulate p-n junction by De Mari in 1968 [7,8] and Read diode oscillator by Scharfetter and Gummel [9] in 1969. In the 1969 publication, Scharfetter and Gummel reported a methodology for stable upwind discretization of the transport equations [9]. This method is still almost universally used and is responsible for making device simulation a computationally feasible design activity.

In 1968, the first 2D numerical solution of Poisson's equation for a MOSFET was reported by Loeb et al. [10] and, in parallel, by Schroeder and Muller [11]. In 1969, the 2D simulation results on planar devices were published by Barron on MOSFETs [12] and Kennedy and O'Brien on junction field-effect transistors (JFETs) [13] by solving the Poisson equation and one continuity equation. In the same year, Slotboom reported the 2D simulation results on BJTs using the full two-carrier system [14], and in 1971 Reiser reported the first 2D transient simulation results of metal-semiconductor field-effect transistors (MESFETs) [15].

During the 1970s, the major numerical techniques for 2D device simulation [16,17], including the first finite element analysis of the semiconductor equations [18,19], were reported. The finite element method can be considered a precursor to the development of the more general-purpose tools commonly used in the 1980s. In the late 1970s, some of the first publicly available device CAD tools were released including the first version of the CADDETH programs from Hitachi [20] to simulate single carrier field-effect transistor (FET) structures, the SEDAN program from Stanford University [21] to simulate 1D bipolar device phenomena, and a special-purpose MOSFETs simulation program MINIMOS from Vienna [22].

Though the work on a general-purpose, non-planar multidimensional device CAD program began in the mid-1970s, the first device simulation program successfully used for bipolar device analysis was FIELDAY [23] from IBM. Besides FIELDAY, the other first generation non-planar codes include the GEMINI program in 1980 [24] which solved only Poisson's equation and PISCES-I in 1982 [25] which solved Poisson and a single continuity equation in the steady-state, both from Stanford University.

The real impact of a general-purpose device CAD tool came in the mid-1980s with the development of programs like PISCES-II at Stanford University [26,27], DEVICE at AT&T Bell Laboratories [28,29], BAMBI at Vienna [30], and HFIELDS [31] at the University of Bologna. Each of these programs worked for two carriers, allowed arbitrary device non-planarities, and included a more comprehensive set of materials, physical models, and simulation capabilities than the prior state-of-the-art. However, the primary advantages of these tools were improved computational methods such as the discretization and grid generation, non-linear and linear solution techniques that made device simulation practical for device designers. It is worth noting that PISCES-II has been commercialized by Technology Modeling Associates (TMA), now Synopsys [32], and Silvaco [33] and is the source of widely used device simulators MEDICI and ATLAS, respectively.

Major advances in the numerical solution of BTE [34] also began in the late 1970s. In 1979, the initial 2D MC device simulation results were reported by Warriner [35]; in 1982, the PDE for energy transport was first treated numerically in 2D by Cook and Frey [36]; in 1984, Fukuma and Uebbing used the energy balance model to predict velocity overshoot in silicon MOSFETs [37] and the hot carrier post-processors with lucky electron based approach were first implemented by Siemens [38]; in 1985, the first realistic 2D simulations of hetero-structure devices were reported from the University of Illinois [39]; in 1986, Laux and Warren reported the coupled 2D Schrödinger-Poisson solver and directly introduced quantum mechanics in device CAD [40]; in 1988, an MC post-processor for silicon MOSFETs was developed at Bologna [41]; and the first general-purpose code incorporating hydrodynamic solutions was implemented in HFIELDS [42].

With the increasing complexities of IC devices due to continuous downscaling of feature size, the 3D numerical analysis became critical. The first paper using 3D device simulation of MOSFET narrow channel effects was published in 1980 [43]. The FIELDAY was the first of many programs extended into 3D in 1981 [44] by extending the grid uniformly in the depth plane. Also, in 1981 and 1982, the 3D device simulation results of narrow channel effects were published using the simulator WATMOS [45,46] which used a finite difference scheme numerical solution of Poisson's equation. Following the approach of FIELDAY, almost every 2D simulation program has been extended to 3D by extending the grid uniformly in the depth plane [47–52]. The FIELDAY program solved Poisson's equation and both carrier continuity equations. The program was later enhanced to include the hydrodynamic energy-balance equations, Fermi-Dirac carrier statistics, lattice energy equation, and incomplete ionization [53]. Thus, both drift diffusion and hydrodynamic simulations were possible using FIELDAY.

In 1985, Hitachi announced CADDETH as a 3D device simulator designed to run on a supercomputer [49]. CADETH solves both Poisson's equation and two current continuity equations using conjugate gradient-based methods for non-symmetric linear systems; distinguishes between three different materials, namely semiconductors, insulators, and metals; and implemented advanced physical models [49]. Another 3D simulator was developed by Toshiba in 1985, called TOPMOST [54,55]. TOPMOST was designed to analyze MOS structures and solve the semiconductor equations for the drift-diffusion case. Both 1D and 2D simulations could be performed by pseudo-1D and pseudo-2D device models by reducing the number of points in the omitted directions. TOPMOST was used to study the effect of the gate structure on the output characteristics [54] and subthreshold swing in 3D MOSFETs [55].

In 1987 Vienna announced 3D device simulator, MINIMOS Version 5 [56]. MINIMOS-5 is one of the first 3D device simulators [56–59] for MOSFET structures, SOI transistors, and gallium arsenide MESFETs. MINIMOS-6, released in 1994 [60], supports transient analysis and MC modeling to replace the drift diffusion approximation in critical device areas. The fundamental semiconductor equations, consisting of Poisson's equation and two carrier continuity equations, are solved numerically in 2D or 3D space. MINIMOS is able to simulate planar and non-planar device structures along with AC small signal analysis and transient simulations.

The reported 3D device CAD tools from the industry include SMART [61] from Matsushita in 1987, PADRE [62] from Bell Laboratories, SITAR [63] from Siemens in 1988, MAGENTA [64] from Microelectronics and Computer Technology Corporation in 1989, and SIERRA [51] from Texas Instruments in 1989. SIERRA solves Poisson's equation and the carrier continuity equations for static, AC small signal, and transient cases. Based on SIERRA, TMA developed their first 3D device simulator known as DAVINCI in 1991 [32]. DAVINCI has been used to investigate the effects of radiation on DRAM cells and narrow channel effects in MOS structures. In January 1998, TMA merged with Avant!, and in 2001 Avant! merged with Synopsys, and DAVINCI has been the starting basis of the TAURUS 3D DEVICE program from Synopsys [32].

Some of the other reported 3D device CAD tools include HFIELDS-3D [65–67] from the University of Bologna in 1989, SECOND [68–70] from ETH Zürich and STRIDE [71] from Stanford University in 1991, FLOODS [72] from the University of Florida in 1994, and DESSIS [73] in 1996. The development of DESSIS began in 1992 in collaboration with Bologna, ETH Zürich, Bosch, and ST Microelectronics [73]. DESSIS is created by merging the device simulators HFIELDS from Bologna and SIMUL from ETH and circuit simulator BONSIM from Bosch to enable efficient mixed-mode IC circuit/device analysis [73]. In 1993, Integrated Systems Engineering (ISE) was founded and took over the simulator. In 2004 ISE merged with Synopsys which took over DESSIS and is the basis of commercial Sentaurus device simulator from Synopsys [32].

With the evolution of IC technology and devices, TCAD is continuously evolving. Another notable device CAD tool includes MINIMOS-NT [74] reported in 1997. MINIMOS-NT is a generic 2D device simulator that allows the modeling of high electron mobility transistors. A new method is used which divides the device region into several sub-domains, referred to as segments, each segment with its specific physical models. The segmentation of device region allows appropriate physical models to be used where required, such as a hydrodynamic model in the channel region and drift diffusion solution for the non-critical region of the device. This increases the overall efficiency of device simulation with the desired accuracy. Also, with the ongoing reduction of feature sizes, atomistic simulations become reasonable to study the effects of random discrete doping in the channel region [75–77].

The device CAD tools are continuously evolving, and mathematical models describing the performance of advanced devices are continuously implemented, especially in the commercial TCAD tools to model today's complex IC devices.

1.2.2 History of Process CAD

After the invention of ICs in 1958, the IC industry was dominated by BJT technology through the 1960s. However, in the 1970s MOSFET technology began to overtake BJT technology in terms of the functional complexity and level of integration. Since the 1980s, complementary MOS (CMOS) technology with its cost-effective technology solution has become the pervasive technology for ICs. With aggressive scaling of MOSFETs in the mid-1970s, transistor dimensions soon reached the point at which first-order assumptions about the physical effects and dopant distributions began to break down. For the MOSFETs, the intrinsic device problem such as output conductance, velocity saturation, and subthreshold behavior all received substantial interest and effort. TCAD became critical to understand the many interrelated process and device effects in MOSFETs.

By the mid-1970s, the critical role of processing technology in establishing device characteristics was evident. Many important interrelated process and device effects were identified by means of computer coupled analysis tools. Industry leaders such as IBM and Texas Instruments had aggressive efforts to model process physics and relate these models to device characteristics and circuit statistics. A unified process and device simulator, the SITCAP program was developed at Katholieke University, Leuven, Belgium [78]. This program inputs process specifications such as processing times and temperatures along with simple mask geometries to output *I–V* and *C–V* curves, along with selected SPICE model parameters. And, a process analysis program CASPER was jointly developed at Lehigh University (Bethlehem, Pennsylvania) and AT&T Bell Laboratories (Allentown, Pennsylvania) [79].

In 1977, the first version of 1D process simulator, SUPREM, was developed and released by Stanford [80]. Since then, the process models in SUPREM have developed substantially and evolved due to ongoing efforts at Stanford and at industrial and other research laboratories worldwide. Versions II and III of SUPREM were released in 1978 and 1983, respectively [81,82]. The release of a process simulator had a tremendous impact on the accuracy to which device simulations could be performed. Process simulation has continued to improve dramatically since the release of SUPREM. And the notable events in the evolution of process CAD include the development of the full 2D simulators such as SUPRA from Stanford [83] in 1982 and BICEPS from AT&T Bell Laboratories [84] in 1983. In 1986, the most advanced 2D process simulation program SUPREM-IV was developed at Stanford [85]. The SUPREM-IV included more physically based models including point defect calculations and stress dependent oxidation required for modeling ultra-small device structures. A 2D SUPREM-IV process simulation program has been commercialized by Crosslight [86], Sillvaco [33], and TMA [32] as CSUPREM, ATHENA, and TSUPREM4, respectively.

In 1992, ISE (now Synopsys [32]) developed 1D and 2D process simulators TESIM and DIOS, respectively. DIOS was a widely accepted tool prior to the introduction of the Sentaurus process by Synopsys in 2005.

The development of 3D process CAD tools began in the mid-1980s for accurate modeling of ultra-small geometry device technology [87,88]. In 1993, FLOOPS was released from the University of Florida [89] and became the source of Synopsys 3D process simulator, Sentaurus process that was released in 2005 [32]. Other 3D process simulators include PROPHET from AT&T Bell Laboratories [90] released in 1994 and Taurus 3D process from TMA released by Avant! in 1998 [32].

The history of commercial TCAD began with the foundation of TMA in 1979. TMA was the first commercial supplier of TCAD software that was the result of the TCAD research program at Stanford University under the supervision of Professor Dutton and Professor Plummer. Currently, the major sources of commercial TCAD tools are Silvaco [33] and Synopsys [32]. Synopsys TCAD tools include Taurus process/device TSUPREM4/MEDICI for 2D TCAD and Sentaurus process/device for both 2D and 3D TCAD. Silvaco TCAD tools include ATHENA for 2D process simulation, ATLAS for 2D device simulation, and Victory process/device for 3D simulation.

1.3 Motivation for TCAD

The major motivation for the use of TCAD in the semiconductor industry is the cost-effective and efficient development of IC fabrication technology using device CAD to analyze device performance and process CAD to input realistic structural information from process flow to device CAD [1,5].

1.3.1 Motivation for Device CAD

The motivation for the use of device CAD in IC device analysis is the optimization of device performance for specific applications. This optimization is a complex task due to the complexities of the equations describing semiconductor device performance. For semiconductor devices, particle conservation is modeled by several cross-coupled non-linear PDEs: the interaction of charged particles due to electric fields is modeled by Poisson's equation, and the particle concentrations relating to particle fluxes and generation and recombination are modeled by continuity equations. Also, electron and hole concentrations are exponentially related to potentials through Boltzmann, Fermi-Dirac, or other exponentially determined probability distribution functions. These equations are difficult to solve by hand, making computer aided analysis a desirable alternative. Again, by the introduction of the SPICE program from Berkeley in 1975, the circuit simulator became a useful design tool, essentially replacing the bread-boarding of prototypes [91]. However, for accurate circuit analysis, compact device models, commonly known as SPICE models, are required. For the generation of SPICE models the device CAD became a necessity. Thus, the widespread use of circuit simulation also motivated the development and use of device CAD for IC device analysis.

Figure 1.5 illustrates the use of device CAD to generate compact models for circuit analysis. Compact models such as BSIM4 [92] provide an excellent framework to analyze different modes of MOSFET circuit behavior. However, in order for the BSIM4 model to be useful in practice, reliable values for the model parameters must be generated. Device CAD tools are useful for linking fabrication conditions to BSIM4 parameters. For example, the device simulator uses inputs such as device geometry, doping profiles, and bias conditions to generate data files for device characteristics such as capacitance (C) and current (I) versus voltage. The simulated data files are then used to extract BSIM4 model parameters for circuit analysis.

Another major motivation for using device CAD is to study the feasibility of realizing concept devices in manufacturing. As the MOSFET devices are scaled down, device performance is severely degraded by short channel effect (SCE), drain-induced barrier lowering (DIBL), quantum-mechanical (QM) effects, and so on. The use of device CAD is critical to minimize these physical effects and improve the device performance by optimizing the device structure. Figure 1.6 shows a nanoscale "double-halo" MOSFET device structure designed to suppress SCE, reduce DIBL and QM effects, and improve device performance [93–97].

The double-halo MOSFET structure as shown in Figure 1.6 includes a polysilicon gate, a gate oxide (T_{OX}), vertically and laterally non-uniform channel doping profiles, shallow source-drain extensions (SDEs), deep source-drain (DSD) regions, and two halo profiles: first around SDE and the second around DSD regions to control DIBL from SDE and DSD junctions, respectively. A

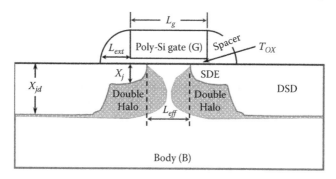

FIGURE 1.6
An idealized double-halo MOSFET device structure showing the basic technology elements: L_g and L_{eff} are the drawn and effective channel lengths, respectively, T_{OX} is the gate oxide thickness, L_{ext} is the spacer width, and X_j and X_{jd} are the junction depths of the source-drain extension (SDE) and deep source-drain (DSD) regions, respectively.

super-steep retrograde (SSR) or "low-high" channel doping profile with a low impurity concentration at the silicon/silicon-dioxide (Si/SiO_2) interface and a higher peak concentration at a finite depth below the interface is used to provide the non-uniform vertical channel profile [98]. The SSR profile is optimized to achieve the target threshold voltage (V_{th}) for the long and wide devices. The SSR profile also provides superior V_{th} control. The halo doping profiles are optimized to achieve the target leakage current for the nominal devices. In reality, the low-high channel doping profiles are achieved using multiple buried layers under an undoped epitaxial layer of appropriate thickness [99]. Because, in reality, the ion-implanted profiles are non-linear, more sophisticated 2D/3D device analysis tools are required for device optimization. Thus, the device CAD use is critical to optimize the halo doping profiles in conjunction with other key technology parameters such as T_{Ox} and SDE junction depth (X_J) to realize the nanoscale double-halo MOSFETs in manufacturing.

In optimizing the double-halo device performance, it is critical to analyze the effect of scaling key technology parameters such as X_j and T_{Ox} on device performance. Figure 1.7 shows 2D device simulation results of drain current (I_{ds}) versus source-drain voltage (V_{ds}) with a given gate-to-source voltage (V_{gs}). Two similar technologies are compared: one optimized for $T_{Ox} = 1.5$ nm, and the other with $T_{Ox} = 1.0$ nm. For both cases, the devices are optimized to achieve the same off-state leakage current, $I_{off} \cong 10$ nA/μm. From Figure 1.7, we find that for small values of V_{ds}, the curves show almost identical values of I_{ds}. At higher V_{ds}, the device with the lower T_{Ox} shows substantially more current handling capability. Thus, both the technologist and the circuit designers are anxious to understand and control the dependencies of key technology parameters to realize the optimum circuit performance of double-halo CMOS technology. Therefore, the motivation to use device CAD is to understand the dependence of key building blocks of device structure on device and circuit performance.

1.3.2 Motivation for Process CAD

In the previous section, we have established the motivation for device CAD in determining circuit design parameters with reference to MOSFET devices. Now, let us discuss certain critical parameters in MOSFETs that depend directly on the quantitative features of the doping profiles within the device. Because doping profiles are determined by process variables such as ion implantation energy, total implanted dose, and drive-in time/temperature cycles, the dependence of device parameters on process variables provides motivation for process simulation.

For MOSFET devices, V_{th} and other device parameters are directly related to the distribution of the channel doping profile within the device structure. In order to produce accurate compact model parameters such as V_{TH0},

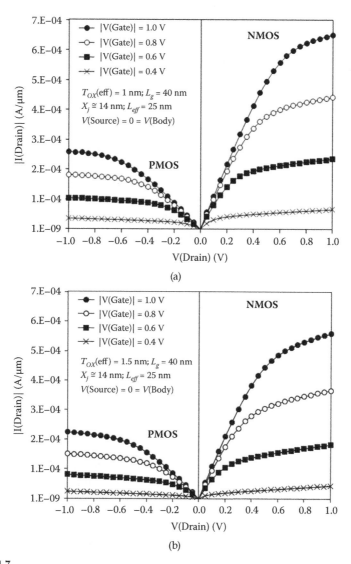

FIGURE 1.7

I–V characteristics of double-halo MOSFETs for (a) T_{OX} (*eff*) = 1 nm and (b) T_{OX} (*eff*) = 1.5 nm; the simulation data are obtained for 40 nm devices with L_{eff} = 25 nm and optimized for $|I_{off}|$ = 10 nA/μm at $|V(Drain)|$ = 1 V; $V(Gate) = V(Source) = V(Body) = 0$). (From S. Saha, Scaling considerations for high performance 25 nm metal-oxide-semiconductor field-effect transistors, *J. Vac. Sci. Tech. B*, vol. 19, no. 6, pp. 2240–2246, November 2001. With permission.)

modeling V_{th} for large devices, *K*1 and *K*2 describing body effect, *U*0 describing inversion layer carrier mobility, *PDIBL*1 and *PDIBL*2 describing DIBL, and so on, the device simulator must have an exact description of the channel doping profile in all dimensions. This accurate description of channel doping profile can be generated using process CAD.

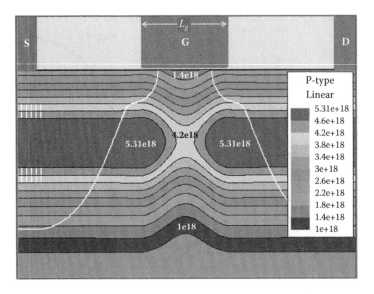

FIGURE 1.8
Simulated 2D-doping contours of a typical double-halo nMOSFET device with laterally and vertically non-uniform p-type channel doping generated using device CAD MEDICI; 2D cross-section shows S, G, and D are the source, gate, and drain terminals, respectively, and the outline of SDE and DSD junctions. (From S. Saha, Device characteristics of sub-20-nm silicon nanotransistors, in *Proc. SPIE Conf. on Design and Process Integration for Microelectronic Manufacturing*, vol. 5042, pp. 172–179, July 2003. With permission.)

Figure 1.8 shows the 2D doping profile of the 25 nm double-halo MOSFET structure shown in Figure 1.6. A number of details of the technology are apparent from the cross sections shown. First, the 2D halo doping contours from the source and drain regions approach a peak at a certain depth from the Si/SiO_2 interface under the gate. The halo diffusion from the source and the drain regions has enhanced the p-type doping concentration at the center of the device under the gate region. With scaling T_{OX}, the doping distribution within the device changes both laterally and vertically. Calibrated process, especially, diffusion models are essential for accurate generation of a 2D doping profile within the active region of the simulation structure. Therefore, process simulation is critical to reproduce the doping distributions within the structure for accurate device simulation and compact model parameter extraction for circuit analysis.

Thus, the motivation to use process CAD is to couple the relevant fabrication information into the device CAD. The process CAD captures the critical aspects of the target fabrication process to accurately determine device models for circuit analysis and predict the limitations in device performance. For most device parameters, the exact description of doping profile is needed to obtain agreement between the simulated and measured data.

1.4 TCAD Flow for IC Process and Device Simulation

TCAD flow for IC process and device analysis includes the (1) generation of simulation structure to numerically solve the PDEs that model IC processing and device performance, (2) verification of the robustness of the simulation structure and sensitivity of simulation results on grid space, (3) calibration of physical models for accuracy and predictability, and (4) coupled process and device simulation.

1.4.1 Generation of Simulation Structure

The first step in the numerical approach is the discretization of the simulation domain in time and space. In this step, the device cross section is partitioned into sub-domains or small cells, each of which is evaluated at discrete time intervals. Then in each of these cells, the PDEs are approximated by algebraic equations that include only the values of the continuous dependent variables at discrete points in the domain and the knowledge of the structure of the functions approximating the dependent variables within each cell. In case of process modeling, the time and space must be partitioned in such a way so that the concentrations of the various impurities present are constant over each individual cell during each time increment along with the diffusivity and other physical parameters. Finally, the solution is computed at each discrete point, known as the mesh or grid, within the domain. The grid spacing must be sufficiently dense so that all the relevant features of the doping profile are accurately represented. The increments of time must be short enough to model important physical effects. On the other hand, it is important not to use excessively small intervals to avoid time-consuming and expensive numerical solutions. The detailed discretization technique is discussed in the literature [100].

The layout of a mesh or grid in a simulation structure is a very important aspect of the numerical solution of PDEs, as it directly determines how well the discrete model represents the actual problem. There are a number of considerations regarding grid selection. First, the points must be allocated to accurately approximate any physical quantities of interest including potentials, concentrations, fields, and currents, as well as any irregularities in the geometry of the domain. Second, because the overall computation time depends on the total number of grid points, grid points must be optimized for computational efficiency. Finally, the finer grid must be allocated in the active regions of device operation under the biasing conditions [3,101].

Solution variables such as potentials, doping, charge, and recombination appear in the PDEs. Therefore, high grid densities must be allocated in regions where any of these quantities undergo rapid changes (e.g., p-n junctions). Conversely, the spacing between points could be relaxed in areas where values remain relatively constant without adding any significant

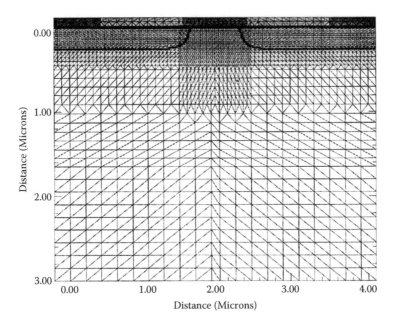

Distance (Microns)

FIGURE 1.9
A typical optimized mesh of an LDD MOSFET simulation structure; grid is denser in the gate oxide and channel region as well as at the source-drain junction. (From S. Saha, MOSFET test structures for two-dimensional device simulation, *Solid-State Electron.*, vol. 38, no. 1, pp. 69–73, January 1995. With permission.)

contribution to the overall error (e.g., quasi-neutral regions deep inside the device). In addition, the simulation structure must be robust so that the simulated device performance is independent of grid density, and the robustness of the simulation domain; that is, sensitivity of simulation results on grid must be checked [101] after the structure generation. Figure 1.9 shows a typical robust mesh of a MOSFET simulation structure.

1.4.2 Verification of the Robustness of Simulation Structure

In the previous subsection we discussed that an appropriate simulation domain is required to emulate the impurity distribution within the device as accurately as possible. Inaccurate mesh may cause fluctuation in the simulation results. Therefore, it is critical to check the robustness of the simulation structure by varying the grid space to generate grid-independent simulation results [101].

Figure 1.10 shows the sensitivity of simulated *I–V* characteristics of nMOS-FETs on grid allocation. Figure 1.10(a) shows the sensitivity of *I–V* data on vertical grid space, Y.GRID in the channel at the SiO_2/Si interface. It is seen from Figure 1.10(a) that for Y.GRID $\leq$ 200 Å, the magnitude of I_{ds} attains a maximum value and the electrical characteristics become insensitive to the

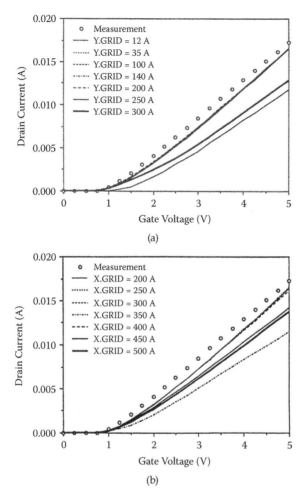

FIGURE 1.10
Sensitivity of grid density on MOSFET device performance; (a) sensitivity of vertical grid, Y.GRID on I_{ds} and (b) sensitivity of lateral grid, X.GRID on I_{ds}; here $L_g = 0.8 \ \mu m$, $W = 40 \ \mu m$, and $T_{OX} = 15$ nm; all data are obtained at V(Drain) = 5 V and V(Body) = 0 = V(Source). (From S. Saha, MOSFET test structures for two-dimensional device simulation, *Solid-State Electron.*, vol. 38, no. 1, pp. 69–73, January 1995. With permission.)

values of Y.GRID at the surface. Figure 1.10(a) also shows that the magnitudes of the measured and simulated data are closer for Y.GRID ≤ 200 Å. Thus, Y.GRID = 12 Å at the surface near the Si/SiO$_2$ interface generates robust test structures for device simulation and accurately models critical physical effects such as inversion layer quantization in the MOSFET channel [98].

Figure 1.10(b) shows the sensitivity of *I–V* data on the lateral grid space, X.GRID in the channel region. It is also seen from Figure 1.10(b) that for

X.GRID ≤ 300 Å, the magnitude of I_{ds} attains a maximum value, and the device characteristics are independent of X.GRID. It is obvious from Figure 1.10(b) that the simulated and measured data are in close agreement for X.GRID ≤ 300 Å. Thus, X.GRID ≤ 200 Å can be used to reduce the uncertainty in the simulated electrical characteristics due to incorrect grid allocations and generate robust test structures for device simulation.

For simulation accuracy and computational efficiency, fine grid is allocated only in the regions of most physical importance as shown in Figure 1.9. The procedure to verify the robustness of the generated mesh by studying the electrical behavior as a function of grid space minimizes the probable errors in the simulation data due to incorrect grid allocation and therefore provides accurate calibration of the fundamental material parameters for device simulation.

1.4.3 Calibration of Physical Models

The accuracy of simulation results relates to the underlying physics, numerical discretization issues, and the proper characterization of the models implemented in the process and device CAD tools. Over the years, the capabilities of TCAD point tools have significantly improved. However, due to the deficiencies in physical models, the simulation results are not 100% predictable and reliable. Although the development and implementation of advanced physical models have been moving ahead with a steady-state growth pattern, the characterization of the existing models is critical to make TCAD accurate. Therefore, the predictability issue must be addressed by proper calibration of the process and device models implemented in TCAD point tools. The calibration is a complex and time-consuming task. In the following section, a brief overview of physical model calibration is presented.

The physical process models implemented in the process CAD tool must be calibrated for the target technology under development. The calibration methodology includes designing short loop wafer fabrication experiments to calibrate all critical as-implanted as well as final doping profiles of all relevant impurities in the simulation structures. For a typical CMOS technology this includes 1D doping profiles for (1) channel, (2) SDE, and (3) DSD regions along the cutlines shown in Figure 1.11.

In order to calibrate the physical device models implemented in device CAD tools, a set of appropriate device characteristics are measured from the fabricated wafers of the target technology. For a typical CMOS technology this includes I_{ds} versus V_{gs}, I_{ds} versus V_{ds}, substrate current, I_{sub} versus V_{gs}, and so on to characterize the physical device models [102,103]. Figures 1.12(a) and 1.12(b) show a basic flow for process and device model calibration of a typical CMOS technology. As shown in Figure 1.12(a) coupled process and device CAD is used to calibrate 2D doping profiles.

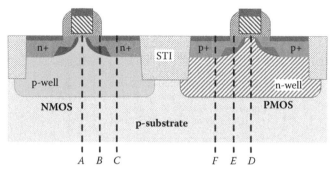

2D-CMOS cross-section to obtain 1D doping profiles along the cut lines:

$A/D \Rightarrow$ NMOS / PMOS *channel*;

$B/E \Rightarrow$ NMOS / PMOS *SDE*;

$C/F \Rightarrow$ NMOS / PMOS *DSD*.

FIGURE 1.11
A typical 2D-CMOS cross-section showing the cutlines along the depth of the simulation structure to obtain 1D doping profiles for process model calibration. (STI represents the shallow trench isolation.)

For greater accuracy and predictability of simulation results, two levels of calibration are performed, namely (1) local calibration and (2) global calibration [1]. A brief outline of the calibration levels is presented below:

Local calibration: The physical models implemented in the process and device CAD tools can be characterized with the measured data of the target technology for predictive process and device simulation. During the IC fabrication process development cycle, a large number of measured data are generated. This experimental database of the target technology can be used to calibrate the process and device models and correlate simulation data with the measured data of the target technology. The calibrated TCAD models and simulation files are transferred to manufacturing at the end of the technology development (TD) cycle. These models can be used with a higher degree of accuracy (1) to evaluate the effects of process control variables on device and product performance and (2) for manufacturing process control [104–107]. Therefore, the calibrated TCAD models of a particular process technology can be used for the predictive simulation of that technology.

Global calibration: In order to develop a new IC fabrication technology, the TCAD point tools must be accurate and must accurately predict new physical phenomena, as may emerge during the development phase. Unfortunately, due to the lack of accurate physical models, effective calibration methodologies must be developed for the successful application of TCAD in designing a new technology. For the

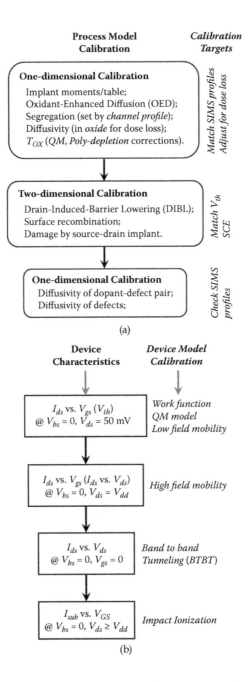

FIGURE 1.12
A simplified process and device model calibration flow for a typical CMOS technology: (a) process model calibration using coupled process and device CAD; (b) device model calibration using device CAD with measured device characteristics. In (a), SIMS represents secondary ion mass spectrometry.

predictive application of TCAD, the numerical models must be initially characterized with a known previous generation of technology. These models can be applied to several similar technologies with the available experimental database. An iterative method of model updates to correlate the simulation and the measurement data must be continued up to an acceptable limit of tolerance of the simulation data with respect to the measurement data. This iterative method of model characterization with different known technologies and constant updating of the models provides a global calibration of the numerical models, which can be applied for next-generation IC fabrication process development with a higher degree of confidence. In addition, the calibration database must include the data from a wider range of anticipated process conditions for the next-generation process technology and device characteristics. Finally, a limited number of wafers of the target new technology should be processed concurrently for a local update of the globally calibrated models. These calibrated physical models and the model parameters will provide a higher degree of accuracy for simulation results and can be applied successfully in developing next-generation IC fabrication technology.

Calibration database: The calibration database used in simulation must include the measured doping profiles and device characteristics. The doping profiles must be obtained under a wide range of processing conditions covering the anticipated process variations of the target technology. The device characteristics must be obtained under various biasing conditions of circuit operations. The database must also be updated to include any new physical phenomena observed in the fabrication facility. In order to develop the next-generation IC fabrication technology, the database must also include the measured data of the anticipated doping materials under the anticipated processing conditions of the target technology. In reality, the individual process module for a new technology is developed prior to the start of the cycle for TD. Therefore, working with the unit process or the module development group, the calibration database can be prepared to include the data for the development of a target technology. In addition, the measurement data for an advanced technology with respect to the newer fabrication equipment can be obtained in collaboration with the equipment vendors. Thus, the calibration database must have experimental data for all the possible effects of the existing and the target next-generation technology.

1.4.4 Coupled Process and Device Simulation

Process CAD is used for unit process development such as time and temperature required for growing the MOSFET gate oxide, shallow trench isolation (STI) module for CMOS technology, and so on. Similarly, the device CAD is used to analyze the effect of process variables on device operation. However, the coupled process and device simulation allows IC designers to directly investigate the effect of process specifications on electrical variations in devices and circuits. The key task of process CAD is to capture the features that accurately reflect the performance limitations of a given fabrication technology.

Figure 1.13 presents an overview of the simulation flow that is used to link process specifications to circuit performance. For the most part, the user input is simply a description of how the actual fabrication sequence progresses. The process simulation generates a set of data with the structural information like device geometry, doping distribution, and so on. The output of process simulation is then used as the input of device CAD to perform device simulation to obtain *I–V, C–V* data. These simulation data are used to generate SPICE models by well-established parameter extraction techniques, and

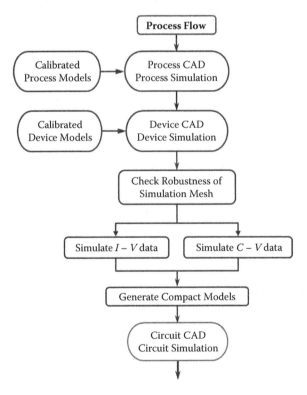

FIGURE 1.13
Link IC process flow to circuit performance using coupled process and device CAD.

the SPICE models are used for circuit simulation. At each step, the emphasis is on focusing a broad spectrum of inputs into a coherent set of outputs that will be of maximum utility in the next step. Thus, the coupled process and device CAD can be considered as a "virtual factory" simulating ICs from the process flow analogous to wafer fabrication facility manufacturing ICs from the target process flow as shown in Figure 1.13 [104].

Through well-defined data exchange formats, the various levels of CAD are linked quite efficiently with or without simulation framework. The linkage of process to device CAD occurs through the exchange of both topographic information and arrays of data representing dopant distributions.

1.5 TCAD Application

In the semiconductor industry the major use of TCAD is in advanced device research to study new device concepts, TD, and technology transfer process [1,105–107]. In the following subsections we discuss some examples of these TCAD usages.

1.5.1 TCAD in Device Research

TCAD is used to study the feasibility of new device concepts or exploratory devices for the next-generation IC fabrication TD. For exploratory devices, the fabrication process is yet to be determined. Therefore, only device CAD is used to design and optimize the device structure and device performance. In this case, the device structure is generated using analytical doping profiles. Figure 1.14 shows a typical flow to use device CAD in the feasibility study of new concept devices.

In Figure 1.14, the basic simulation flow includes device architecture by analytical doping profile, use of device CAD to simulate device characteristics, verification of the robustness of the simulation structure, and optimization of the device structure to achieve the target performance objectives. After device optimization, perform device simulation to generate *I–V* and *C–V* data, format device simulation data, and generate compact models for circuit analysis. In Figure 1.14, note that calibrated device models are required for accurate prediction of device performance. After achieving the target device performance by iterative device simulation, appropriate fabrication process is designed for technology optimization using coupled process and device CAD flow. To illustrate the use of device CAD in device development for the next-generation fabrication technology, let us discuss the architecture and performance of nanoscale double-halo MOSFETs and split-gate (SG) flash memory cells.

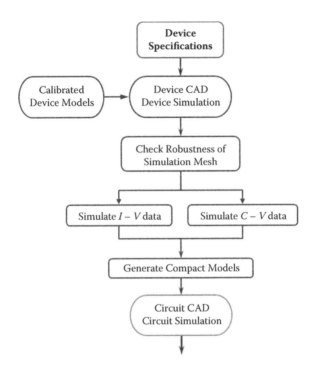

FIGURE 1.14
A typical device CAD-based simulation flow to study new device concepts and generate electrical device characteristics for circuit analysis.

1.5.1.1 Double-Halo MOSFET Devices

It is well known that the conventional scaling of MOSFET devices in the nanoscale regime requires a reduction of the gate oxide thickness (T_{OX}) and an increase in the channel doping concentration (NCH) to control SCE, DIBL, and leakage. However, the combination of such high NCH and ultra-thin *Tox* is likely to cause severe performance degradation due to higher V_{th}. In order to suppress SCE and control DIBL, typically, a single halo profile is used around SDE. However, as the MOSFET devices approach their ultimate dimension near the 10 nm regime, both SDE and DSD regions near the source-end of the channel contribute to DIBL, resulting in higher I_{off} and degradation in the subthreshold swing. In order to control DIBL from both SDE and DSD, we can use two halo doping profiles, one around SDE and other around DSD. Then use device CAD to optimize these halo doping profiles to control V_{th}, SCE, DIBL, and I_{off} in the nanoscale MOSFETs. Therefore, our objective is to design high performance nanoscale MOSFETs with low leakage, fast switching, controlled SCE, and reduced DIBL and QM effects [93–97]. In order to achieve our objective an ideal structure shown in Figure 1.6 is used to optimize the device performance.

In device architecture, the n/p MOS device structure shown in Figure 1.6 includes an n+/p+ polysilicon gate, a gate oxide, vertically and laterally non-uniform channel doping profile, shallow n+/p+ SDE, deep n+/p+ DSD, and two halo doping profiles: the first around SDE and the second around DSD regions. An SSR doping profile with a low impurity concentration at the Si/SiO$_2$ interface and a higher peak concentration at a finite depth below the interface is used to provide the non-uniform vertical channel profile [97, 98]. The SSR profile is optimized to achieve a target V_{th} for the long channel devices. The SSR profile also provides superior V_{th} control caused by dopant fluctuations [108].

The non-uniform analytical lateral channel doping profile is achieved using two Gaussian halo profiles during drain-profile engineering [109]. The first halo doping profile is a heavily doped shallow profile defined on both sides of the gate region aligning at the gate edge. The peak concentration of the shallow halo profile is defined immediately below the SDE junction (X_j) to reduce DIBL due to SDE. The second halo profile is a lightly doped deep profile defined on both sides of the gate region by an offset distance from the gate edge. The peak concentration of the second halo doping is placed immediately below the DSD junction (X_{jd}) to control DIBL due to DSD. The detailed fabrication procedure is described in [95]. After device architecture, device CAD is used to study the device performance of the double-halo MOSFETs.

A hydrodynamic model for semiconductors with full energy balance solution along with an analytical QM model [110] is used for device simulation [111]. The halo profiles along with the SSR channel doping profile are optimized to the target $I_{off} = 10\ nA/\mu m$. The 2D doping distribution of an optimized double-halo nMOSFET device is shown in Figure 1.8. And, Figures 1.15(a) and 1.15(b) show the simulated device performance of double-halo nMOSFET and pMOSFET devices with effective channel length, $L_{eff} = 25$ nm.

Device simulation results in Figure 1.15 show excellent device performance, thus achieving the target objective of this study. Thus, device CAD can be used to optimize exploratory devices and assess the feasibility of these devices for next-generation IC fabrication technology.

After device optimization using device CAD, an initial guess process flow is created to reproduce the analytical doping profiles used in device simulation. Then the process CAD is used to perform process simulation and optimize the process flow. Figure 1.16 shows only the section of process flow used to integrate the double-halo MOSFETs in advanced CMOS technology. The detailed process flow is described in [95].

In Figure 1.16, the shallow halo doping profile is implanted after the gate definition with peak concentration immediately below Xj. Then about a 2-nm wide offset spacer is used to implant the deep halo and SDE profiles. Because the depth of the shallow halo is ~Xj, SDE regions are encroached by about 2 nm with the halo doping near the Si/SiO$_2$ interface within the channel. Thus, the shallow doping profiles only enhance the channel doping by about 4 nm of $L_{eff} \cong 25$ nm near the Si/SiO2 interface. On the other hand, the deep halo

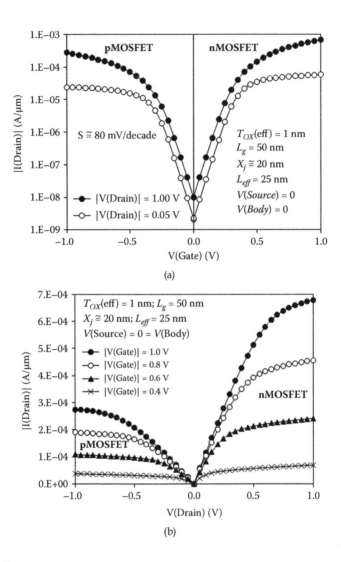

FIGURE 1.15
Device characteristics of double-halo MOSFETs obtained by device CAD: (a) I_{ds} versus V_{gs} and (b) I_{ds} versus V_{ds}; the simulation data are obtained for 50 nm devices with $L_{eff} = 25$ nm and optimized for $|I_{off}| = 10\ nA/\mu m$ at $|V(\text{Drain})| = 1$ V; $V(\text{Gate}) = V(\text{Source}) = V(\text{Body}) = 0$. (From S. Saha, Design considerations for 25 nm MOSFET devices, *Solid-State Electron.*, vol. 45, no. 10, pp. 1851–1857, October 2001. With permission.)

doping profiles with depth ~ DSD junction depth diffuse laterally into the channel region to enhance the channel doping at a finite depth below the Si/SiO2 interface. Thus, the combination of double-halo profiles provides the non-uniform lateral channel doping while maintaining a lower channel doping concentration near the surface due to the SSR channel doping profile. The halo implant dose and energy are optimized to achieve the target value of I_{off}

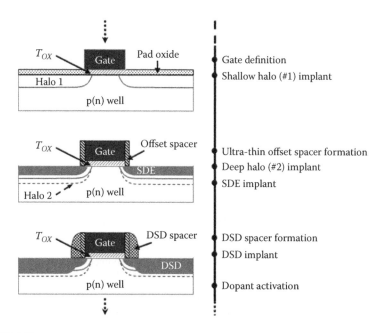

FIGURE 1.16
Part of the process flow showing the integration of two halo profiles in CMOS technology to fabricate nanoscale double-halo MOSFET devices. (From S. Saha, Device characteristics of sub-20-nm silicon nanotransistors, in *Proc. SPIE Conf. on Design and Process Integration for Microelectronic Manufacturing*, vol. 5042, pp. 172–179, July 2003. With permission.)

for the nominal devices of the target technology. The source-drain regions are optimized to achieve an improved device behavior, and the peak impurity concentrations for SDE and DSD profiles used are 2.5×10^{20} cm^{-3} and 3.7×10^{20} cm^{-3}, respectively [93,94].

1.5.1.2 Sub-90 nm Split-Gate Flash Memory Cells

Our second example of device CAD use in advanced device research and development is to study the feasibility of scaling split-gate (SG) NOR-type flash memory cells below 90 nm. A typical SG cell with top coupling gate (CG), called SG-TCG cell [112], is shown in Figure 1.17. The structure includes a select gate or word-line (WL) with gate oxide (T_{ox}), a floating gate (FG) with tunneling oxide ($T_{ox,tun}$), a CG with inter-poly oxide ($T_{ox,IPO}$), an n+ source as the "Bitline" (BL), and an n+ drain as the "Sourceline" (SL). The FG is completely isolated within the gate dielectric and acts as a potential well to store charge by programming the cell using source-side injection (SSI) of hot carriers. The erase operation is performed by poly to poly Fowler-Nordheim tunneling of carriers from the FG to WL. The sharp FG-tip near the WL edge in Figure 1.17 improves the erasing efficiency of the programmed cells. Because the cell structure in Figure 1.17 consists of two MOSFET devices in series, it is difficult to scale these cells in the sub-90 nm regime due to several constraints [113,114].

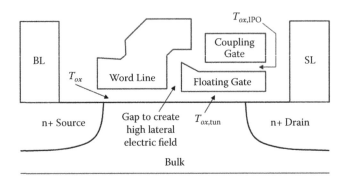

FIGURE 1.17
An idealized SG-TCG flash memory cell structure used for scaling in the sub-90 nm regime: here, BL = bitline, T_{ox} = WL-transistor gate oxide thickness, $T_{ox,tun}$ = tunneling oxide thickness, $T_{ox,IPO}$ = inter-poly oxide thickness and SL = Sourceline. (From S.K. Saha, Non-linear coupling voltage of split-gate flash memory cells with additional top coupling gate, *IET Circuits, Devices & Systems*, vol. 6, no. 3, pp. 204–210, May 2012. With permission.)

The major scaling constraints of NOR-type SG-TCG cells are (1) the high value of SL programming voltage, $V_{SL} \equiv V_{sp} \gg 3.2$ V required for an efficient hot-electron programming [113] and V_{sp} > floating gate (FG) transistor saturation voltage ($V_{SL,sat}$) required to mitigate the risk of supply voltage fluctuations [112]; (2) excessive program-inhibit leakage current, $I_{off(BLI)}$ causing inhibited cells susceptible to soft-write error [113,114]; (3) high program cell leakage current, I_{r0}, causing ineffective sensing of the write and erase states by the sense amplifiers [113,114]; and (4) degradation of program/erase (P/E) coupling ratio causing degradation in P/E efficiency [113,114].

Our objective is to design high performance sub-90 nm NOR-type SG-TCG cells within the above described scaling constraints. First, the requirement for $V_{sp} \gg 3.2$ V makes the scaled SG-TCG cells susceptible to punchthrough at the operating conditions, causing degradation in the cell reliability. Therefore, in order to improve the punchthrough voltage, the channel doping concentration must be increased which in turn decreases the SL p-n junction breakdown voltage, BV_j. Thus, we need to optimize the channel doping profile and use graded SL/BL p-n junctions to improve the overall cell breakdown voltage (BV) to achieve the target V_{sp} for the scaled SG-TCG cells. In this case, three channel doping profiles are used to optimize cell performance for the target V_{sp} value.

The next requirements are tolerable leakage currents $I_{off(BLI)}$ and I_{r0} at the target SL voltage (V_{SL}) for the sub-90 nm SG-TCG cells. $I_{off(BLI)}$ is the off-state leakage current of WL-MOSFETs, whereas I_{r0} is the off-state leakage current of FG MOSFETs. Thus, $I_{off(BLI)}$ can be optimized to the target value required for the target V_{sp} using shallow BL-extension (BLE) and optimizing BL-halo to reduce SCE and DIBL. Similarly, to achieve the target I_{r0} shallow SL junction along with a lightly doped FG-channel profile is used as shown in Figure 1.17. The target values of leakage currents at the required $V_{SL} = V_{sp}$ are obtained using device CAD. In this case, we can use device CAD to determine the maximum tolerable parasitic leakage current for $BV > V_{sp}$. In this example, the device simulation data show that $I_{r0} \sim 200$ pA/cell at $V(BL) = 0.8$ V and $V(FG) = 0$ with WL device on and $I_{off(BLI)} \sim 200$ nA/kbit at $V(BL) = 1.8$ V and $V(WL) = 0$ with FG device on are required to maintain $V_{sp} = 6.5$ V $< BV$. The detailed optimization technique is reported in the references [113,114].

The final requirement to design sub-90 nm SG-TCG cells is to account for the degradation of the coupling ratio in the scaled devices. The addition of top CG improves the programming coupling ratio, whereas the shallow BLE without overlap under the WL transistor improves the erase coupling ratio.

Thus, the use of shallow SL junction will control SCE and DIBL and improve scalability of SG-TCG cells. In addition, a shallow BL-junction will control DIBL and improve scalability of the cells. In this example, a shallow BLE and deep BL regions along with BL-halo are used to optimize the cell to the desired performance objective [112–114]. The final simulation structure of the optimized SG-TCG cell is shown in Figure 1.18.

Figure 1.19 shows device simulation results of the optimized SG-TCG cells. Figure 1.19 shows that $V_{th}(WL)$ and cell read current (I_{r1}) as function of WL transistor channel length. In Figure 1.19(a), $V_{th}(WL)$ is extracted from the extrapolated $I_{ds} - V_{gs}$ plots of WL-devices at $V_{ds} = 50$ mV with overdrive $V(FG) = 2.5$ V. And $I_{r1} = I_{cell}$ is obtained at the read condition, $V(WL) = 2.5$ V, $V(BL) = 0.8$ V, $V(FG) = 1.8$ V $= V(CG)$, and $V(SL) = 0$. The device simulation data show acceptable read current $I_{r1} \approx 22.2$ μA for 65 nm cells optimized for $V_{sp} = 6.5$ V. Figure 1.19(b) shows the

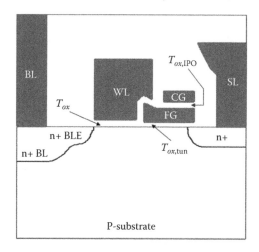

FIGURE 1.18
The final sub-90 nm SG-TCG flash memory cells structure generated using device CAD tool MEDICI: here, BL = bitline, WL = word line, FG = floating gate, CG = coupling gate, T_{ox} = WL-transistor gate oxide thickness, $T_{ox,tun}$ = tunneling oxide thickness, $T_{ox,IPO}$ = inter-poly oxide thickness, SL = sourceline, and BLE = shallow BL extension. (From S.K. Saha, Non-linear coupling voltage of split-gate flash memory cells with additional top coupling gate, *IET Circuit, Devices & Systems*, vol. 6, no. 3, pp. 204–210, May 2012.)

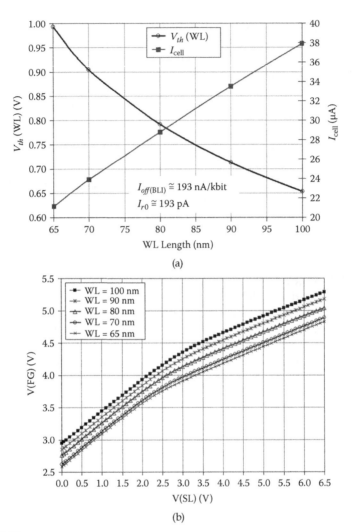

FIGURE 1.19
Simulated device performance of SG-TCG cells: (a) V_{th} and I_{cell} versus WL-transistor channel length and (b) programming coupling voltage $V(FG)$ as a function of programming voltage $V(SL)$ obtained at function of programming voltage $V(SL)$ obtained at $V(CG) = 10$ V. (From S.K. Saha, Design considerations for sub-90 nm split-gate Flash memory cells, *IEEE Trans. Electron. Devices*, vol. 54, no. 11, pp. 3049–3055, November 2007. With permission.)

simulated $V(FG)$ versus $V(SL)$ plots of the SG-TCG cells at $V(CG) = 10$ V. Similarly, Figure 1.20 shows an improvement in both programming coupling ratio due to CG and the new device architecture with shallow SL and BLE junctions.

    The device simulation study clearly shows the feasibility of SG-TCG cell scaling near the 65-nm regime. After the device optimization, an initial guess process flow is created as shown in Figure 1.21 for complete technology optimization using coupled process and device CAD as shown in Figure 1.13.

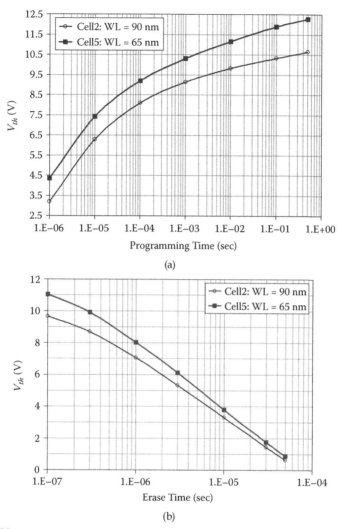

FIGURE 1.20
Programming/erase characteristics of SG-TCG cells: (a) programming at $V(WL) = 1.2$ V, $V(SL)$ = 6.5 V, $V(BL) = 0.3$ V, and $V(CG) = 10$ V; and (b) erase; at $V(WL) = 10$ V and $V(SL) = 0 = V(BL)$; from simulation data the time to program, T2P $\approx$ 30 μs and time to erase, T2E $\approx$ 40 μs. (From S.K. Saha, Design considerations for sub-90 nm split-gate Flash memory cells, *IEEE Trans. Electron. Devices*, vol. 54, no. 11, pp. 3049–3055, November 2007. With permission.)

The device-simulation results show that the sub-90-nm SG-TCG flash-memory cells can be achieved with tolerable $I_{r0} < 200$ pA, $I_{OFF(BLI)} < 200$ nA/kb, acceptable I_{r1}, and excellent T2P $\approx$ 30 μs and T2E $\approx$ 40 μs. Thus, the present design methodology demonstrates the feasibility of high performance sub-90-nm split-gate flash-memory cells down to 65 nm with $I_{r1} \approx 235$ μA/μm along with efficient P/E characteristics. From the results of device CAD, a process flow is developed for coupled process and device simulation.

... ...

P-well 1 implant: peak concentration depth ~ 0.4 μm
P-well 2 implant: peak concentration depth ~ SL-junction depth
P-well 3 implant: defines WL V_{th}

FG V_{th}-adjust implant
FG-transistor gate oxide growth: thickness ~ 9 nm
n+ FG-poly definition

Inter-poly (FG – CG) oxide deposition: thickness ~ 12 nm
n+ CG-poly definition

SL-offset spacer deposition
Ultra-shallow SL implant
n+ SL-poly formation

WL-transistor gate oxide growth: thickness ~ 13-nm
n+ WL-poly definition

BL-halo-offset spacer deposition
BL-halo implant
BL-offset spacer deposition: thickness ~ BLE junction depth
Ultra-shallow BLE implant
BL-spacer deposition
Deep BL implant and anneal

Contact formation
... ...

FIGURE 1.21
The major technology steps to integrate sub-90 nm split-gate NOR-flash memory cells in a standard CMOS technology. (From S.K. Saha, Design considerations for sub-90 nm split-gate Flash memory cells, *IEEE Trans. Electron. Devices*, vol. 54, no. 11, pp. 3049–3055, November 2007. With permission.)

1.5.2 TCAD in Fabrication Technology Development (TD)

The coupled process and device CAD flow as shown in Figure 1.13 is used in IC fabrication TD. Typically, in the semiconductor industry, the next-generation IC fabrication technology is derived by scaling the current technology. In this case, the current generation fabrication process flow with appropriate modifications is used as the initial guess to optimize the process recipe of the next-generation technology. As shown in Figure 1.13, the initial guess process recipe is iteratively optimized using coupled process and device CAD to the target achievable device specifications; then the compact model parameters are extracted from the simulated device characteristics; and finally, circuit CAD is used to analyze the circuit performance. Figure 1.13 also shows that for accurate prediction of device and circuit performance, calibrated physical models are essential. Also, the calibration database with experimental doping profiles under various processing conditions and device characteristics are essential to validate the simulation results in each step of the optimization process.

Besides the conventional TD approach described above, the initial guess process recipe can be efficiently obtained by "reverse modeling" as shown in Figure 1.1. As we discussed in Section 1.3, the ultimate motivation for TCAD is to extract circuit design parameters to enable designers to predict circuit performance of the target technology. In the TCAD flow in Figure 1.13, the circuit design parameters are generated by optimizing the initial guess process recipe of the current generation technology using coupled process and device CAD. However, for exploratory devices, the current generation technology does not exist; therefore, the initial process flow and recipe are obtained by reverse simulation flow (e.g., circuit CAD to process CAD), as shown in Figure 1.1. Because the ultimate goal of the new technology generation is to predict circuit performance, circuit/product specific process recipe will ensure the target circuit performance. The basic simulation flow for the product-specific IC process design is shown in Figure 1.22. The initial guess process recipe is generated by reverse engineering from the target product specifications in three sequential steps: (1) generation of device models using circuit simulation as shown in Figure 1.22(a); (2) generation of doping profiles

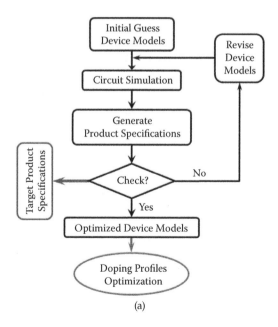

(a)

FIGURE 1.22
Flowchart for the generation of initial guess process recipe using three-step approach: (a) optimization of device models to the target product specifications; (b) optimization of process profiles to the target device models; (c) optimization of process recipe to the target process profiles. (From S. Saha, Technology CAD for integrated circuit fabrication technology development and technology transfer, *Proc. SPIE*, vol. 5042, pp. 63–74, July 2003. With permission.) (*continued*)

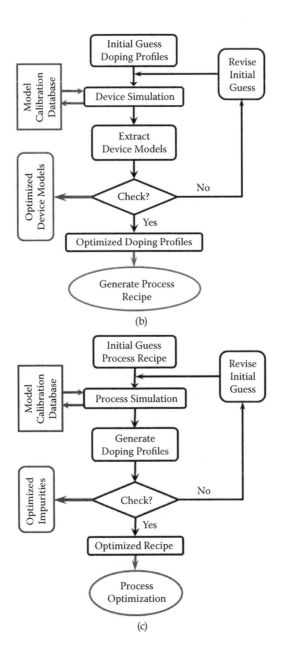

FIGURE 1.22
(*continued*) Flowchart for the generation of initial guess process recipe using three-step approach: (a) optimization of device models to the target product specifications; (b) optimization of process profiles from the target device models; (c) optimization of process recipe to the target process profiles. (From S. Saha, Technology CAD for integrated circuit fabrication technology development and technology transfer, *Proc. SPIE*, vol. 5042, pp. 63–74, July 2003. With permission.)

within the device using device CAD as shown in Figure 1.22(b); and (3) generation of process recipe using process CAD as shown in Figure 1.22(c). This initial guess product-specific process recipe can then be further optimized using coupled process and device CAD as shown in Figure 1.13 to generate the final process flow and device specifications.

1.6 Benefit of TCAD in TD Projects

TCAD is widely used in advanced IC fabrication TD to optimize IC devices and fabrication processes through critical analysis and detailed understanding of process, device, and circuit simulation data. The TCAD tools that accurately predict the process and device characteristics of the anticipated wafer fabrication technology have shown great promise in nanoscale device and advanced IC manufacturing TD. Therefore, TCAD has become indispensable to reduce the development cycle time and cost of advance IC fabrication technology and products [1–5]. TCAD tools offer manufacturing and development engineers vast improvements in flexibility, innovation, efficiency, and customer responsiveness [1].

In Section 1.3, we established the motivation and importance of TCAD in IC device and process architecture. It is obvious that the knowledge gained by TCAD is impossible to quantify. However, for industrial applications it is essential to quantify the benefit of TCAD in reducing TD cycle time and cost [1,5]. By conducting a survey of TCAD users, International Technology Roadmap for Semiconductors (ITRS) reported about 32% reduction in development time and about 30% in development cost by the appropriate use of TCAD [2]. Though this demonstrates high relevance and potential for the industrial use of TCAD, a survey-based benefit assessment is not adequate to undertake a TCAD-based development project in the industry. Therefore, a quantitative model to assess the benefit of TCAD use in a development project is crucial for project managers.

In 1999, the basic formulations of a quantitative model to compute the benefit of TCAD use in IC development were reported [1] and the detailed analytical model to compute the benefit of TCAD was reported in 2010 [5]. In these reports, a set of realistic assumptions is used to derive a set of mathematical expressions to compute the cycle time and cost benefit of a TCAD-based project compared to that of the trial-and-error experimentation or "conventional" approach. The key assumption of this model is that the TCAD tools accurately predict the process and device characteristics of the target wafer fabrication technology. In order to develop a realistic analytical model to compute the benefit of TCAD over the conventional approach, a typical IC fabrication TD project is divided into three phases (ϕ): phase 1, generation of initial guess process recipe; phase 2, process optimization to generate process and device specifications; and phase 3, evaluation of process manufacturability. According to this model, the reduction in the development cycle time, Δt, and cost, ΔC, on a project by TCAD use compared to the conventional approach is described by

$$\Delta t \cong Ft_{conv} \tag{1.1}$$

$$\Delta C \geq C_{wfr}\left(\frac{F}{1-\rho}+(1+ROI)\right)\Delta n - C_{tcad} \tag{1.2}$$

$$\Delta C \geq C_{wfr}[F+(1+ROI)(1-\rho)]A \cdot m \cdot n_p - C_{tcad} \tag{1.3}$$

where the model parameters in Equations (1.1)–(1.3) are as follows: F is the development cycle time reduction factor by a TCAD-based project compared to a conventional poject; t_{conv} is the conventional TD time without using TCAD; C_{wfr} is the cost of processing a single wafer-lot in the fab; ρ is the fraction of the conventional TD wafers used in a TCAD-based project; ROI is the return on investment from IC sale; Δn is the reduction in the number of wafer-lots in the fab by using TCAD; C_{tcad} is the total cost of implementation of CAD infrastructure; m is the total number of iterations required to optimize a technology in each phase of a TD project and is assumed to be the same for both conventional and TCAD-based processes; n_p is the number of process-control variables p (e.g., ion implant) and defines the complexity of an IC fabrication technology; and $A = n_x n_y \phi/w_n$ is the model parameter that depends on the number of project phases ϕ, the split conditions of implants (e.g., energy n_x and dose n_y), and the number of wafers in a wafer-lot w_n.

By using the appropriate values of the parameters in Equations (1.1) and (1.2) or (1.1) and (1.3) for the TCAD tools that accurately predict the process and device characteristics of the target wafer fabrication technology, the above analytical model provides a reduction in the TD cycle time of about 67% with multimillion dollar cost savings compared to the conventional approach [1,5].

1.7 Summary

This chapter presents the mission and scope of extended technology CAD in IC process modeling and device performance analysis. A brief history of the evaluation of device and process CAD during the past four decades leading to the commercialization of TCAD software is described. The motivation for TCAD use is outlined with a few typical examples. A brief outline of TCAD flow including generation of robust simulation structure and physical model calibration for accurate IC process and device simulation is presented. A few examples of typical TCAD application, especially, in studying the feasibility of exploratory devices for the next-generation fabrication technology are discussed. Finally, an analytical cost model to compute the benefit of TCAD use in saving IC TD cycle time and cost over the conventional trial-and-error-experimentation based TD project is discussed.

References

1. S.K. Saha, Managing technology CAD for competitive advantage: An efficient approach for integrated circuit fabrication technology development, *IEEE Trans. Eng. Manage.*, vol. 46, no. 2, pp. 221–229, May 1999.
2. *International Technology Roadmap for Semiconductors*: *Modeling and Simulation*, 2011, http://www.itrs.net/Links/2011ITRS/2011Chapters/2011Modeling.pdf.
3. R.W. Dutton and Z. Yu, *Computer Simulation of IC Processes and Devices*. Boston: Kluwer Academic, 1993.
4. R.W. Dutton and A.J. Strojwas, Perspective on technology and technology-driven CAD, *IEEE Trans. Computer Aided Design*, vol.19, no. 12, pp. 1544–1560, December 2000.
5. S.K. Saha, Modelling the effectiveness of computer aided development projects in the semiconductor industry, *Int. J. Eng. Manage. Econ.*, vol. 1, no. 2/3, pp. 162–178, November 2010.
6. H.K. Gummel, A self-consistent iterative scheme for one-dimensional steady state transistor calculations, *IEEE Trans. Electron Dev.*, vol. 11, no. 10, pp. 455–465, October 1964.
7. A. De Mari, An accurate numerical steady-state one-dimensional solution of the p-n junction, *Solid-St. Electron.*, vol. 11, no. 1, pp. 33–58, January 1968.
8. A. De Mari, An accurate numerical one-dimensional solution of the p-n junction under arbitrary transient conditions, *Solid-St. Electron.*, vol. 11, no. 11, pp. 1021–1053, November 1968.
9. D.L. Scharfetter and H.K. Gummel, Large-signal analysis of a silicon read diode oscillator, *IEEE Trans. Electron Dev.*, vol. 16, no. 1, pp. 66–77, January 1969.
10. H.W. Loeb, R. Andrew, and W. Love, Application of 2-dimensional solutions of the Shockley-Poisson equation to inversion-layer M.O.S.T. devices, *Electron. Lett.*, vol. 4, no. 17, pp. 352–354, August 1968.
11. J.E. Schroeder and R.S. Muller, IGFET analysis through numerical solution of Poisson's equation, *IEEE Trans. Electron Devices*, vol. 15, no. 12, pp. 954–961, December 1968.
12. M.B. Barron, *Computer Aided Analysis of Insulated Gate Field Effect Transistors*, Ph.D. thesis, Stanford University, California, November 1969.
13. D.P. Kennedy and R.R. O'Brien, Electric current saturation in a junction field-effect transistor, *Solid-St. Electron.*, vol. 12, no. 10, pp. 829–830, October 1969.
14. J.W. Slotboom, Iterative scheme for 1- and 2-dimensional d.c.-transistor simulation, *Electron. Lett.*, vol. 5, no. 26, pp. 677–678, December 1969.
15. M. Reiser, Difference methods for the solution of time-dependent semiconductor flow equations, *Electron. Lett.*, vol. 7, no. 12, pp. 353–355, June 1971.

16. M.S. Mock, A two-dimensional mathematical model of the insulated-gate field-effect transistor, *Solid-St. Electron.*, vol. 16, no. 5, pp. 601–609, May 1973.

17. M.A. Green and J. Shewchun, Minority carrier effects upon the small signal and steady-state properties of Schottky diodes, *Solid-St. Electron.*, vol. 16, no. 10, pp. 1141–1150, October 1973.

18. J.J. Barnes, *A Two-Dimensional Simulation of MESFETs*, Ph.D. Thesis, University of Michigan, Ann Arbor, May 1976.

19. J.J. Barnes and R.J. Lomax, Finite-element methods in semiconductor device simulation, *IEEE Trans. Electron. Dev.*, vol. 24, no. 8, pp. 1082–1089, August 1977.

20. T. Toyabe, K. Yamaguchi, S. Asai, and M.S. Mock, A numerical model of avalanche breakdown in MOSFETs, *IEEE Trans. Electron Dev.*, vol. 25, no. 7, pp. 825–832, July 1978.

21. D.C. D'Avanzo, M. Vanzi, and R.W. Dutton, One-dimensional semiconductor device analysis, Tech Report, No. G-201-5, Stanford Electronics Laboratories, Stanford University, California, October 1979.

22. S. Selberherr, W. Fichtner, and H.W. Potzl, MINIMOS—A program package to facilitate MOS device design and analysis, in *Proc. NASCODE-I*, pp. 275–279, June 1979.

23. P.E. Cottrell and E.M. Buturla, Two-dimensional static and transient simulation of mobile carrier transport in a semiconductor, in *Proc. NASCODE* I, pp. 31–64, June 1979.

24. J.A. Greenfield and R.W. Dutton, Nonplanar VLSI device analysis using the solution of Poisson's equation, *IEEE Trans. Electron Dev.*, vol. 27, no. 8, pp. 1520–1532, August 1980.

25. C.H. Price, *Two-Dimensional Numerical Simulation of Semiconductor Devices*, Ph.D. Thesis, Stanford University, California, May 1982.

26. M.R. Pinto, C.S. Rafferty, and R.W. Dutton, *PISCES II: Poisson and Continuity Equation Solver*, Stanford University, California, September 1984.

27. M.R. Pinto, C.S. Rafferty, and R.W. Dutton, *PISCES IIB: Supplementary Report*, Stanford University, September 1984.

28. R.E. Bank, W.M. Coughran, Jr., W. Fichtner, E.H. Grosse, D.J. Rose, and R.K. Smith, Transient simulation of silicon devices and circuits, *IEEE Trans. Computer Aided Design of ICs & Sys.*, vol. 4, no. 4, pp. 436–451, October 1985.

29. R.E. Bank, W.M. Coughran, Jr., W. Fichtner, D.J. Rose, and R.K. Smith, Computational aspects of semiconductor device simulation, in *Process and Device Modeling*, ed. W.L. Engl, pp. 229–264, North-Holland: Amsterdam, 1986.

30. A.F. Franz and G.A. Franz, BAMBI—A design model for power MOSFETs, *IEEE Trans. Comp. Aided Design of ICs*, vol. 4, no. 3, pp. 177–189, July 1985.

31. G. Baccarani, R. Guerrieri, P. Ciampolini, and M. Rudan, HFIELDS: A highly flexible 2D semiconductor device analysis program, *Proc. NASCODE-IV*, pp. 3–12, June 1985.

32. Synopsys, Inc., Home page, http://www.synopsys.com/.

33. Silvaco, Inc., Home page, http://www.silvaco.com/.

34. T. Kurosawa, Monte Carlo calculation of hot-electron problems, *J. Phys. Soc. Japan Suppl.*, vol. 21, pp. 424–426, 1966.

35. R.A. Warriner, Computer simulations of gallium arsenide field-effect transistors using the Monte Carlo method, *Solid-State Electron Devices*, vol. 1, no. 4, pp. 105–110, July 1979.

36. R. K. Cook and J. Frey, Two-dimensional numerical simulation of energy transport effects in Si and GaAs MESFETs, *IEEE Trans. Electron Dev.*, vol. 29, no. 6, pp. 970–977, June 1982.

37. M. Fukuma and R.H. Uebbing, Two-dimensional MOSFET simulation with energy transport phenomena, in *IEDM Tech. Dig.*, pp. 621–624, 1984.

38. C. Werner, R. Kuhnert, and L. Risch, Optimization of lightly doped drain MOSFETs using a new quasiballistic tool, in *IEDM Tech. Dig.*, pp. 770–773, 1984.

39. D.J. Widiger, I.C. Kiziyalli, K. Hess, and J.J. Coleman, Two-dimensional transient simulation of an idealized high-electron mobility transistor, *IEEE Trans. Electron Dev.*, vol. 32, no. 6, pp. 1092–1102, June 1985.

40. S.E. Laux and A.C. Warren, Self-consistent calculation of electron states in narrow channels, in *IEDM Tech. Dig.*, pp. 567–570, 1986.

41. E. Sangiorgi, B. Ricco, and F. Venturi, MOS²—An efficient Monte Carlo simulator for MOS devices, *IEEE Trans. Comp. Aided Design of ICs & Sys.*, vol. 7, no. 2, pp. 259–271, February 1988.

42. A. Forghieri, F. Guerrieri, P. Ciampolini, A. Gnudi, M. Rudan, and G. Baccarani, A new discretization strategy of the semiconductor equations comprising momentum and energy balance, *IEEE Trans. Comp. Aided Design of ICs*, vol. 7, no. 2, pp. 231–242, February 1988.

43. E.M. Buturla, P.E. Cottrell, B.M. Grossman, M.B. Lawlor, C.T. McMullen, and K.A. Salsburg, Three dimensional finite element simulation of semiconductor devices, *IEEE Intl. Solid State Circuits Conf. Dig.*, pp. 76–77, February 1980.

44. E. Buturla, P. Cottrell, B. Grossman, and K. Salsburg, Finite-element analysis of semiconductor devices: The FIELDAY program, *IBM J. Res. & Dev.*, vol. 25, no. 4, pp. 218–231, July 1981.

45. S.G. Chamberlain and A. Husain, Three-dimensional simulation of VLSI MOSFETs, in *IEDM Tech. Dig.*, pp. 592–595, 1981.

46. A. Husain and S.G. Chamberlain, Three-dimensional simulation of VLSI MOSFETs: The three-dimensional simulation program WATMOS, *IEEE J. Solid-State Circuits*, vol. SC-17, no. 2, pp. 261–268, April 1982.

47. W. Fichtner, R.L. Johnston, and D.J. Rose, Three-dimensional numerical simulation of small-size MOSFETs, *IEEE Trans. Electron Dev.*, vol. 28, no. 10 pp. 1215–1216, October 1981.

48. N. Shigyo, M. Konaka, and R. Dang, Three-dimensional simulation of inverse narrow-channel effect, *Electron. Lett.*, vol. 18, no. 6, pp. 274–275, March 1982.

49. T. Toyabe, H. Masuda, Y. Aoki, H. Shukuri, and T. Hagiwara, Three-dimensional device simulator CADDETH with highly convergent matrix solution algorithms, *IEEE Trans. Electron Dev.*, vol. 32, no. 10, pp. 2038–2044, October 1985.

50. M. Thurner and S. Selberherr, Comparison of long and short channel MOSFETs carried out by 3D MINIMOS, *Proc. ESSDERC*, pp. 409–412, 1987.

51. J.H. Chern, J.T. Maeda, L.A. Arledge, Jr., and P. Yang, SIERRA: A 3D device simulator for reliability modeling, *IEEE Trans. Comp. Aided Design of ICs*, vol. 8, no. 5, pp. 516–527, May 1989.

52. P. Ciampolini, A. Pierantoni, M. Melanotte, C. Cecchetti, C. Lombardi, and G. Baccarani, Realistic device simulation in three dimensions, *IEDM Tech. Dig.*, pp. 131–134, 1989.

53. E. Buturla, J. Johnson, S. Furkay, and P. Cottrell, A new 3D device simulation formulation, in *Proc. Numerical Analysis of Semiconductor Devices and Integrated Circuits*, pp. 291–295, 1989.

54. N. Shigyo and R. Dang, Analysis of an anomalous subthreshold current in a fully recessed oxide MOSFET using a three-dimensional device simulator, *IEEE Trans. Electron Devices*, vol. 32, no. 2, pp. 441–445, February 1985.

55. N. Shigyo, S. Fukuda, T. Wada, K. Hieda, T. Hamamoto, H. Watanabe, K. Sunouchi, and H. Tango, Three-dimensional analysis of subthreshold swing and transconductance for fully recessed oxide (trench) isolated 1/4-μm-width MOSFETs, *IEEE Trans. Electron Devices*, vol. ED-35, no. 7, pp. 945–951, July 1988.

56. M. Thurner and S. Selberherr, The extension of MINIMOS to a three-dimensional simulation program, in *Proc. Numerical Analysis of Semiconductor Devices and Integrated Circuits*, pp. 327–332, 1987.

57. M. Thurner, P. Lindorfer, and S. Selberherr, Numerical treatment of nonrectangular field-oxide for 3-D MOSFET simulation, *IEEE Trans. Computer Aided Design*, vol. 9, no. 11, pp. 1189–1197, 1990.

58. M. Thurner and S. Selberherr, 3D MOSFET device effects due to field oxide, in *Proc. European Solid-State Device Research Conf.*, pp. 245–248, 1988.

59. M. Thurner and S. Selberherr, Three-dimensional effects due to the field oxide in MOS devices analyzed with MINIMOS 5, *IEEE Trans. Computer Aided Design*, vol. CAD-9, no. 8, pp. 856–867, 1990.

60. C. Fischer, P. Habaš, O. Heinreichsberger, H. Kosina, P. Lindorfer, P. Pichler, H. Pötzl, C. Sala, A. Schütz, S. Selberherr, M. Stiftinger, and M. Thurner, *MINIMOS 6 User's Guide*. Institut für Mikroelektronik, Technische Universität Wien, 1994. http://www.iue.tuwien.ac.at/software.

61. S. Odanaka, M. Wakabayashi, H. Umimoto, A. Hiroki, K. Ohe, K. Moriyama, H. Iwasaki, and H. Esaki, SMART: Three-dimensional process/device simulator integrated on a super-computer, in *Proc. Intl. Symp. Circuits and Systems*, pp. 534–537, 1987.

62. M.R. Pinto and R.K. Smith, *PADRE*, AT&T Bell Laboratories Internal Memorandum, 1987.

63. W. Bergner and R. Kircher, SITAR—An efficient 3D-simulator for optimization of nonplanar trench structures, in *Proc. SISPAD 3 Conf.*, pp. 165–175, 1988.

64. T. Linton and P. Blakey, A fast, general three-dimensional device simulator and its application in a submicron EPROM design study, *IEEE Trans. Computer Aided Design*, vol. 8, no. 5, pp. 508–515, 1989.

65. P. Ciampolini, A. Forghieri, A. Pierantoni, A. Gnudi, M. Rudan, and G. Baccarani, Adaptive mesh generation preserving the quality of the initial grid, *IEEE Trans. Computer Aided Design*, vol. 8, no. 5, pp. 490–500, 1989.

66. P. Ciampolini, A. Pierantoni, and G. Baccarani, Efficient 3-D simulation of complex structures, *IEEE Trans. Computer Aided Design*, vol. 10, no. 9, pp. 1141–1149, September 1991.

67. P. Ciampolini, A. Pierantoni, M. Melanotte, C. Cecchetti, C. Lombardi, and G. Baccarani, Realistic device simulation in three dimensions, in *IEDM Tech. Dig.*, pp. 131–134, 1989.

68. G. Heiser, *Design and Implementation of a Three-Dimensional, General Purpose Semiconductor Device Simulator*. Hartung-Gorre Verlag, 1991.

69. G. Heiser, C. Pommerell, J. Weis, and W. Fichtner, Three-dimensional numerical semiconductor device simulation: Algorithms, architectures, results, *IEEE Trans. Computer Aided Design of ICs & Sys.*, vol. 10, no. 10, pp. 1218–1230, October 1991.

70. M. Noell, S. Poon, M. Orlowski, and G. Heiser, Study of 3-D effects in BOX isolation technologies, in *Proc. Simulation of Semiconductor Devices and Processes*, vol. 4, pp. 331–340, 1991.

71. K.-C. Wu, G.R. Chin, and R.W. Dutton, A STRIDE towards practical 3-D device simulation—Numerical and visualization considerations, *IEEE Trans. Computer Aided Design*, vol. 10, no. 9, pp. 1132–1140, September 1991.

72. M. Liang and M. Law, An object-oriented approach to device simulation-FLOODS, *IEEE Trans. Computer Aided Design of IC Design & Sys.*, vol. 13, no. 10, pp. 1235–1240, 1994.

73. G. Baccarani, M. Rudan, M. lorenzini, W. Fichtner, J. Litsios, A. Schenk, P. Van Staa, L. Kaeser, A. Kampmann, A. Marmiroli, C. Sala, and E. Ravanelli, Device simulation for smart integrated systems (DESSIS), in *Proc. ICECS*, pp. 752–755, 1996.

74. T. Simlinger, H. Brech, T. Grave, and S. Selberherr, Simulation of submicron double-heterojunction high electron mobility transistors with MINIMOS-NT, *IEEE Trans. Electron. Devices*, vol. 44, no. 5, pp. 700–707, May 1997.

75. A. Asenov, Random dopant induced threshold voltage lowering and fluctuations in sub-0.1μm MOSFETs: A 3-D atomistic simulation study, *IEEE Trans. Electron Devices*, vol. 45, no. 12, pp. 2505–2513, December 1998.

76. A. Asenov, A. Brown, J. Davies, and S. Saini, Hierarchical approach to atomistic 3-D MOSFET simulation, *IEEE Trans. Computer Aided Design of ICs and Sys.*, vol. 18, no. 11, pp. 1558–1565, November 1999.

77. A. Asenov and S. Saini, Polysilicon gate enhancement of the random dopant induced threshold voltage fluctuations in sub-100 nm MOSFET's with ultrathin gate oxide, *IEEE Trans. Electron Devices*, vol. 47, no. 4, pp. 805–812, April 2000.

78. H.J. DeMan and R. Mertens, SITCAP—A simulator of bipolar transistors for computer aided circuit analysis programs, in *IEEE ISSCC Technical Dig.*, pp. 104–105, February, 14–16, 1973.

79. P.H. Langer and J.J. Goldstein, Impurity redistribution during silicon epitaxial growth and semiconductor device processing, *J. Electrochemical Soc.*, vol. 121, no. 4, pp. 563–571, April 1974.

80. D.A. Antoniadis, S.E. Hansen, R.W. Dutton, and A.G. Gonzalez, SUPREM I—A program for IC process modeling and simulation, Stanford *Technical Report*, No. 5019-1, May 1977.

81. D.A. Antoniadis, S.E. Hansen, and R.W. Dutton, SUPREM II—A program for IC process modeling and simulation, Stanford *Technical Report*, No. 5019-2, June 1978.

82. C.P. Ho, J.D. Plummer, S.E. Hansen, and R.W. Dutton, VLSI process modeling—SUPREM III, *IEEE Trans. Electron. Dev.*, vol. 30, no. 11, pp. 1438–1453, November 1983.

83. D.J. Chin, M. Kump, H.G. Lee, and R.W. Dutton, Process design using coupled 2D process and device simulators, *IEEE Trans. Electron Dev.*, vol. 29, no. 2, pp. 336–340, February 1982.

84. B.R. Penumalli, A comprehensive two-dimensional VLSI process simulation program, BICEPS, *IEEE Trans. Electron Dev.*, vol. 30, no. 9, pp. 986–992, September 1983.

85. M.E. Law, C.S. Rafferty, and R.W. Dutton, *SUPREM-IV Users Manual*, Stanford Electronics Laboratories, Stanford University, California, July 1986.

86. Crosslight Software, Inc., Home page, http://www.crosslight.com/.

87. S. Onga and K. Taniguchi, A three-dimensional process simulator and its application to submicron VLSIs, in *Proc. Sump. VLSI Tech.*, pp. 68–69, 1985.

88. S. Odanaka, H. Umimoto, M. Wakabayashi, and H. Esaki, SMART-P: Rigorous three-dimensional process simulator on a supercomputer, *IEEE Trans. Comp. Aided Design*, vol. 7, no. 6, pp. 675–683, June 1988.

89. M. Law, *FLOODS/FLOOPS Manual*. University of Florida, Department of Electrical Engineering, 2000. http://www.tec.ufl.edu.

90. C.S. Rafferty, PROPHET user's manual, *AT&T Tech. Memo.*, May 1, 1991.

91. L.W. Nagel, *SPICE2—A computer program to simulate semiconductor circuits*, University of California, Berkeley, Memo ERL-M250, Electronic Research Laboratory, May 1975.

92. *BSIM4.6.0 MOSFET Model—User's Manual.* http://www-device.eecs.berkeley.edu/bsim/Files/BSIM4/BSIM460/doc/BSIM460_Manual.pdf.

93. S. Saha, Design considerations for 25 nm MOSFET devices, *Solid-State Electron.*, vol. 45, no. 10, pp. 1851–1857, October 2001.

94. S. Saha, Scaling considerations for high performance 25 nm metal-oxide-semi-conductor field-effect transistors, *J. Vac. Sci. Tech. B*, vol. 19, no. 6, pp. 2240–2246, November 2001.

95. S. K. Saha, Transistors having optimized source-drain structures and methods for making the same, *US Patent 6,344,405*, February 5, 2002.

96. S. Saha, Device characteristics of sub-20-nm silicon nanotransistors, in *Proc. SPIE Conf. on Design and Process Integration for Microelectronic Manufacturing*, vol. 5042, pp. 172–179, July 2003.

97. S.K. Saha, Method for forming channel-region doping profile for semiconductor device, *US Patent 6,323,520*, November 27, 2001.

98. S. Saha, Effects of inversion layer quantization on channel profile engineering for nMOSFETs with 0.1 μm channel lengths, *Solid-State Electron.*, vol. 42, no. 11, pp. 1985–1991, November 1998.

99. K. Fujita et al., Advanced channel engineering achieving aggressive reduction of VT variation for ultra-low-power applications, in *IEDM Tech. Dig.*, pp. 749–752, 2011.

100. S. Selberherr, *Analysis and Simulation of Semiconductor Devices*, Springer-Verlag, Vienna, 1984.

101. S. Saha, MOSFET test structures for two-dimensional device simulation, *Solid-State Electron.*, vol. 38, no. 1, pp. 69–73, January 1995.

102. S. Saha, C.S. Yeh, and B. Gadepally, Impact ionization rate of electrons for accurate simulation of substrate current in submicron devices, *Solid-State Electron.*, vol. 36, no. 10, pp. 1429–1432, October 1993.

103. S. Saha, Extraction of substrate current model parameters from device simulation, *Solid State Electron.*, vol. 37, no. 10, pp. 1786–1788, October 1994.

104. S.K. Saha, Organizational visions of virtual manufacturing: Sociotechnical aspects of adopting technology computer aided design based manufacturing process, in *Proc. IEMC'96*, pp. 570–575, 1996.

105. S. Saha, Technology CAD for integrated circuit fabrication technology development and technology transfer, *Proc. SPIE*, vol. 5042, pp. 63–74, July 2003.

106. S.K. Saha, Improving the efficiency and effectiveness of IC manufacturing technology development, in *Technology and Innovation Management*, D.F. Kocaoglu, T.R. Anderson, D.Z. Milosevic, K. Niwa, and H. Tschirky (eds.) Portland, OR: PICMET 1999, pp. 540–547, 1999.

107. S.K. Saha, Improving the efficiency and effectiveness of technology transfer process in the semiconductor industry, in *Proc. IEMC'98*, pp. 207–212, 1998.

108. S.K. Saha, Modeling process variability in scaled CMOS technology, *IEEE Design & Test of Computers,* vol. 27, no. 2, pp. 8–16, March–April 2010.

109. S.K. Saha, Drain profile engineering for MOSFET devices with channel lengths below 100 nm, in *Proc. SPIE Conference on Microelectronic Device Technology,* vol. 3881, pp. 195–204, September 1999.

110. M. J. van Dort, P. H. Woerlee, and A. J. Walker, A simple model for quantisation effects in heavily-doped silicon MOSFETs at inversion conditions, *Solid-State Electron.,* vol. 37, no. 3, pp. 411–414, March 1994.

111. TMA MEDICI, Version 1998.4, Avant! Corporation, Fremont, California, 1998.

112. S.K. Saha, Non-linear coupling voltage of split-gate flash memory cells with additional top coupling gate, *IET Circuits, Devices & Systems,* vol. 6, no. 3, pp. 204–210, May 2012.

113. S.K. Saha, Design considerations for sub-90 nm split-gate Flash memory cells, *IEEE Trans. Electron. Devices,* vol. 54, no. 11, pp. 3049–3055, November 2007.

114. S.K. Saha, Scaling considerations for sub-90 nm split-gate Flash memory cells, *IET Circuits, Devices & Systems,* vol. 2, no. 1, pp. 144–150, February 2008.

2

Basic Semiconductor and Metal-Oxide-Semiconductor (MOS) Physics

Swapnadip De

CONTENTS

2.1 Introduction

The invention of the transistor in 1947 started the exponential growth of an industry that is now, some decades later, a several hundred billion dollar industry. The first bipolar transistor was announced in December 1947 by William Shockley, John Bardeen, and Walter Brattain at Bell Labs. The first metal-oxide-semiconductor (MOS) transistor and the first integrated circuits were demonstrated in the early 1960s. From that time on the development in the field of microelectronics was impressive. The integration density grew exponentially. Not only has the integration density been steadily growing, but pressure on the industry to deliver in short time-to-market has also been increasing, leaving minimal research and development times for new technology nodes. This has led to intense efforts in the field of numerical simulation of the semiconductor manufacturing process and the resulting device structure, called technology computer aided design (TCAD). TCAD can reduce the number of test cycles with real semiconductor devices and drastically increase the possibilities to vary process parameters as doping concentrations, device geometries, materials, and their composition to a minimum. Here, TCAD gives the opportunity to analyze the effect of process variation within hours instead of weeks for real processing.

The fundamentals of semiconductors are typically found in textbooks discussing quantum mechanics, electro-magnetics, solid-state physics, and statistical thermodynamics. The purpose of this chapter is to review the physical concepts, which are needed to understand the fundamentals of semiconductor devices. We start from the concepts of energy bands, energy band gaps, and the density of states in an energy band. We then discuss the important concept of effective mass. The analysis of most semiconductor devices requires some knowledge of Gauss's law and Poisson's equation, which are explained briefly. We then look at transport in semiconductors through the semi-classical Boltzmann transport equation (BTE). Two carrier transport mechanisms—the drift of carriers in an electric field and the diffusion of carriers due to a carrier density gradient—will be discussed. Recombination mechanisms and the continuity equations are then combined into the diffusion equation. We then present the drift-diffusion model, which combines all the essential elements discussed in this chapter. The rapid developments in semiconductor technology over the past 20 years have caused huge interest in device modeling. The need to understand the detailed operation of very large scale integrated (VLSI) devices and compound semiconductor devices has meant that device modeling now plays a crucial role in modern technology. The main focus is on the phonon transition or Shockley-Read-Hall mechanism. It is of special interest for modeling the carrier generation and recombination at silicon/dielectric interface traps that are caused by negative bias temperature instability. Mobility modeling is also discussed in brief. The operation and types of the MOSFET (metal-oxide-semiconductor field-effect transistor) or MOS transistor are discussed in detail. The steady down-scaling of MOS transistor dimensions over the past two decades obeying Moore's law is discussed along with the short-channel effects in detail.

2.2 Band Formation Theory of Semiconductor

The concept of electronic energy bands provides the basis for the classification of solids as good conductors, insulators, or semiconductors. Quantum physics describes the state of electrons in an atom according to the four-fold scheme of quantum numbers. The quantum number system describes the allowable states the electrons may assume in an atom. Individual electrons may be described by the combination of quantum numbers they possess. Electrons may change their status, given the presence of available spaces for them to fit, and available energy. Because shell level is closely related to the amount of energy that an electron possesses, "leaps" between shell (and even subshell) levels require the transfer of energy. If an electron is to move into a higher-order shell, it requires that an additional energy be given to the electron from an external source. Conversely, an electron "leaping" into a lower shell gives up some of its energy, the expended energy manifesting as heat and sound released upon impact.

Leaps between different shells require a substantial exchange of energy, while leaps between subshells or between orbitals require lesser exchanges. When atoms combine to form substances, the outermost shells, subshells, and orbitals merge, providing a greater number of available energy levels for electrons to assume. When large numbers of atoms exist in close proximity to each other, these available energy levels form a nearly continuous band wherein electrons may transit.

The width of these bands and their proximity to existing electrons determine how mobile those electrons will be when exposed to an electric field. In metallic substances, as in Figure 2.1, empty bands overlap with bands containing electrons, meaning that electrons may move to what would normally be a higher-level state with little or no additional energy imparted. Thus, the outer electrons are said to be "free," and ready to move at the beckoning of an electric field [1].

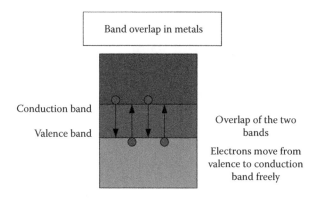

FIGURE 2.1
Energy band structure of metal.

Band overlap will not occur in all substances, no matter how many atoms are in close proximity to each other. In some substances, as in Figure 2.2, a substantial gap remains between the highest band containing electrons (the so-called valence band) and the next band, which is empty (the so-called conduction band). As a result, valence electrons are "bound" to their constituent atoms and cannot become mobile within the substance without a significant amount of imparted energy [2–4]. These substances are electrical insulators.

Materials that fall within the category of semiconductors have a narrow gap between the valence and the conduction bands, as shown in Figure 2.3. Thus the amount of energy required to motivate a valence electron into the conduction band where it becomes mobile is quite low.

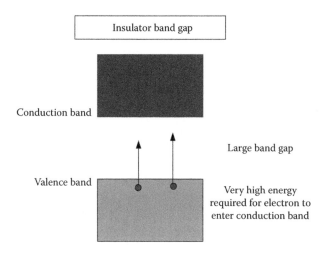

FIGURE 2.2
Energy band structure in insulator.

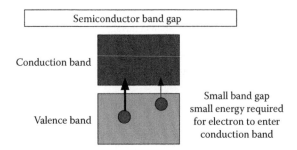

FIGURE 2.3
Energy band structure in semiconductor.

At low temperatures, there is little thermal energy available to push valence electrons across this gap, and the semiconductor material acts as an insulator. At higher temperatures, the ambient thermal energy becomes sufficient to force electrons across the gap, and the material will conduct electricity.

The free electron model of a metal is used to explain the photoelectric effect. This model assumes that electrons are free to move within the metal but are confined to the metal by potential barriers, as illustrated by Figure 2.4. The minimum energy needed to extract an electron from the metal equals $q\Phi_M$ = work function. This model is used for analyzing metals but does not work well for semiconductors because the effect of the periodic potential due to the atoms in the crystal has been ignored. In semiconductors, the energy levels are grouped into bands. The behavior of electrons at the top and bottom of such a band is similar to that of a free electron [5]. However, the electrons are affected by the presence of the periodic potential. The combined effect of the periodic potential is included by adjusting the mass of the electron to a different value. This mass will be referred to as the effective mass.

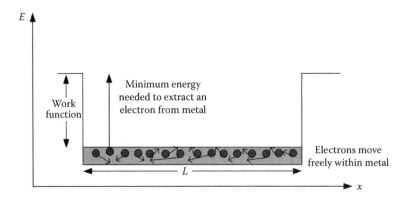

FIGURE 2.4
The free electron model of a metal.

2.2.1 Band Formation in Silicon

A schematic representation of an isolated Silicon (Si) atom is shown in Figure 2.5(a). In Si crystal each isolated Si atom has an electronic structure $1s^2 2s^2 2p^6 3s^2 3p^2$. Ten out of fourteen Silicon atom electrons occupy energy levels close to the nucleus. The four remaining valence electrons are weakly

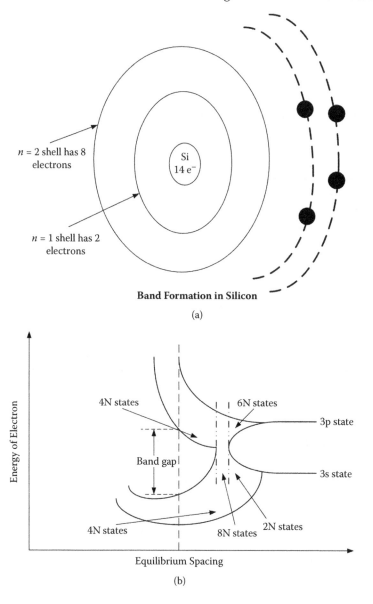

FIGURE 2.5
Energy band formation in silicon.

bound, and these electrons are involved in chemical reactions [6]. They determine the chemical and electrical properties of Silicon (i.e., 14 electrons can be distributed in 18 states—two $1s$, two $2s$, six $2p$, two $3s$, and six $3p$ states). If we consider N atoms, there will be $2N$, $2N$, $6N$, $2N$, and $6N$ states of type $1s$, $2s$, $2p$, $3s$, and $3p$, respectively. Figure 2.5(b) shows the band splitting of the silicon. Because the first two shells are completely filled and tightly bound to the nucleus, we have to consider $n = 3$ level for valence electron. The $3s$ states corresponding to $n = 3$ and $l = 0$ contain two quantum states per atom. This state will contain two electrons at $T = 0$ K. The $3p$ states corresponding to $n = 3$ and $l = 1$ contain six quantum states per atom. This state will contain two remaining electrons of the individual Si atom (i.e., the band of '$3s$-$3p$' levels contain $8N$ available states). When the inter-atomic spacing decreases, these energy levels split into bands, beginning with the outer ($n = 3$) shell. The $3s$ and $3p$ bands merge into a single band composed of a mixture of energy levels. At the equilibrium inter-atomic distance, the bands again split, having four quantum states in the lower band and four quantum states in the upper band. At $T = 0$ K electrons will reside at the lowest energy state, thus the lower band (valence band will be full) and the energy states in the upper band (conduction band) will be empty. The band-gap energy E_g is the width of the energy between the top of the valance band and the bottom of the conduction band.

In other words, for inter-atomic spacing, this band splits into two bands separated by an energy gap E_g. The upper band (the conduction band) contains $4N$ states, as does the lower band (the valence band). The energy gap contains no allowed energy levels for electrons to occupy, thus it is called *forbidden band*, E_g. The lower bands ($1s$, $2s$, $2p$) are fully occupied. But $4N$ electrons originally in $n = 3$ shells ($2N$ in $3s$ and $2N$ in $3p$ states) must occupy states in the valence band or the conduction band in the crystal. At 0 K the electrons will occupy the lowest energy states available to them. In the case of Si crystal, there are exactly $4N$ states in the valence band available to the $4N$ electrons. Thus at 0 K every state in the valence band is totally filled, while the conduction band is empty.

2.2.2 Band Structure in Compound Semiconductor

A particularly interesting and useful feature of the III–IV compounds is the ability to vary the mixture of elements. For example, in the ternary compound AlGaAs, it is possible to vary the composition of the ternary alloy by choosing the fraction of Al or Ga atoms. It is common to represent the composition by assigning subscripts to the various elements. For example $Al_xGa_{1-x}As$ refers to a ternary alloy in which a fraction of X of Al atoms and $(1 - X)$ of Ga atoms are present. X can vary from 0 to 1, thus varying the optical and electronic properties. The composition $Al_{0.3}Ga_{0.7}As$ has 30% of aluminum and 70% Ga atoms [7]. GaAs is a direct band-gap semiconductor

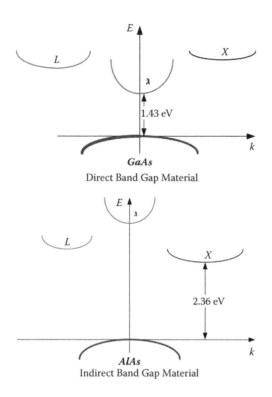

FIGURE 2.6
Energy band diagram in GaAs and AlAs.

material with a band gap of 1.43 eV. The direct ($k = 0$) conduction band minimum is denoted as λ. The lowest-lying indirect minimum is denoted as L and the other as X. In GaAs, as shown in Figure 2.6 (top), there are two higher-lying indirect minima, but these are sufficiently far above λ and few electrons reside there.

In AlAs shown in Figure 2.6 (bottom), the direct transition minimum is much higher than the indirect minimum and so this material is an example of indirect band-gap semiconductor with a band gap of 2.16 eV at room temperature. In the ternary compounds, all of these conduction band minima move up relative to the valence band, as the composition X varies from 0(GaAs) to 1(AlAs). However, the indirect minimum moves up less than the others when the compositions are above 38 percent Al. Also here this indirect minimum is actually the lowest-lying conduction band. AlGaAs is a direct band-gap semiconductor when $X = 0$ to $X = 0.38$ and is an indirect semiconductor for higher Al mole fractions. $GaAs_{1-x}P_x$ is generally similar to AlGaAs, which is also a direct band-gap semiconductor up to $X = 0.45$. This material is often used in light-emitting diode (LED) fabrication. Light emission is most efficient for direct materials, in which the electron can drop to the valence bands without changing k and therefore the momentum. By changing X, we can change the color of the light by changing the band gap [8].

2.3 Concept of Effective Mass

The effective mass of a semiconductor is obtained by fitting the actual E-k diagram around the conduction band minimum or the valence band maximum by a paraboloid shown in Figure 2.7. While this concept is simple enough, the issue turns out to be substantially more complex due to the occasional anisotropy of the minima and the maxima. A single electron is assumed to travel through a perfectly periodic lattice. The parameter m^* called the effective mass takes into account all the internal forces in the lattice. Electrons are considered as classical particles whose motions are governed by classical mechanics when all the internal forces are taken care of through the concept of the effective mass.

Now classical laws like $F = m^*a$ are applied, where a is now directly related to external force. We know the electron momentum is $p = mv = \hbar k$, where the symbols have their usual significances. Thus,

$$E = \frac{1}{2}mv^2 = \frac{1}{2}\frac{p^2}{m} = \frac{1}{2}\frac{\hbar^2}{m}k^2 \tag{2.1}$$

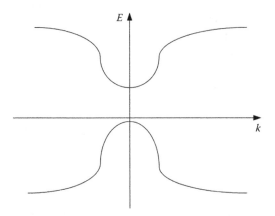

FIGURE 2.7
Example of an E-K diagram.

The electron energy is parabolic with wave vector k.

$$\frac{dE}{dk} = \frac{\hbar^2 k}{m} = \frac{\hbar p}{m}$$

$$\frac{1}{\hbar}\frac{dE}{dk} = \frac{p}{m} = v \tag{2.2}$$

where v is the velocity of particle. $\frac{dE}{dk}$ is related to the velocity of the particle. The electron mass is inversely related to $\frac{d^2E}{dk^2}$. So

$$\frac{d^2E}{dk^2} = \frac{\hbar^2}{m}$$

$$\frac{1}{\hbar^2}\frac{d^2E}{dk^2} = \frac{1}{m} \tag{2.3}$$

For a free electron the mass is a constant, so $\frac{d^2E}{dk^2}$ is a constant. Also

$$k = \frac{\sqrt{2mE}}{\hbar}$$

For a free electron total energy E is equal to the kinetic energy. So,

$$\sqrt{\frac{2mE}{\hbar^2}} = \sqrt{\frac{2m\left(\frac{1}{2}mv^2\right)}{\hbar^2}} = \frac{p}{\hbar} = k \tag{2.4}$$

$$E = \frac{p^2}{2m} = \frac{k^2\hbar^2}{2m}$$

The E-k relationship is parabolic. The effective mass is a parameter that relates the quantum mechanical results to the classical force equations.

2.4 Basic Semiconductor Equations

The analyses of most semiconductor devices include the calculation of the electrostatic potential within the device as a function of the charge distribution [9]. Electromagnetic and electrostatic theories are used to obtain the potential. Short descriptions of the necessary tools, namely Gauss's law, Poisson's equation, and the Boltzmann transport equation, are given below.

2.4.1 Gauss's Law

Gauss's law gives a relationship between the charge density, $\rho(x)$, and the electric field, $E(x)$. Considering no time-dependent magnetic fields, the one-dimensional equation is given by

$$\frac{dE(x)}{dx} = \frac{\rho(x)}{\varepsilon} \tag{2.5}$$

Integrating the above equation, the electric field for 1D charge distribution is given as

$$E(x_2) - E(x_1) = \int_{x_1}^{x_2} \frac{\rho(x)}{\varepsilon} \, dx \tag{2.6}$$

In three dimensions, application of Gauss's law gives the divergence of the electric field:

$$\nabla E(x, y, z) = \frac{\rho(x, y, z)}{\varepsilon} \tag{2.7}$$

2.4.2 Poisson's Equation

The electric field is defined as the negative gradient of the electrostatic potential, $\Psi(x)$, or in one dimension, as the negative derivative of the electrostatic potential:

$$\frac{d\Psi(x)}{dx} = -E(x) \tag{2.8}$$

The electric field starts from a higher-potential region and points toward a lower-potential region.

Integration of the electric field gives the expression of potential as

$$\Psi(x_2) - \Psi(x_1) = -\int_{x_1}^{x_2} E(x) \, dx \tag{2.9}$$

Putting the expression of the electric field from Equation (2.8) into Equation (2.5), the relation between the charge density and the potential is obtained as

$$\frac{d^2\Psi(x)}{dx^2} = -\frac{\rho(x)}{\varepsilon} \tag{2.10}$$

This equation is referred to as Poisson's equation.

In three dimensions, the potential gradient is given by

$$\nabla\Psi(x,y,z) = -E(x,y,z) \tag{2.11}$$

Combining Equations (2.11) and (2.7), the general form of Poisson's equation is given by

$$\nabla^2\Psi(x,y,z) = -\frac{\rho(x,y,z)}{\varepsilon} \tag{2.12}$$

2.4.3 Boltzmann Transport Equation

The Boltzmann transport equation (BTE) is a semi-classical approach to carrier transport. It describes the evolution of the trajectory of a particle by using a combination of Newtonian mechanics and quantum probabilistic scattering rates. The former account for the classical motion of the particle, and the latter for dissipative processes from one energy state to another. Unlike quantum transport, the energy eigen-states are not determined during the solution but are pre-computed by an independent method. The BTE can be derived from the Liouville-von Neumann transport equation under simplifying assumptions and ignoring all phase coherence [10,11]. A good introduction to the BTE, its physical parameters, and its application for device simulation can be found in [10].

For deriving the BTE a phase space is considered about the points (x, y, z, p_x, p_y, p_z) as in [12], where (p_x, p_y, p_z) are the angular components of p-type orbitals. The particles entering the considered phase space in time dt are equal to the number of particles in $(x - v_x dt, y - v_y dt, z - v_z dt, p_x - F_x dt, p_y - F_y dt, p_z - F_z dt)$ at some earlier dt time, where F is the external force on a particle and (F_x, F_y, F_z) are the respective directional components of the force. Let the number of particles per unit volume in the phase space be represented by $f(x, y, z, p_x, p_y, p_z)$. The change in this distribution function in time dt as a result of the external force and the movement of particles is given from [12] as $df = f(x - v_x dt, y - v_y dt, z - v_z dt, p_x - F_x dt, p_y - F_y dt, p_z - F_z dt) - f(x, y, z, p_x, p_y, p_z)$.

The time derivative of this function from Taylor series is given by $\frac{df}{dt} = -(F\nabla_p f + v\nabla_r f)$. However, if the scattering or collision of particles in the phase space due to particles from outside is considered, the time derivative takes the form

$$\frac{df}{dt} = -(F\nabla_p f + v\nabla_r f) + s(r,p,t) + \left.\frac{\partial f}{\partial t}\right|_{collision} \tag{2.13}$$

where $\left.\frac{\partial f}{\partial t}\right|_{collision}$ is the rate of change of distribution function due to collisions from other particles outside the phase space; $s(r,p,t)$ is the term accounting for the generation-recombination processes; $s(r,p,t)$ stands for the probability of having generation-recombination at a position r at time t having particle momentum p. If particle momentum is replaced by crystal momentum the time derivative of f is given by

$$\frac{\partial f(r,k,t)}{\partial t} + \frac{1}{\hbar} F \nabla_k f(r,k,t) + \frac{1}{\hbar} \nabla_k E(k) \nabla_r f(r,k,t) = s(r,k,t) + \frac{\partial f}{\partial t}\bigg|_{collision} \qquad (2.14)$$

This equation is the Boltzmann transport equation or continuity equation in 6D-phase space.

Because the BTE does not include phase information, it is simpler to solve than quantum transport. There are several methods of solution of BTE. Among the earlier approaches was the Legendre polynomial expansion [13]. Such methods did not achieve much success because the drastic approximations used to simplify the problem and to obtain analytical solutions were valid only in the simplest cases and not for any practical devices. Other earlier methods were based on an iterative integration technique that worked well only for low-field transport [14]. In the 1960s, a method based on the Monte Carlo technique was suggested as a means to solve the BTE. It has achieved the most success among all other methods so far, due to its ease of programming, ease of including a variety of physical effects in the same framework, simple numerical algorithms, and low memory requirements (review in [15]). The Monte Carlo technique can simulate transport in complicated device geometries with complicated band structures [16,17]. However, the Monte Carlo technique suffers from several fundamental disadvantages (i.e., statistical noise in low-bias near-equilibrium conditions). These conditions involve events that occur at exponentially decreasing probabilities and cannot be detected by a stochastic method that has a well-known convergence only for nearly uniform distributions. Some methods have been suggested to "enhance" the exponential tails of distribution functions so that they can be detected and that has provided some respite [18]. However, a stochastic method inherently has lesser accuracy than a direct numerical method with controlled discretization error. Among other significant methods to solve the BTE, the Cellular Automata methods, the Scattering matrix method, and the Spherical Harmonic method must be mentioned.

2.5 Carrier Transport

Current in a semiconductor is defined as the rate of flow of charge carriers. The flow of charge carriers called *carrier drift* is due to an externally applied electric field. The carriers move from a higher carrier density region to a lower density region. This movement of carriers called *diffusion* is due to the thermal energy. The total current in a semiconductor is equal to the sum of the drift and the diffusion currents.

When an electric field is applied to a semiconductor, the electrostatic force causes the carriers to first accelerate due to the electrostatic field. Then due to collisions with impurities a constant average velocity, v, is reached. Mobility is defined as the average drift velocity per applied electric field. Saturation velocity is reached at high electric fields. Carriers along the semiconductor surface are subjected to surface scattering as a result of which the mobility degrades. Due to variation in doping density, a density gradient is created in the semiconductor due to which diffusion of carriers takes place.

Both drift and diffusion mechanisms are related because the same particles and scattering mechanisms are involved. This leads to the Einstein relation which is a relationship between the mobility and the diffusion constant.

2.5.1 Carrier Drift

The drifting of a carrier in a semiconductor on application of an externally applied electric field, E, is shown in Figure 2.8. Due to the applied bias the carriers move with an average velocity, v [1].

The current flowing through the semiconductor can be expressed as the total charge divided by the time taken to travel from one electrode to the other—that is,

$$I = Q/t_r = \frac{Q}{L/v} \tag{2.15}$$

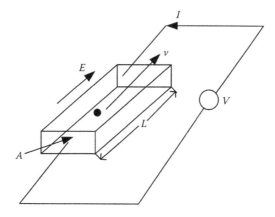

FIGURE 2.8
Drift of a carrier due to an applied electric field.

where t_r is the transit time of carrier, moving with velocity, v, covering the distance L. The current density, J, can be expressed in terms of the charge density ρ as

$$J = I/A = \frac{Q}{AL}v = v\rho \tag{2.16}$$

Considering negatively charged electrons, the current density is given by

$$J = -qnv \tag{2.17}$$

Considering positively charged holes, it is given by

$$J = qpv \tag{2.18}$$

where n and p are the semiconductor electron and hole density.

Due to scattering, the carriers move around the semiconductor randomly with a constantly changing path instead of a straight-line path along the electric field. This occurs when no electric field is applied externally and is due to the thermal carriers. Electrons in a non-degenerate electron gas have a thermal energy of $kT/2$ per particle per degree of freedom [1,19]. The typical thermal velocity is around 10^7 cm/s at room temperature, which is greater than the drift velocity in semiconductors. The movement of carriers in the semiconductor in the presence and absence of an electric field is shown in Figure 2.9.

When no external field is applied, the carriers move randomly with rapidly changing directions. On application of an external electric field, the holes move in the direction of the applied field, while the electrons move in the opposite direction.

The force on a carrier can be obtained from Newton's law.

$$F = ma = m\frac{d\langle v \rangle}{dt} \tag{2.19}$$

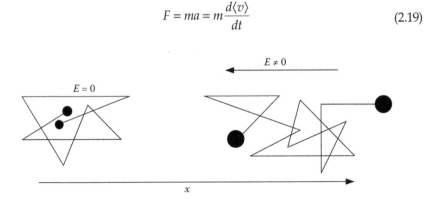

FIGURE 2.9
Random motion of carriers in a semiconductor with and without an applied electric field.

The force equals the difference between the electrostatic force and the scattering force. The scattering force equals the ratio of the momentum of the carriers to the average time between collisions, τ—that is,

$$F = qE - m\frac{\langle v \rangle}{\tau} \qquad (2.20)$$

where q is the charge of a carrier particle.

Equating Equations (2.19) and (2.20):

$$qE = m\frac{d\langle v \rangle}{dt} + m\frac{\langle v \rangle}{\tau} \qquad (2.21)$$

The average particle velocity can be obtained from Equation (2.21). In steady state, the carrier particles after acceleration reach a constant velocity. In such a condition, the velocity of the particle is proportional to the externally applied electric field. The mobility is defined as the average velocity per applied field.

$$\mu = \frac{\langle v \rangle}{E} = \frac{q\tau}{m} \qquad (2.22)$$

Mobility of a semiconductor particle is small when the mass is large and the time between collisions is small. In terms of mobility, the drift current density for electrons may be expressed as

$$J_n = qn\mu_n E \qquad (2.23)$$

Similarly, the drift current density for holes may be expressed as

$$J_p = qp\mu_p E \qquad (2.24)$$

considering the mass, m, of the semiconductor particle. But the effective mass, m^*, rather than the free particle mass, m, must be considered for taking into account the effect of the periodic potential of the atoms:

$$\mu = \frac{q\tau}{m^*} \qquad (2.25)$$

2.5.2 Diffusion Current

Semiconductor devices fall into two broad categories: majority carrier devices and minority carrier devices. In the majority carrier devices, current flow is dominated by electric field driven current. In minority carrier devices, the current flow is dominated by the diffusion effects. Whenever there is a gradient in the concentration of mobile particles, the particles diffuse from the regions of high concentration to the regions of low concentration. In addition to drift, this is the alternate mechanism that can lead to current flow. Suppose a drop of ink falls into a glass of water. Introducing a high local concentration of ink molecules, the drop begins to "diffuse"—that is, the ink molecules tend to flow from a region of high concentration to regions of low concentration. This mechanism is called *diffusion*. A similar phenomenon occurs if charge carriers are dropped into a semiconductor so as to create a non-uniform density. In the absence of an electric field, the carriers move toward regions of low concentration, thereby carrying an electric current so long as the non-uniformity is sustained. Diffusion is therefore distinctly different from drift.

The derivation is based on the idea that carriers at non-zero temperature (Kelvin) have an additional thermal energy equal to $kT/2$ per degree of freedom. Thermal energy drives the diffusion process. At $T = 0$ K there is no diffusion. Because thermal energy is random, the average value needs to be considered for deriving the diffusion current for a one-dimensional semiconductor [1].

Let the average values of the thermal velocity be v_{th}, the collision time τ, and the mean free path l. The thermal velocity is the average velocity of the semiconductor carriers in the positive or the negative direction. The collision time is the time taken by the carriers to move with the same velocity before a collision occurs with another carrier. The mean free path is the average length a carrier moves between collisions. From this basic concept the thermal velocity is given by

$$v_{th} = l/\tau \tag{2.26}$$

In order to find an expression for diffusion current density, a variable carrier density $n(x)$ is considered in Figure 2.10. The carrier densities of two

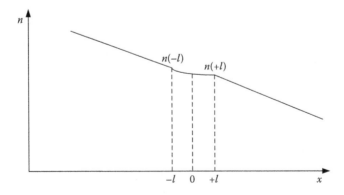

FIGURE 2.10
Carrier density profile used to derive the diffusion current expression.

points $x = -l$ and $x = l$ are considered as in [1], which are one mean free path away from $x = 0$. The flux due to the semiconductor carriers at $x = 0$ due to carriers moving from $x = -l$ is given by

$$\eta_{left-right} = \frac{1}{2}v_{th}n(-l) \tag{2.27}$$

The "1/2" term is because only half of the carriers move to the left while the other half moves to the right. The flux due to the semiconductor carriers at $x = 0$ due to carriers moving from $x = +l$ is given by

$$\eta_{right-left} = \frac{1}{2}v_{th}n(+l) \tag{2.28}$$

The total flux of carriers at $x = 0$ may be obtained by subtracting the flux due to carriers moving from right to left from the flux of carriers moving from left to right:

$$\eta = \eta_{left-right} - \eta_{right-left} = \frac{1}{2}v_{th}\{n(-l) - n(+l)\} \tag{2.29}$$

Considering small mean free path, the carrier density derivative may be obtained as

$$\eta = -lv_{th}\frac{n(+l) - n(-l)}{2l} = -v_{th}l\frac{dn}{dx} \tag{2.30}$$

The diffusion current density for electrons may be obtained by multiplying the flux with the charge of an electron:

$$J_n = -q\eta = qv_{th}l\frac{dn}{dx} \tag{2.31}$$

Let the diffusion constant, D_n, be equal to the product of thermal velocity, v_{th}, and the mean free path, l:

$$J_n = qD_n\frac{dn}{dx} \tag{2.32}$$

For holes the diffusion current density is given as

$$J_p = -qD_p\frac{dp}{dx} \tag{2.33}$$

2.5.3 Total Drift-Diffusion Current

Based on the concepts derived in the previous sections, we can now establish the drift-diffusion equations [3,20]. The total hole current density in a semiconductor is composed of the sum of the drift and the diffusion components of current. Similarly, the total electron current density in a semiconductor is composed of the sum of the drift and the diffusion components of current. For electrons the total current density is given as

$$J_n = qD_n \frac{dn}{dx} + qn\mu_n E \tag{2.34}$$

Similarly for holes,

$$J_p = -qD_p \frac{dp}{dx} + qp\mu_p E \tag{2.35}$$

The total current is the sum of the electron and hole current densities multiplied by the area, A, perpendicular to the current direction:

$$I_{tot} = A(J_p + J_n) \tag{2.36}$$

2.5.4 Einstein Relation

In a non-uniformly doped semiconductor the doping concentration decreases with increase in x. As a result, diffusion of majority carriers takes place from a high concentration to a low concentration region along the $+x$ direction. The flow of electrons leaves behind positively charged donor ions. The separation of positive and negative charges creates an electric field in the direction opposite to the diffusion process. At equilibrium the induced electric field prevents further diffusion.

When taking the electric field along the x direction, the energy bands are as shown in Figure 2.11. The potential energy increases in the direction of the electric field. The electrostatic potential, which varies in the opposite direction as it is defined in terms of positive charges, is given by

$$V(x) = -\frac{E(x)}{q} \tag{2.37}$$

From the definition of electric field

$$E(x) = -\frac{dV(x)}{dx}$$

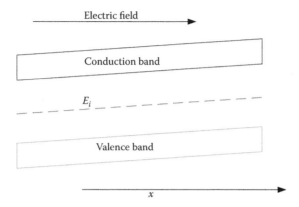

FIGURE 2.11
Energy band diagram of a semiconductor in the presence of an electric field $E(x)$.

Considering E_i as a convenient reference, the electron potential energy may be related to $E(x)$ as

$$E(x) = -\frac{dV(x)}{dx} = -\frac{d}{dx}\left[\frac{E_i}{(-q)}\right] = \frac{1}{q}\frac{dE_i}{dx} \qquad (2.38)$$

Since the band diagram indicates electron energies, we know that the slope of this band must be such that electrons drift downhill in the field. Therefore E points uphill in the band diagram. At equilibrium no current flows. Putting

$$J_p = qp\mu_p E - qD_p\frac{dp}{dx} = 0 \qquad (2.39)$$

we get

$$E(x) = \frac{D_p}{\mu_p}\frac{1}{p(x)}\frac{dp(x)}{dx} \qquad (2.40)$$

Also,

$$p_0 = n_i e^{(E_i - E_F)/kT}$$

so

$$E(x) = \frac{D_p}{\mu_p}\frac{1}{KT}\left(\frac{dE_i}{dx} - \frac{dE_F}{dx}\right) \qquad (2.41)$$

The equilibrium Fermi level does not vary with x. So

$$\frac{dE_F}{dx} = 0$$

and

$$\frac{dE_i}{dx} = qE(x)$$

Thus the equation takes the form $\frac{D}{\mu} = \frac{kT}{q}$. The relationship between drift parameter (μ) and diffusion parameter (D) is given by the Einstein relationship.

2.6 Carrier Recombination and Generation

Recombination is a process by which the electrons occupy the empty states associated with holes. The carriers as a result disappear. The energy released is the difference between the initial state energy and the final state energy of the electrons. The energy is emitted in the form of a photon for radiative recombination. For non-radiative recombination, it is transmitted to one or more phonons, and for Auger recombination it is transferred as kinetic energy to another electron [1,2,21]. The different processes are shown in Figure 2.12.

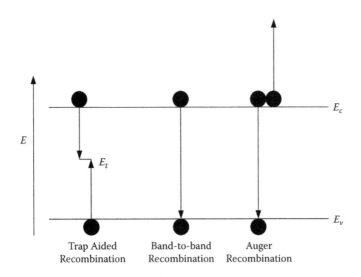

FIGURE 2.12
Carrier recombination mechanisms in semiconductors.

In band-to-band recombination an electron moves from the conduction band into the valence band state associated with the hole. In case of direct band-gap semiconductors, it is a case of radiative transition. In trap-assisted recombination, the electrons fall into the band gaps caused by some defects or some foreign atoms. The electrons occupying those band gaps move into an empty valence band state, thus completing the recombination process. So this is a two-step transition of an electron from the conduction to the valence band called Shockley-Read-Hall (SRH) recombination. Auger recombination is a process of recombination of an electron and a hole with the resultant energy given to another electron or hole. Auger recombination is different from band-to-band recombination due to this third electron or hole. All the recombination processes when reversed cause carrier generation instead of recombination. A single expression can be used to describe both generation and recombination processes. Generation of carriers by light absorption is a process that does not have recombination associated with it. This process is referred to as *ionization*. Impact ionization also belongs to this category. The different generation mechanisms are shown in Figure 2.13.

If the photon energy is large, an electron from the valence band may move into the conduction band generating an electron-hole pair as a result. This photon energy must be larger than the band-gap energy for electron-hole pair generation. Kinetic energy $(E_{ph} - E_g)$ is added to the electron and the hole due to absorption of the photon.

Carrier generation due to high-energy charged particles is similar except that the available energy may be far greater than the band-gap energy causing multiple electron-hole pairs generation. The high-energy particle gradually loses its energy and eventually stops [5,22].

Impact ionization is the counterpart of Auger recombination. It is caused by an electron/hole with energy, much greater/smaller than the conduction/valence band edge. The process is shown in Figure 2.14.

The excess energy is transmitted to generate an electron-hole pair in band transition [23]. Avalanche multiplication is caused in semiconductor diodes

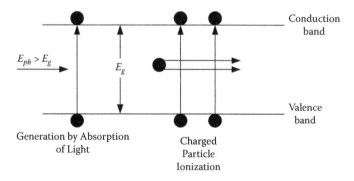

FIGURE 2.13
Carrier generation due to light absorption and ionization due to high-energy particles.

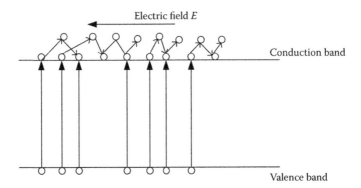

FIGURE 2.14
Impact ionization and avalanche multiplication of electrons and holes in the presence of a large electric field.

under high reverse bias as a result of this generation process. The accelerated carriers gain kinetic energy that is given off to an electron in the valence band, causing an electron-hole pair. The two electrons created in the process can create two more electrons causing an avalanche multiplication effect. Both electrons as well as holes take part in avalanche multiplication.

2.7 Continuity Equation and Solution

The continuity equation describes a basic concept, namely that a change in carrier density over time is due to the difference between the incoming and outgoing flux of carriers plus the generation and minus the recombination. If a volume of space is considered in which charge transport and recombination are taking place, we have the simple equality as in Figure 2.15(a). As a result of consideration of particle current, Net rate of particle flow = Particle flow rate due to current − Particle loss rate due to recombination + Particle gain due to generation.

Thus a continuity equation is based on the conservation of mobile charges [24]:

$$\frac{\partial n}{\partial t} = \frac{1}{q}\frac{\partial J_n}{\partial x} - R_n + G_n$$

$$\frac{\partial p}{\partial t} = -\frac{1}{q}\frac{\partial J_p}{\partial x} - R_p + G_p$$

(2.42)

where G_n and G_p are the electron and hole generation rates, R_n and R_p are the electron and the hole recombination rates, and $\frac{\partial J_n}{\partial x}$ and $\frac{\partial J_p}{\partial x}$ are the net flux of mobile charges in and out of x.

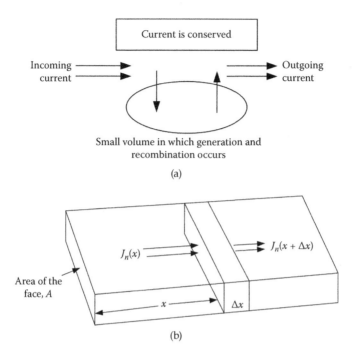

FIGURE 2.15
(a) A conceptual description of the continuity equation. (b) Geometry used to develop the current continuity equation.

A solution to these equations can be obtained by substituting the expression for the electron and hole current. This then yields two partial differential equations as a function of the electron density, the hole density, and the electric field. The electric field is obtained from Gauss's law as in Figure 2.15(b).

Let us now collect the various terms in this continuity equation. If δn is the excess carrier density in the region, the recombination rate R in the volume $A \cdot \Delta x$ may be written approximately as

$$R = \frac{\delta n}{\tau_n} \cdot A \cdot \Delta x \tag{2.43}$$

where τ_n is the electron recombination time per excess particle due to both the radiative and the non-radiative components. The particle flow rate into the same volume due to the current density $J_n(x)$ is given by the difference of particle current coming into the region and the particle current leaving the region:

$$\left[\frac{J_n(x)}{(-e)} - \frac{J_n(x + \Delta x)}{(-e)} \right] A \cong \frac{1}{e} \frac{\partial J_n(x)}{\partial x} \Delta x \cdot A$$

If G is the generation rate per unit volume, the generation rate in the volume $A \cdot \Delta x$ is $GA\Delta x$. $\delta n/\tau n$ is the net recombination rate of electrons and $U = G - R$—that is, Rate of electron buildup (U) = Increase in electron concentration in $\Delta x A$ per unit time (G) – recombination rate (R). The rates of electron buildup in volume $A \cdot \Delta x$ is then

$$A \cdot \Delta x \left[\frac{\partial n(x,t)}{\partial t} \equiv \frac{\partial \delta n}{\partial t} = \frac{1}{e}\frac{\partial J_n(x)}{\partial x} - \frac{\delta n}{\tau_n} \right]$$

As Δx approaches zero, we can write equations in the derivative form for electrons and holes as

$$\frac{\partial \delta n}{\partial t} = \frac{1}{e}\frac{\partial J_n(x)}{\partial x} - \frac{\delta n}{\tau_n}$$

$$\frac{\partial \delta p}{\partial t} = -\frac{1}{e}\frac{\partial J_p(x)}{\partial x} - \frac{\delta p}{\tau_p}$$

(2.44)

Using these expressions, the diffusion currents are

$$J_n(diff) = eD_n \frac{\partial \delta n}{\partial x}$$

$$J_p(diff) = -eD_p \frac{\partial \delta p}{\partial x}$$

(2.45)

The time-dependent continuity equation for electrons and holes, valid separately:

$$\frac{\partial \delta n}{\partial t} = D_n \frac{\partial^2 \delta n}{\partial x^2} - \frac{\delta n}{\tau_n}$$

$$\frac{\partial \delta p}{\partial t} = D_p \frac{\partial^2 \delta p}{\partial x^2} - \frac{\delta p}{\tau_p}$$

(2.46)

These equations are used to study the steady-state charge profile in p-n diodes and bipolar transistors. In steady state,

$$\frac{\partial^2 \delta n}{\partial x^2} = \frac{\delta n}{D_n \tau_n} = \frac{\delta n}{L_n^2}$$

$$\frac{\partial^2 \delta p}{\partial x^2} = \frac{\delta p}{D_p \tau_p} = \frac{\delta p}{L_p^2}$$

(2.47)

Here is the diffusion length for electrons, and L_p is the diffusion length for holes. Considering the case where an excess electron density $\delta n(0)$ is maintained at $x = 0$, at point L in the semiconductor, the excess carrier density is maintained at $\delta(L)$. The general solution of the above second-order differential equation is

$$\delta n(x) = A_1 e^{x/L_n} + A_2 e^{-x/L_n} \qquad (2.48)$$

When $L \gg L_n$ and $\delta_n(L) = 0$, the semiconductor is much longer than L_n, for example in the case of the long p-n diode. A_1 and A_2 can be found from the boundary conditions. For a large value of x, $\delta_n = 0$ at $x = \infty$ and so $A_1 = 0$. Similarly $\delta_n = 0 = \delta_n(0)$ at $x = 0$ giving $A_2 = \delta_n(0)$. The solution of the equation is given by

$$\delta n_p(x) = \delta n_p(0) e^{-x/L_n} \qquad (2.49)$$

It is seen from the above equation that the carrier density decays exponentially in the semiconductor.

However, when $L \ll L_n$, the carrier density is linear from one boundary value to the other because over a short distance exponential can be approximated as linear. When excess carriers are injected into a thick semiconductor sample, both diffusion and recombination take place. L_n represents the distance over which the injected carrier density falls to $1/e$ of its original value. It also represents the average distance an electron diffuses before recombination.

The probability that an electron survives up to a distance x without recombination is given by

$$\frac{\delta n_p(x)}{\delta n_p(0)} = e^{-x/L_n}$$

The steady-state distribution of excess holes causes diffusion and a hole current in the direction of decreasing concentration.

$$J_p(x) = -q D_p \frac{\delta_p}{\delta x} = q \frac{D_p}{L_p} \delta p(x) \qquad (2.50)$$

It will be useful for the current calculation of the p-n junction where the injection of minority carriers across a junction will lead to exponential distribution.

2.8 Mobility and Scattering

The relationship between the velocity of electrons and the applied electric field is complex. When the electric field is low, the relationship is in a simple form. The distance versus time trajectory of an electron is shown in Figure 2.16. Considering d as the distance traveled in time t, the electron motion is described by

$$d = vt$$
$$v = \mu E$$

The velocity of electron v is proportional to the electric field applied, and μ is the mobility. For a large electric field the relation between the velocity and the applied field is not so simple [2–4] and will be discussed later.

When no electric field is applied externally, the occupation of a state with momentum $+\hbar k$ is the same as that with momentum $-\hbar k$. So no current flows in this case as the momentum gets canceled out. Figure 2.17(a) shows the distribution function in momentum space. When an electric field is applied, the electron distribution shifts, as shown schematically in Figure 2.17(a), and there is a net momentum of the electrons. Current flows as a result.

For perfect and rigid crystal, no scattering of the electron takes place. On application of an external electric field E, the electron behaves as a "free" electron in the absence of scattering. However, there are always imperfections due to which electrons scatter. The process is shown in Figure 2.17(b). The average behavior of the electrons represents the transport properties of the electrons.

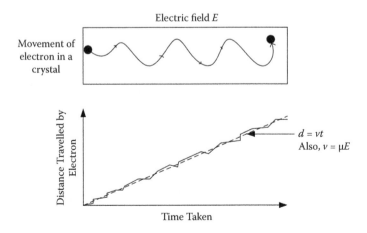

FIGURE 2.16
A typical electron trajectory in a sample, and the distance versus time plot.

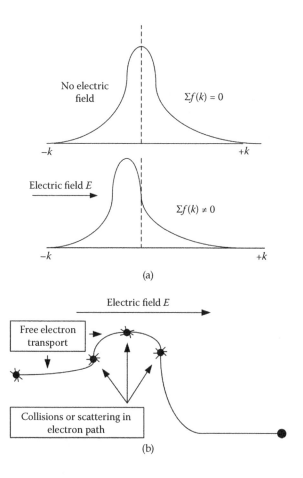

FIGURE 2.17

(a) Schematic of the electron momentum distribution function in the presence and absence of an electric field. (b) Schematic view of an electron moving under an electric field in a semiconductor.

In the absence of scattering, the electron transport is very simple although scattering dominates transport in semiconductor devices. Due to scattering, electrons in a semiconductor do not achieve constant acceleration. However, they can be considered as classical particles moving at a constant average drift velocity.

The electron suffers a scattering as it moves. In between scattering, the electron moves according to the "free" electron equation of motion. The scattering of carriers is mainly due to lattice vibrations caused by thermal energy. Carriers scatter from various imperfections in crystal, alloy disorder, and interface imperfections.

On application of an electric field E, drift current flows in a semiconductor. The force experienced by an electron is given by $F = -qE = m_e \frac{dv}{dt}$. The velocity of electrons will saturate at a constant value in steady state. In fact, $F = m_e \frac{dv}{dt} + m_e \frac{v}{\tau}$, where the latter term represents the impeded force due to scattering which describes collision caused by lattice vibrations and crystal imperfections. At steady state ($dv/dt = 0$), the drift velocity is

$$V_d = \frac{F\tau}{m_e} = \frac{-qE\tau}{m_e} = -\mu E \tag{2.51}$$

where

$$\mu = \frac{q\tau}{m_e} \tag{2.52}$$

is the mobility of the electron (or hole), and τ is the mean free time between collisions.

2.9 Different Distribution Laws

There are three distribution laws that govern the particle distribution among available energy states.

1. *Maxwell-Boltzmann probability function*: Particles are distinguishable with no limit to the number of particles in each energy state.
2. *Bose-Einstein function*: Particles are indistinguishable and no limit to the number of particles in each energy state.
3. *Fermi-Dirac probability function*: Particles are indistinguishable and only one particle is permitted in each quantum state.

The Fermi factor that expresses the probability that a state at a given energy level is occupied by an electron has a value between 0 and 1. A probability of 0 means that the state is unoccupied and a probability of 1 means that the state is occupied. A probability of ½ means that the chance of the state being occupied is 50%.

2.9.1 Fermi-Dirac Distribution

The Fermi-Dirac distribution function, also called the Fermi function, gives the probability of occupancy of energy levels by Fermions. Fermions are half-integer spin particles obeying the Pauli exclusion principle. As the Fermions are added to an energy band, they will fill the available states in an energy band. The states with the lowest energy are filled first, followed by the next higher ones. At absolute zero temperature ($T = 0$ K), the energy levels are all filled up to a maximum energy called the *Fermi level*. No states above the Fermi level are filled. At higher temperature, the transition between completely filled and completely empty states is gradual. The Fermi function provides the probability that energy level at energy, E, in thermal equilibrium with a large system, is occupied by an electron. The system is characterized by its temperature, T, and its Fermi energy, E_F.

The Fermi-Dirac distribution function $f(E)$ gives the probability that an electron has an energy E at a temperature T [25]. This is given by

$$f(E) = \frac{1}{1 + \exp[(E - E_F)/KT]} \tag{2.53}$$

where k is the Boltzmann constant, T is the temperature in Kelvin, and E_F is the Fermi level.

It is seen that at $T = 0$ K if $E = E_F$, $f(E) = 1/2$. Similarly at $T = 0$ K if $E > E_F$ $f(E) = 1/1 + \exp(\infty) = 1/1 + \infty = 0$. Also at $T = 0$ K if $E < E_F$ $f(E) = 1/1 + \exp(-\infty) = 1/1 + 0 = 1$. The condition $E < E_F$, $f(E) = 1$ shows that below the Fermi level all the states are occupied and the probability of finding an electron on a level with energy greater than Fermi energy is zero. It is therefore the energy of the highest filled level. At other temperatures $E = E_F$ and $f(E) = 1/2$. So the Fermi level is that energy at which the probability of a state being occupied is ½. At $T = 0°$K, no covalent bonds are broken and all the lower energy states up to $E = E_F$ are completely occupied. At any other temperature covalent bonds are continuously broken, which generates electron-hole pairs. The recombination of electrons and holes also takes place at the same time. Due to this, the probability of the states being occupied below the Fermi level is slightly less than one, and there is a small probability that a few states are occupied just above the Fermi level. As shown in Figure 2.18, as the energy $E = E_F$, the probability of occupancy of a state by an electron decreases exponentially as E increases in accordance with the above equation. $f(E)$ gives the probability that a state at a given energy level is occupied by an electron. Then probability that a state is not occupied by an electron, or equivalently, the probability that there exists a hole in the valence band will be given by

$$f_p(E) = 1 - f(E) = 1 - \frac{1}{1 + \exp[(E - E_F)/KT]} = \frac{1}{1 + \exp[(E_F - E)/KT]} \tag{2.54}$$

In general the position of E_F is dependent on temperature. The occupation probability is at 0.5 at the Fermi energy. We may write $f(E) = N(E)/g(E)$, where $N(E)$ is the number of carriers per unit volume per unit energy, and $g(E)$ is the number of quantum states per unit volume per unit energy.

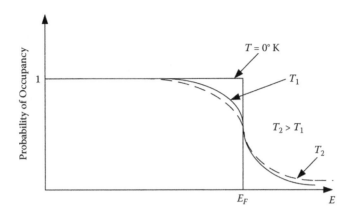

FIGURE 2.18
The Fermi function for electrons.

Therefore, $f(E)$ gives us the probability that a quantum state of energy E may be occupied.

The probability of energy above E_F being occupied increases as temperature increases, and the probability of a state below E_F being empty increases as temperature increases. We may note that the probability of a state being occupied above E_F is the same as the probability of a state below E_F being empty. The curve is symmetrical.

The Bose-Einstein distribution function is applicable to particles with integer spin called *Bosons* and includes photons, phonons, and a large number of atoms. Bosons do not obey the Pauli Exclusion Principle so that any number can occupy a single energy level. The Bose-Einstein distribution function is given by

$$f_{be}(E) = \frac{1}{\exp[(E - E_F)/KT] - 1} \tag{2.55}$$

This distribution function is only defined for $E > E_F$.

The Maxwell-Boltzmann distribution function applies to non-interacting distinguishable particles [9]. This function is also called the *classical distribution function* because it provides the probability of occupancy for non-interacting particles at low densities. The Maxwell-Boltzmann distribution function is given by

$$f_{mb}(E) = \frac{1}{\exp[(E - E_F)/KT]} \tag{2.56}$$

All three functions are almost equal for large energies. The Fermi-Dirac distribution reaches a maximum of 100% for energies, which are a few kT below the Fermi energy, while the Bose-Einstein distribution diverges at the Fermi energy and has no validity for energies below the Fermi energy.

2.10 Semiconductor Device Modeling

2.10.1 Introduction

Rapid developments in semiconductor technology over the past 20 years have caused huge interest in device modeling. The need to understand the detailed operation of very large scale integrated (VLSI) devices and compound semiconductor devices has meant that device modeling now plays a crucial role in modern technology.

As the size of the semiconductor decreases and the complexity of the physical structure increases, the nature of the device characteristics depart from those obtained from many of the classically held modeling concepts. The difficulty encountered in performing measurements on these devices means that greater emphasis must be put on theoretical results. Modeling also allows new device structures to be rigorously investigated prior to fabrication.

Semiconductor device models can be classified into two categories: physical device models and equivalent circuit models. Physical device models account for the physics of device operation, and equivalent circuit models are based on electrical circuit analogies representing electrical behavior. The latter approach is generally limited in its applicability because of the frequency dependence and the non-linear behavior of most devices with respect to signal level. The advantage of this technique is that it is easy to implement. In contrast, physical device models can provide greater insight into the detailed operation of semiconductor devices but usually require lengthy analysis. The analysis requirements for physical models are typically satisfied using numerical techniques implemented on computers. With the advent of fast and powerful computing resources, this makes physical modeling techniques a very attractive proposition for the physicist.

Physical device models are solved using either bulk carrier transport equations, solutions to the Boltzmann transport equation, or quantum transport concepts. Historically, bulk transport solutions have satisfied most device models, while Boltzmann and quantum transport solutions have provided strong insight into the detailed device physics. However, the trend toward very small geometry devices, operating in the hot electron range, means that non-equilibrium transport conditions must be accounted for. The understanding of physical boundary conditions and device-circuit interaction is steadily improving, allowing more intricate models to be developed.

2.10.2 Shockley-Read-Hall (SRH) Generation/Recombination Model

The generation/recombination process by phonon emission is shown in Figure 2.13. A lattice defect energy E_t, as in Figure 2.12, is used in this trap-assisted mechanism. Energy is transmitted/received by phonon during the recombination/generation process. This effect is known as Shockley-Read-Hall (SRH) generation/recombination effect. The four sub-processes are described below as in [1,9]:

1. *Capture of electron*: A conduction band electron is captured by a vacant trap in the semiconductor band gap. $(E_c - E_t)$ is the energy released to the crystal lattice in the form of phonon emission.

2. *Capture of hole*: The electron in the trap now neutralizes a hole in the valence band. Because the direction of motion of the hole is opposite to that of the electron, the hole is captured by the trap in the semiconductor lattice band gap and energy of $(E_t - E_v)$ is generated in the form of a phonon.

3. *Emission of hole*: A valence band electron moves to the trap in the lattice band gap, thereby leaving a hole in the valence band. So it can be said that a hole moves from the trap to the valence band, as a result of which generating $(E_t - E_v)$ energy in the form of phonon.

4. *Emission of electron*: Here an electron moves from the trap level to the conduction band as a result of which requiring $(E_c - E_t)$ amount of energy.

It is seen that both the generation and the recombination processes are two-step processes. Sub-processes 1 and 2 lead to recombination of electron-hole pairs, the excess energy equal to band-gap energy is transferred to the crystal lattice in the form of phonons. Sub-processes 3 and 4 lead to generation of electron-hole pairs where energy needs to be supplied by the lattice.

For finding an expression of the total recombination rate R_{tot}, rates for the four sub-processes need to be determined. In this case traps are assumed to be of the acceptor type, which are neutral when empty and negatively charged when occupied by an electron. Similarly, the derivation for donor traps (neutral when occupied by electron and positive when empty) can be done.

Let the capture rate of an electron be $v_{e,capture}$, proportional to the electron concentration in the conduction band n, the empty traps concentration n_t^0, and a proportionality constant $k_{e,capture}$. With the energy-dependent distribution function for electrons $f_e(E)$ and the density-of-states $g_e(E)$ we get

$$dv_{e,capture} = k_{e,capture}(E)n_t^0 g_e(E) f_e(E)dE \qquad (2.57)$$

The total conduction band electrons is given by

$$n = \int_{E_c}^{\infty} g_e(E) f_e(E) \, dE \tag{2.58}$$

Let the hole capture rate be $v_{h,capture}$, hole concentration in the valence band p, the filled traps concentration n_t^{fill}, with proportionality constant $k_{h,capture}$. We get

$$dv_{h,capture} = k_{h,capture}(E) n_t^{fill} f_h(E) g_h(E) \, dE \tag{2.59}$$

$f_h(E)$ is the distribution function for holes, and $g_h(E)$ is the hole density of state. Total holes in the valence band is

$$p = \int_{E_v}^{\infty} g_h(E) f_h(E) \, dE \tag{2.60}$$

The hole emission rate is $v_{h,emission}$, and the proportionality constant is $k_{h,emission}$:

$$dv_{h,emission} = n_t^0 k_{h,emission}(E)\{1 - f_h(E)\} g_h(E) \, dE \tag{2.61}$$

Electron emission rate $v_{e,emission}$ is proportional to the concentration of filled traps and the proportionality constant $k_{e,emission}$:

$$dv_{e,emission} = n_t^{fill} k_{e,emission}(E)\{1 - f_e(E)\} g_e(E) \, dE \tag{2.62}$$

The total trap concentration n_t is

$$n_t = n_t^{fill} + n_t^0 \tag{2.63}$$

and the fraction of occupied traps $f_{occupied}$ is given by [9,26]

$$f_{occupied} = \frac{n_t^{fill}}{n_t}$$

$$1 - f_{occupied} = \frac{n_t^0}{n_t} \tag{2.64}$$

The net recombination rate for electrons becomes

$$dR_{e.tot} = [k_{e,capture}(E) n_t^0 f_e(E) - k_{e,emission}(E) n_t^{fill}\{1 - f_e(E)\}] g_e(E) \, dE \tag{2.65}$$

In thermal equilibrium $np = n_0 p_0 = n_i^2$, which means that the respective capture and emission rates for electrons and holes are equal:

$$v_{e,emission} = v_{e,capture}, \; v_{h,emission} = v_{h,capture} \tag{2.66}$$

so

$$\frac{k_{e,emission}(E)}{k_{e,capture}(E)} = \frac{1 - f_{occupied}}{f_{occupied}} \frac{f_e(E)}{1 - f_e(E)} \tag{2.67}$$

In thermal equilibrium, $f_{occupied}$ is given by Fermi-Dirac statistics [9]:

$$f_{occupied}(E) = \frac{1}{1 + e^{\left(\frac{E - E_F}{kT}\right)}} \tag{2.68}$$

Hence,

$$\frac{k_{e,emission}(E)}{k_{e,capture}(E)} = e^{\left(\frac{E_t - E_F}{kT}\right)} e^{-\left(\frac{E - E_F}{kT}\right)} = e^{\left(\frac{E_t - E}{kT}\right)} \tag{2.69}$$

The net recombination rate is modified as

$$dR_{e,tot} = [n_t^0 f_e(E) - \frac{k_{e,emission}(E)}{k_{e,capture}(E)} n_t^{fill} \{1 - f_e(E)\}] k_{e,capture}(E) g_e(E) dE \tag{2.70}$$

$$= \left[1 - \frac{k_{e,emission}(E)}{k_{e,capture}(E)} \left(\frac{f_{occupied}}{1 - f_{occupied}} \right) \left\{ \frac{1 - f_e(E)}{f_e(E)} \right\} \right] (1 - f_{occupied}) f_e(E) k_{e,capture}(E) g_e(E) n_t \, dE \tag{2.71}$$

$$= \left\{ 1 - e^{\left(\frac{E_{Ft} - E_F}{kT}\right)} \right\} (1 - f_{occupied}) f_e(E) k_{e,capture}(E) g_e(E) n_t \, dE \tag{2.72}$$

with the trap's quasi Fermi energy E_{Ft}.

The total electron recombination rate is given by

$$R_{e,tot} = [\{1 - e^{\left(\frac{E_{Ft} - E_F}{kT}\right)}\}(1 - f_{occupied}) n_t] \int_{E_c}^{\infty} f_e(E) k_{e,capture}(E) g_e(E) dE \tag{2.73}$$

Let a capture cross section $\alpha_e(E)$ be introduced to rewrite $k_{e,capture}$ as

$$k_{e,capture}(E) = \alpha_e(E)v^e_{thermal} \tag{2.74}$$

with the thermal velocity for electrons

$$v^e_{thermal} = \sqrt{\frac{3kT}{m}} \tag{2.75}$$

For non-degenerate semiconductors near equilibrium the electron recombination is

$$R_{e,tot} = \left[n - N_{c,effective}e^{-\left(\frac{E_c-E_F}{kT}\right)}e^{\left(\frac{E_{Ft}-E_F}{kT}\right)} \right](1 - f_{occupied})K_{e,capture} \tag{2.76}$$

where $N_{c,effective}$ is the effective density of states for electrons and

$$K_{e,capture} = n_t v^e_{thermal} << \alpha_e(E) >>$$

Considering

$$n_1 = N_{c,effective}e^{-\left(\frac{E_c-E_t}{kT}\right)} \tag{2.77}$$

and

$$p_1 = N_{v,effective}e^{\left(\frac{E_v-E_t}{kT}\right)}$$

we get

$$R_{e,tot} = \{n(1 - f_{occupied}) - n_1 f_{occupied}\}K_{e,capture}$$

In the stationary case the recombination rates for electrons and holes are equal,

$$R_{e,tot} = R_{h,tot} = R_{tot} \tag{2.78}$$

Therefore, we can calculate $f_{occupied}$ as

$$f_{occupied} = \frac{k_{e,capture}n + k_{h,capture}p_1}{k_{e,capture}(n+n_1) + k_{h,capture}(p+p_1)} \tag{2.79}$$

Using this expression the total recombination rate is obtained as [3,9]

$$R_{tot} = k_{e,capture}k_{h,capture}n_t \frac{np - n_1 p_1}{k_{e,capture}(n+n_1) + k_{h,capture}(p+p_1)} \tag{2.80}$$

It is common to introduce carrier lifetimes for electrons and holes $T_{electron}$ and T_{hole} as

$$T_{electron} = \frac{1}{k_{e,capture} n_t}$$

and

$$T_{hole} = \frac{1}{k_{h,capture} n_t}$$

By using the capture cross sections for electrons and holes, $\alpha_e(E)$ and $\alpha_h(E)$, and the thermal velocities $v_{thermal}^e$ and $v_{thermal}^h$,

$$T_{electron} = \frac{1}{n_t v_{thermal}^e \alpha_e(E)}$$

$$T_{hole} = \frac{1}{n_t v_{thermal}^h \alpha_h(E)}$$

We come to the final formulation of the Shockley-Read-Hall model for the carrier generation-recombination model:

$$R_{tot} = n_t \frac{np - n_i^2}{T_{hole}(n + n_1) + T_{electron}(p + p_1)} \tag{2.81}$$

2.10.3 Simple Recombination-Generation Model

In this model the recombination-generation rate is proportional to the excess carrier density. If the carrier density equals the thermal equilibrium value no net recombination takes place. The expression for the recombination of electrons in a p-type semiconductor is given by

$$U_n = R_n - G_n = \frac{n_p - n_{p0}}{\tau} \tag{2.82}$$

The expression for the recombination of holes in an n-type semiconductor is given by

$$U_p = R_p - G_p = \frac{p_n - p_{n0}}{\tau} \tag{2.83}$$

where τ is the average time of recombination of excess minority carrier.

The above expressions are valid only for minority carriers in a "quasi-neutral" semiconductor. In steady state the recombination rates of the majority and minority carriers are equal because recombination involves an equal number of holes and electrons. Majority carrier recombination depends on the excess minority carriers.

2.10.4 Impact Ionization Model

Impact ionization is a pure generation process similar to the Auger generation process. When a carrier moves into the conduction or valence band, the energy released is used to move an electron from valence to conduction band causing an electron-hole pair.

The two sub-processes in this case are the electron emission and the hole emission. While in the former a highly energetic electron in conduction band transfers energy to an electron in valence band, in the latter case, a hole in a valence band transfers energy to an electron in the valence band. As a result, the electron from valence band moves to the conduction band. The effect of impact ionization and avalanche multiplication is shown in Figure 2.14.

The current densities for electrons and holes are given as J_e and J_h. The generation rates are modeled proportional to these current densities as [9]

$$g_e = \frac{i_e J_e}{q} \quad \text{and} \quad g_h = \frac{i_h J_h}{q} \tag{2.84}$$

where i_e and i_h are the ionization rates for electrons and holes, respectively. The ionization rates are exponentially dependent on the electric field along the current flow direction. Let $E_{electron}^{critical}$ and $E_{hole}^{critical}$ be the critical electric fields for electrons and holes. Let i_e^∞ and i_h^∞ be the ionization rates at infinite field for electrons and holes, respectively. The ionization rates for electrons and holes are given by

$$i_e = i_e^\infty \exp- \left\{ \frac{E_{electron}^{critical}}{E} \right\}^{j_{electron}} \tag{2.85}$$

and

$$i_h = i_h^\infty \exp- \left\{ \frac{E_{hole}^{critical}}{E} \right\}^{j_{hole}} \tag{2.86}$$

where $j_{electron}$ and j_{hole} are model parameters with values close to 1.

The total impact ionization rate is now found as

$$R_{total,IIR} = -(g_e + g_h) = -\left(\frac{i_e J_e}{q} + \frac{i_h J_h}{q}\right) \tag{2.87}$$

This rate is independent of the electric field but depends on the carrier temperature.

2.10.5 Mobility Modeling

Mobility modeling is divided into four categories: low field mobility, high field mobility, bulk semiconductor mobility, and inversion layer mobility. In low field mobility the carriers are in equilibrium with the lattice and the mobility is very low. This mobility is inversely related to the impurity scattering. The low field mobility models for bulk materials include constant mobility model, Caughey and Thomas model [27], Dorkel-Leturg model [28], Arora model [29], and Klaassen low-field mobility model [30]. The low field mobility is mainly affected by Coulomb scattering. In the high field case the mobility decreases with electric field as the high energy carriers take part in scattering and the mean drift velocity rises slowly with increasing electric field. Finally the drift velocity saturates to a constant value. The bulk mobility model is a three-step process. First the low field mobility is expressed as a function of lattice temperature and doping. Then the saturation velocity is expressed in terms of temperature. And finally the low and high field junction region is described. The mobility models can further be classified into physical based, semi-empirical, and empirical models. As carriers move under the influence of an electric field, the velocity saturates and so the effective mobility reduces because the drift velocity is equal to the product of mobility and the electric field. The field-dependent mobility expressions of Caughey and Thomas [27] are

$$\mu_n(E) = \mu_{n0}\left[1+\left(\frac{\mu_{n0}E}{v_{sat}^n}\right)^{\beta_n}\right]^{-\frac{1}{\beta_n}} \tag{2.88}$$

$$\mu_p(E) = \mu_{p0}\left[1+\left(\frac{\mu_{p0}E}{v_{sat}^p}\right)^{\beta_p}\right]^{-\frac{1}{\beta_p}} \tag{2.89}$$

where μ_{n0} and μ_{p0} are the respective low field electron and hole mobility, and E is the parallel electric field. The saturation velocities are calculated from the temperature-dependent model in [31] as

$$v_{sat}^p = v_{sat}^n = \frac{24 * 10^6}{1 + 0.8 \exp\left(\frac{T}{600}\right)} \, \text{cm/sec} \tag{2.90}$$

2.11 Introduction to MOS Transistor

The most important requirement for a field-effect transistor (FET) is zero or negligible gate leakage current. It is important to isolate the gate from the channel so that no current flows into the gate. One needs some kind of barrier for the electron (or hole) from the gate to the source, the channel, and the drain. In MOSFET, an insulator provides the barrier. It is possible to grow a high-quality and reliable insulator using Silicon (Si). SiO_2 is stable and makes a firm bonding with Silicon. This oxide makes selective diffusion and easy pattern transfer. This has led to the well-established silicon MOSFET technology to become dominant. In MOSFET application of a large gate bias inverts the bands and induces the electrons (or holes) in a channel without the gate leakage. Over the last few years steady progress has been made on using the MOSFET concept with other semiconductors, notably (Gallium Arsenide) GaAs. GaAs n-MOSFETs have channel mobilities much higher than those in n-MOSFET based on the silicon as in [32]. However, GaAs technology does not provide a high-quality oxide. Thus, the widespread use of such devices is still not prevalent because of lack of good-quality oxide.

The Si technology is unique in the sense that a high-quality oxide SiO_2 can be formed on the silicon wafer. The Si-SiO_2 interface perfection is required for a field-effect device. Their higher areal density, better switching characteristics, and lower power dissipation have made them the dominant device in electronic systems.

There are two basic types of MOS transistors: the n-channel and the p-channel. A circuit containing only n-channel devices is produced by an nMOS process. Similarly, a pMOS process fabricates circuits that contain only p-channel transistors.

2.12 Structure and Symbol of MOSFET

The structure of an n-channel MOS transistor is shown in Figure 2.19(a), consisting of two n-type regions embedded in a p-type substrate, connected via metal or polysilicon to external conductors called the *source* and the *drain*.

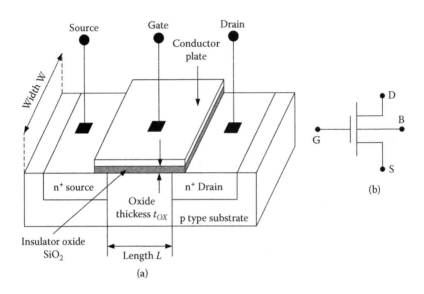

FIGURE 2.19
(a) An n-type NMOS device structure. (b) Symbol of MOSFET.

The symbol of MOSFET is shown in Figure 2.19(b). An n-channel MOS transistor is fabricated in a p-type semiconductor substrate, usually silicon [33]. Two n-type regions are made in the substrate and the current flows between these two regions. The region with the lowest applied potential is called the *source* and that with the highest potential is called the *drain*. On the surface, a thin silicon dioxide (SiO_2) layer is formed and on top of this a conducting polysilicon material called a *gate* is deposited. An electron-rich layer called a *channel* can be created between the source and the drain underneath the gate when a positive gate bias is applied. When appropriate voltages are applied at the source and the drain, electrons can flow from the source into the drain. If the substrate material is n-type and the diffused regions are p-type, a similar structure will represent a p-channel MOS transistor. The gate plate must act as a good conductor and was in fact realized by metals like aluminum in the early generations of the MOS technology. However, it was discovered that the non-crystalline polysilicon exhibits better fabrication and physical properties. The bonding between the silicon substrate and polysilicon gate is better than that between the metal gate and the silicon substrate. Thus the metal M is replaced by a heavily doped polysilicon.

Here L and W denote the length and width of the channel, respectively. In the most common mode of operation of the transistor, the source and the substrate are grounded and the drain is connected to a supply voltage V_{DD} through a load resistor, which is positive for an n-channel transistor and negative for a p-channel transistor. In a MOSFET the channel charge is induced electrostatically by the gate by using it as a capacitor without the need for doping.

2.13 Basic Operation of MOSFET

2.13.1 Operation of MOSFET with Zero Gate Voltage

Let us consider the gate voltage equal to zero while the p-type substrate and the source are grounded ($V_{Sub} = V_S = 0$) [34]. The drain is connected to a positive voltage source. Because the source and the substrate are at the same potential, no current flows in the source-substrate junction. The drain-substrate junction is reverse biased, and except for a small reverse leakage current, no current flows in that junction. These back-to-back diodes prevent current conduction from the source to the drain, as in Figure 2.20(a). The depletion formation is shown in Figure 2.20(b). The MOSFET has a very high resistance between the source and the drain. It is operating in the "cut off" as there is no conducting channel between the source and the drain. Small current flows due to the second-order effect under the weak inversion, called *subthreshold current*.

2.13.2 Operation of MOSFET with a Positive Gate Voltage

In this case a constant positive bias is applied to the gate as in Figure 2.21(a). There is no gate current because the metal electrode is insulated from the silicon. However, the positively biased gate electrode attracts electrons from the semiconductor, and an electron-rich layer forms underneath the gate insulator. In effect, negative charges are induced in the underlying Si by the formation of a depletion region and a thin surface region containing mobile electrons.

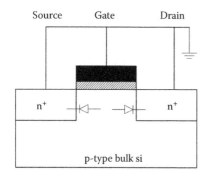

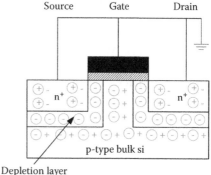

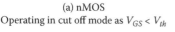

(a) nMOS
Operating in cut off mode as $V_{GS} < V_{th}$

(b) Depletion layer formation

FIGURE 2.20
(a) Two reverse biased p-n diodes representing a MOSFET working in cutoff regime. (b) MOSFET operating in the "cut-off" mode.

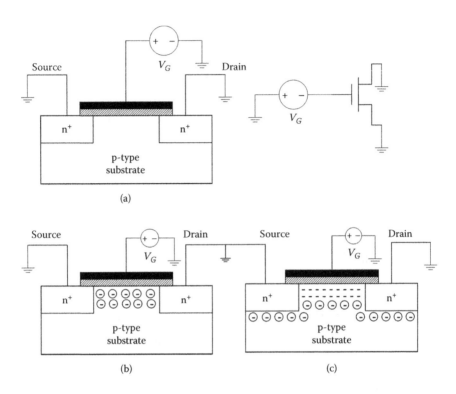

FIGURE 2.21
(a) MOSFET with positive gate bias. (b) Formation of depletion region. (c) Formation of channel.

The positive voltage on the gate causes the free holes to be repelled from the region underneath the gate [35]. The holes are pushed downward into the substrate, leaving behind a depletion region. The depletion region is populated by the bound negative charge associated with the acceptor atoms as in Figure 2.21(b). Also, the positive gate voltage attracts electrons from the source and the drain regions into the channel. When a large number of electrons accumulates near the surface underneath the gate, an n region is created, connecting the source and the drain regions, as in Figure 2.21(c).

The electron-rich layer underneath the gate is called the *channel*. The n-type source and the n-type drain are connected by the electron-rich channel. When a voltage is applied between the drain and the source, current flows between them. The gate bias creates an electric field that can either induce or prevent the formation of an electron-rich region at the surface of the semiconductor. The channel is created by inverting the substrate surface from p to n type. Hence the induced channel is also called an *inversion layer* and is shown in Figure 2.22.

Threshold voltage (V_{TH}) in MOSFET is defined as the minimum gate voltage required to induce the channel. For an n-channel device, positive gate

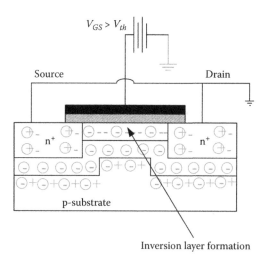

FIGURE 2.22
Inversion layer formation in nMOS for positive gate bias.

voltage greater than V_{TH} is required to induce a conducting channel consisting of electrons. For a p-channel device, a negative gate voltage larger than V_{TH} is required to induce a conducting channel consisting of holes.

2.13.3 Effect of a Small V_{DS}

A small positive voltage V_{DS} between the drain and the source induces a channel. This positive voltage V_{DS} causes electrons to flow from the source to the drain through the induced channel, causing a flow of current. The magnitude of this current I_D flowing from the drain to the source depends on the density of the electrons in the channel, which in turn depends on the magnitude of V_{GS}. For $V_{GS} = V_{TH}$, the channel is just induced and the current is small. When V_{GS} exceeds V_{TH} more electrons are attracted to the channel and the current increases. We observe that the MOSFET is operating as a linear resistor whose value is controlled by V_{GS}, and the channel is uniform when a small V_{DS} is applied. Thus the MOSFET is said to operate in the *linear* or *triode region*.

2.13.4 Operation of MOSFET as V_{DS} Is Increased

Let V_{GS} be kept constant at a voltage greater than V_{TH}. As V_{DS} is increased the voltage drop across the length of the channel increases [36]. As one moves along the channel from the source to the drain, the voltage increases from 0 to V_{DS}. So the voltage between the gate and points along the channel decreases from V_{GS} at the source end to $(V_{GS} - V_{DS})$ at the drain end. Because the voltage is not constant, the channel is no longer of uniform depth; rather it will be

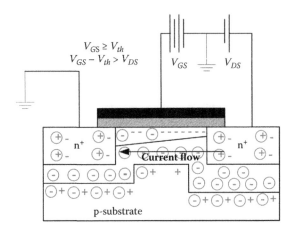

FIGURE 2.23
Current flow through the channel for small V_{DS}.

tapered as shown in Figure 2.23. As V_{DS} is increased, the channel becomes more tapered and its resistance increases correspondingly. Eventually, when V_{DS} is increased to the value that reduces the voltage between the gate and the channel at the drain end to V_{TH}—that is, $V_{GD} = V_{TH}$ or $V_{GS} - V_{DS} = V_{TH}$ or $V_{DS} = V_{GS} - V_{TH}$—the channel depth at the drain end decreases to almost zero, and the channel is said to be pinched off as shown in Figure 2.24. Increasing V_{DS} beyond this has negligible effect on the channel shape, and the current through the channel remains fixed. The drain current saturates at this value, and the MOSFET is said to have entered the saturation region of operation. Because the current is constant in the saturation region, the MOSFET is said to operate as a constant current source whose value depends upon the applied gate voltage. The voltage V_{DS} at which saturation occurs is denoted as $V_{Dsat} = V_{GS} - V_{TH}$. So for every value of V_{GS}, there is a corresponding value

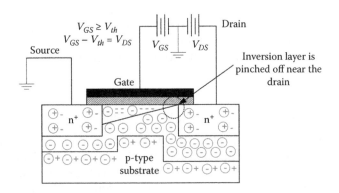

FIGURE 2.24
Pinched off channel, with deeper depletion layer near the drain side.

of V_{Dsat} [37]. When $V_{\text{DS}} \geq V_{\text{Dsat}}$, the device operates in the *saturation region*. The region of the $I_\text{D} - V_{\text{DS}}$ characteristic for $V_{\text{DS}} < V_{\text{Dsat}}$ is called the *triode region*.

2.14 Threshold Voltage of MOSFET

Conduction between the source and the drain takes place for MOSFET under the influence of the source-to-gate voltage. The current flow does not begin sharply, but it is assumed that if the gate voltage exceeds a given value called the *threshold voltage* V_{th}, conduction starts. The first step is to study the band diagram. The analytical expression of this important parameter comes from the study of the *MOS* structure. To understand the operation of the MOSFET we first need to examine the *MOS capacitor*, shown in Figure 2.25. An oxide layer is grown on top of a p-type semiconductor and a metal contact is placed on the oxide. In general, the insulator could be large band-gap material.

The MOS capacitor consists of a metal gate, an insulating oxide layer, and a semiconductor. The thickness of the oxide varies from 5 to 50 nanometers. We first consider the case of a hypothetical metal whose Fermi level is the same as that of silicon [38]. When such a structure is fabricated, the Fermi level of the system is unique, and because the metal has the same Fermi level as the silicon, the band structure is that shown in Figure 2.26. *Work function* is defined as the energy required for moving an electron from the Fermi level to the outside. In this idealized case let $\Phi_m = \Phi_S$ so that there is no difference in the work functions. This condition is referred to as flat band for obvious reasons.

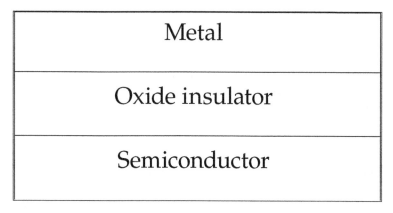

FIGURE 2.25
A MOS capacitor.

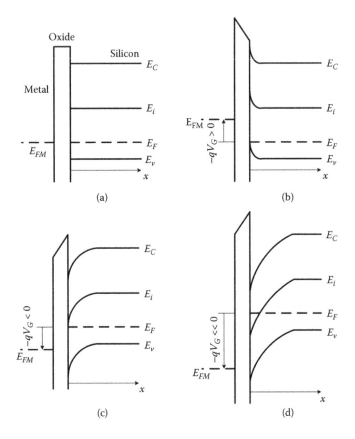

FIGURE 2.26
Energy band diagram of MOSFET: (a) flat band, (b) accumulation, (c) depletion, and (d) inversion.

2.14.1 Accumulation of Holes

When a negative bias is applied between the metal and the semiconductor, a negative surface charge is deposited on the metal at the metal oxide interface, and the structure behaves as a parallel-plate capacitor whose electrodes are the silicon and the metal, with oxide as the insulator [39]. In response an equal net positive charge appears at the surface of the semiconductor at the silicon-oxide interface. This silicon charge whose thickness is approximately 10 nm can also be considered as a surface charge. This hole-rich thin layer is called an *accumulation layer*.

The energy band diagrams are drawn for negative charges, whereas an electrostatic potential diagram is drawn for positive test charges. In metal, application of negative bias reduces the electrostatic potential, and as a result electron energies are raised in the metal relative to the semiconductor. The Fermi level for the metal rises above its equilibrium position by qV, where V

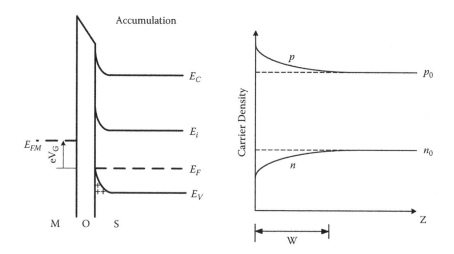

FIGURE 2.27
Energy band diagram for ideal MOS capacitor under the application of applied voltage.

is the applied voltage. The difference between the Fermi level in the metal and the semiconductor is the applied bias.

Because Φ_m and Φ_S do not change with applied voltage, but E_{Fm} moves up relative to E_{FS}, causing a tilt in the oxide conduction band, the energy bands of the semiconductor bend near the interface (the valence bands are bent to come closer to the Fermi level) accommodating the accumulation of holes at the interface as in Figure 2.27. The effect of depositing a negative charge in the gate of a MOS transistor causes *hole accumulation*.

An increase in surface hole concentration implies an increase in $E_i - E_F$ at the surface. Because the Fermi level within the semiconductor remains unchanged as no current flows though the MOS structure with increasing $(E_i - E_F)$, E_i must move up in the energy near the surface. This results in band bending near the surface. From Figure 2.27 it is clear that near the surface the Fermi level lies closer to the valence band, creating a larger hole concentration than that arising from the doping of the p-type semiconductor.

2.14.2 Depletion

When a positive bias is applied to the metal, positive charges are deposited on the metal and a corresponding net negative charge accumulates at the semiconductor surface. Such a negative charge in the p-type material is due to depletion of holes from the surface, leaving behind the uncompensated ionized acceptors. Thus the hole concentration decreases, moving E_i closer to E_F, bending the band down near the semiconductor surface as in Figure 2.28. If the positive voltage is increased, the band bends down more

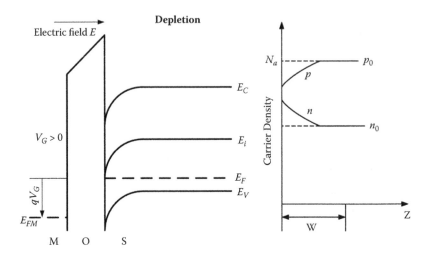

FIGURE 2.28
Effect of applied electric field on the interface charge density in the ideal MOS capacitor: positive gate voltage (V_G) creates a depletion region.

strongly, resulting in inversion that occurs for the higher positive applied positive gate bias [39].

2.14.3 Inversion

If the positive bias on the metal side is increased further, the bands at the semiconductor surface bend down more strongly as shown in Figure 2.29. A large positive voltage can bend E_i below E_F. Thus the conduction band at the oxide-semiconductor region comes close to the Fermi level in the semiconductor. This reverses the mobile charges from holes to electrons at the interface and the electron density increases. If the positive bias is increased until E_C comes close to the electron quasi-Fermi level near the interface, the electron density increases and the semiconductor near the interface has electrical properties of an n-type semiconductor. This n-type surface layer is formed not by doping, but by *inversion* of the original p-type semiconductor due to the applied bias. This inverted layer is separated by the underlying p-type material by a depletion region.

2.14.4 Surface Potential

Figure 2.30 shows the band bending of the semiconductor on the onset of strong inversion [37–39]. It is described by the quantity $\Phi(x)$, which measures the position of the intrinsic Fermi level with respect to the bulk intrinsic Fermi level. The band bending at the oxide-semiconductor interface is described in terms of the potential Φ_S as in Figure 2.30. We notice that $\Phi_S = 0$ is the flat band condition for the MOS (Figure 2.26). When $\Phi_S < 0$, the

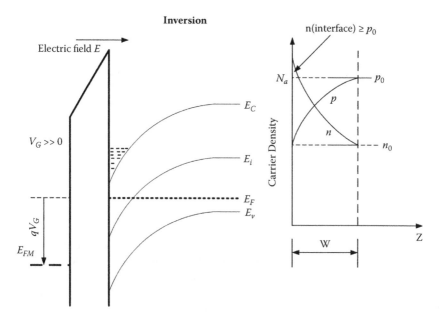

FIGURE 2.29
Band bending with increase of positive bias on the metal side of MOSFET.

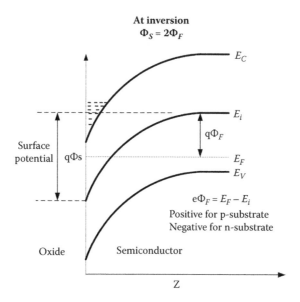

FIGURE 2.30
Band bending of the semiconductor in the inversion mode.

bands bend up at the surface and holes accumulate. Similarly when $\Phi_S > 0$, depletion takes place, and finally when $\Phi_S > 0$ and greater than Φ_F, the bands bend in such a way that E_i lies below E_F, resulting in inversion.

The onset of inversion is a gradual process and is a function of gate bias. The strong inversion occurs when the electron concentration at the interface is equal to the bulk p-type concentration. Strong inversion means the surface will be as strongly n-type as p-type. So the intrinsic level E_i is at a position Φ_F below the Fermi level at the interface. The surface band bending is given in (2.91) as

$$\Phi_S(inv) = 2\Phi_F \qquad (2.91)$$

For an n-MOSFET, the substrate is p-type and Φ_F is positive, and a positive Φ_S is required for inversion. For a p-MOSFET the substrate is n-type and Φ_F is negative, causing inversion.

The charge density of the metal Q_m is balanced by the channel depletion charge Q_d and the inversion charge Q_n. We are interested in calculating the threshold voltage (i.e., the gate voltage needed to cause inversion in the channel).

The total surface charge density is related to the surface field by Gauss's law. This charge Q_S is the total surface charge density at the semiconductor-oxide interface region and includes the induced free charge in inversion and the background ionic charge. The charge Q_S is zero when the bands are flat.

For a larger positive gate voltage the surface potential increases. The hole concentration near the surface decreases while the electron concentration increases, according to the following relationships:

$$p_{(x=0)} = N_a \exp\left(\frac{-q\Phi_S}{kT}\right) \qquad (2.92)$$

and

$$n_{(x=0)} = \frac{n_i^2}{N_a} \exp\left(\frac{q\Phi_S}{kT}\right) \qquad (2.93)$$

since

$$n = n_i \exp\left(\frac{E_F - E_i}{kT}\right) \qquad (2.94)$$

$$p = n_i \exp\left(\frac{E_i - E_{Fi}}{kT}\right) \qquad (2.95)$$

$$np = n_i^2 \qquad (2.96)$$

The electron surface concentration is equal to that of hole ($n(0) = p(0) = n_i$), when E_i coincides with E_F at $x = 0$. This happens when

$$\Phi_S = \Phi_F = \frac{kT}{q}\ln\left(\frac{N_a}{n_i}\right) \tag{2.97}$$

as shown in Figure 2.31.

With further increase in the gate voltage, the electron surface concentration increases up to a point where $n(x = 0)$ becomes equal to $p_{po} = N_a$ which is the original hole concentration in the substrate. This is because the band curvature at the surface ($x = 0$) places E_i at an energy $q\Phi_F$ below E_F. In other words, the band curvature is equal to $2(E_i - E_F)$ or $\Phi_S = 2\Phi_F$.

When this condition is met, the semiconductor surface is said to be in *strong inversion*.

For $\Phi_F \leq \Phi_S \leq 2\Phi_F$, the electron concentration is larger than the hole concentration, and the surface is in *weak inversion*, while for $\Phi_S \geq 2\Phi_F$ it is in strong inversion as shown in Figure 2.32. The inversion layer is rich in electrons and hence is a good conductor. The MOS capacitor consists of two conducting electrodes (the metal gate and the inversion layer at the silicon surface). As in the case of accumulation, the capacitance of the MOS structure is again equal to C_{ox}.

When an inversion layer is formed, electrons are the local majority carriers at the surface. Any subsequent increase in gate voltage increases the electron concentration in the inversion layer and produces a larger inversion charge Q_{inv}. However, the thickness of the inversion layer remains very small. Its actual thickness is similar to that of an accumulation layer.

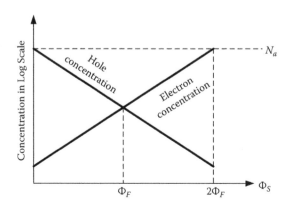

FIGURE 2.31
Hole and electron concentration as a function of surface potential.

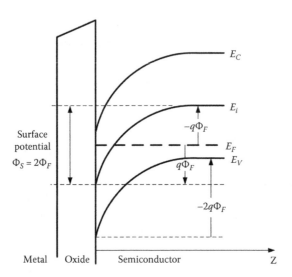

FIGURE 2.32
Band bending under strong inversion at the surface.

The electron charge in an inversion layer can be thought of as a surface charge [40]. As in the case of an accumulation layer, the inversion charge depends exponentially on the *surface potential*:

$$Q_{inv} \alpha \exp\left(\frac{q\Phi_S}{kT}\right) \tag{2.98}$$

When the gate voltage is increased beyond inversion, the surface potential Φ_S increases very slightly above $2\Phi_F$ and one can assume that $\Phi_S = 2\Phi_F$ when an inversion layer is present. Because the semiconductor is p-type, the electrons in the inversion layer are produced by a slow process called *thermal genera-tion* at room temperature. They can also be produced by external generation (if a light source is present). If the semiconductor is in the dark and at cryo-genic temperature, the inversion layer may never form.

In summary the following rules will be used to describe the relationships between the charge on the metal gate and the charge in the accumulation, depletion, and inversion layers as shown in Figure 2.33.

$$-Q_G = Q_{acc} \quad \text{(accumulation)} \tag{2.99}$$

$$-Q_G = Q_d \quad \text{(depletion)} \tag{2.100}$$

$$-Q_G = Q_d + Q_{inv} \quad \text{(inversion)} \tag{2.101}$$

$-Q_G$ = charge at the back-side contact of the sample (dielectric mode)

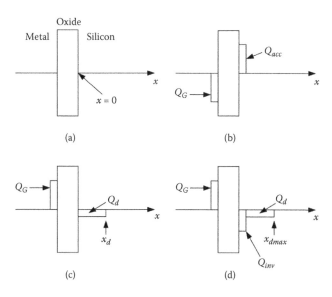

FIGURE 2.33
Charges in the MOS structure: (a) flat band, (b) accumulation, (c) depletion, and (d) inversion.

2.15 Flat-Band Voltage: Effect of Real Surfaces

2.15.1 Equalization of the Fermi Levels

The Fermi level of the metal gate and the silicon are so far considered equal. However, in practice this is not the case. In modern devices the gate material is not actually a metal, but heavily doped polysilicon. The doping concentration for that material is so high ($\approx 10^{20}$ cm^{-3}) that it can be considered as a metal. Let us first consider the metal and the semiconductor separately. The energy needed to extract an electron with an energy E_{FM} from the metal is called the *work function* $q\Phi_m$. The work function in the semiconductor is denoted by $q\Phi_{sc}$. The potential work function difference is $\Phi_{ms} = \Phi_m - \Phi_s$. It is to be noted that Φ_{ms} is negative for both n+ poly-silicon gate and n-type Si substrate and n+ poly-silicon gate and p-type Si substrate.

When a MOS structure is formed, the Fermi levels align, and the charge transfer causes a tilt in the oxide conduction band [40,41]. The bands bend down near the semiconductor surface to accommodate work function difference as shown in Figures 2.34(a),(b),(c). To go back to a flat-band condition a voltage must be applied to the gate. This voltage called *work function difference* is denoted by Φ_{ms}:

$$\Phi_{ms} = \Phi_m - \Phi_{sc} = \frac{E_F - E_{FM}}{q} \tag{2.102}$$

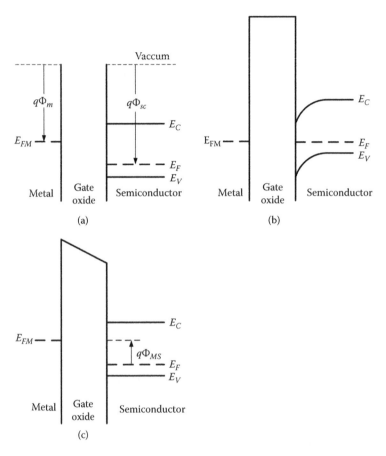

FIGURE 2.34
Energy band diagram of MOSFET when (a) the metal and the semiconductor are taken separately, (b) no bias is applied, and (c) a bias equal to Φ_{ms} is applied to the gate.

2.15.2 Oxide Charges

Oxides grown on silicon contain positive charges due to the presence of contaminating metallic ions or imperfect Si-oxide bonds [40]. These charges can either be fixed or mobile in the oxide. However, for simplicity only the case of fixed charges is considered.

Let us consider an elementary real positive charge Q(Coulomb m^{-2}) at a depth x in the oxide. Let $x = 0$ be defined at the metal/oxide interface as shown in Figure 2.35(a). Negative charges will appear in the metal and the silicon. The sum of these three charges is equal to zero. The charge in the silicon can be removed when an appropriate negative bias is applied to the gate as in Figure 2.35(b). If the charge is closer to the semiconductor, a larger compensation gate bias is required to remove it. In an actual device, charges

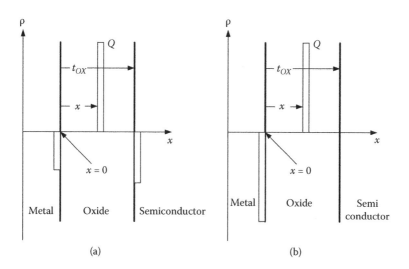

FIGURE 2.35
Single charge in an oxide for (a) $V_G = 0$ and (b) an applied gate voltage.

are distributed throughout the oxide. The compensating voltage is obtained as equal to

$$V_Q = \frac{Q_{OX}}{C_{OX}} \tag{2.103}$$

2.15.3 Interface Traps

A set of charges arise from the interface states at the Si-SiO$_2$ interface. They are created by the sudden termination of the semiconductor crystal lattice at the oxide interface. In the oxidation process, Si reacts with oxygen forming a SiO$_2$ layer. When the oxidation process is suddenly stopped, some ionic Si left near the interface causes perturbation to the periodic crystal structure of the semiconductor and hence some Si-Si bonds are unfulfilled or *dangling*. As a result there are energy states in the band gap at the silicon surface. These states called *interface states* or *interface traps* can be charged positively or negatively, and depending on their nature and their energy with respect to the Fermi level will affect the surface potential.

2.15.4 Flat-Band Voltage

Flat-band voltage is defined as the voltage that must be applied to the gate to bring the semiconductor energy bands to a flat level. It can also be defined as the voltage applied to the gate such that there is no band bending in the semiconductor. Flat-band is achieved by applying a gate voltage that compensates for the following:

- The differences in work functions of the semiconductor and the gate electrode
- The presence of charges in the oxide
- The interface traps

$$V_{FB} = V_Q + \Phi_{ms} + V_i = \Phi_{ms} - \frac{Q_{OX}}{C_{OX}} + \frac{Q_i}{C_{OX}} \tag{2.104}$$

For simplicity the various oxide and interface charges are included in an effective positive charge Q_{it} (C/cm^2) at the interface. Q_{it} includes both Q_i and Q_{OX}. This charge will introduce an effective negative charge in the semiconductor. To compensate for these charges, a bias $V_{it} = \frac{Q_{it}}{C_{OX}}$ must be applied to the gate. Flat band voltage can be given as

$$V_{FB} = V_{it} + \Phi_{ms} = \Phi_{ms} - \frac{Q_{it}}{C_{OX}} \tag{2.105}$$

2.16 Expression of Threshold Voltage

The threshold voltage of a MOSFET is defined as the voltage that must be applied to the gate to form an inversion layer. In a MOS transistor, the gate voltage is equal to the sum of the potential drops in the semiconductor and the oxide given as

$$V_G = \Phi_S + \frac{Q_G}{C_{OX}} \tag{2.106}$$

where Q_G is equal to the positive charge on the gate electrode. An equal amount of negative charge also exists in the semiconductor, composed of ionized impurities in the depletion region, and the free electrons at the oxide/silicon interface at the inversion. If the charge due to the free electrons is assumed to be much smaller than that due to ionized impurities when the inversion layer starts to form, the above equation can be written as

$$V_G = 2\Phi_F - \frac{Q_d}{C_{OX}} = V_{THO} \tag{2.107}$$

V_{THO} is called the *ideal threshold voltage,* and it is measured with respect to the source [40].

The flat-band voltage in Equation (2.105) must be added to the expression of the threshold voltage in (2.107), in order to accurately describe the actual, "non-ideal" threshold voltage given as

$$V_{TH} = V_{FB} + 2\Phi_F - \frac{Q_d}{C_{OX}} = \Phi_{ms} - \frac{Q_{it}}{C_{OX}} + 2\Phi_F - \frac{Q_d}{C_{OX}} \qquad (2.108)$$

For strong inversion the voltage required must be strong enough to first achieve the flat band condition (first two terms of Equation 2.108), then to induce an inverted region ($2\Phi_F$ term of Equation 2.108) and to accommodate the charge in the depletion region (last term of Equation 2.108).

The threshold voltage may be either positive or negative, depending on the doping concentration N_a, the material used to form the gate electrode, etc. If the threshold voltage is negative, the n-channel MOSFET is a depletion-mode device. However, if V_{TH0} is positive, the device is an enhancement-mode MOSFET.

Depletion-mode devices have an inversion layer when the gate voltage is equal to zero. Such devices are sometimes referred to as "normally on." *Enhancement-mode* devices called "normally off" require an applied positive gate bias to create the inversion layer. The value of the threshold voltage can be adjusted by applying a controlled amount of doping impurities in the channel region during device fabrication.

2.17 I–V Characteristics of MOSFET

The derivation of the MOSFET I–V relationship for different conditions needs several approximations. The analysis of an actual three-dimensional MOSFET would be a very complex task without these assumptions. Derivation of closed form I–V equations is not possible without these assumptions. Here *gradual channel approximation (GCA)* is used for deriving the I–V relationship, which will effectively reduce the problem to a one-dimensional current flow problem. This will allow us to devise simple current equations that are in agreement with experimental results. However, GCA has limitations, particularly in the case of short-channel MOSFETs.

2.17.1 Gradual Channel Approximation

A semiconductor bar carrying a current I is considered in Figure 2.36. If Q_d coulombs per meter is the density of charge along the direction of current and v meters per second is the velocity of the charge, then

$$I = Q_d v \qquad (2.109)$$

The total charge that passes through a cross section of the bar per unit time can be measured. With a velocity v, the charge enclosed in v meters of the bar must

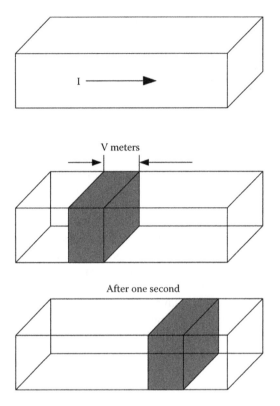

FIGURE 2.36
A semiconductor bar carrying current and snaps of the carriers after 1 second.

flow through the cross section in one second. Because the charge density is Q_d, the total charge in 'v' meters is given by $Q_d v$.

In Figure 2.37(a) a coordinate system for the MOSFET structure is considered taking the x-direction parallel to the surface and the y-direction perpendicular to the surface. The origin of the x-coordinate is at the source end of the channel. The channel voltage with respect to the source end is denoted by $V(x)$. Now let the threshold voltage V_{th} be constant along the entire channel region between $x = 0$ and $x = L$. However, in reality the threshold voltage changes along the channel because the channel voltage is not constant. Let the electric field component E_x along x-coordinate be dominant compared to electric field component E_Y along the y-coordinate. This allows us to reduce the current flow problem in the channel along the x-direction only.

Boundary conditions used are

$$V_{(x=0)} = V_S = 0 \tag{2.110}$$

$$V_{(x=L)} = V_{DS} \tag{2.111}$$

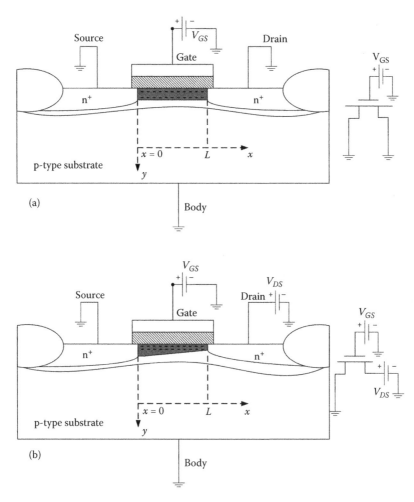

FIGURE 2.37
(a) Channel charge with equal source and drain voltages and (b) with only drain voltage.

Let an n-channel MOSFET whose source and drain are connected to ground be considered. When $V_{GS} = V_{TH}$, the inversion charge density produced by the gate oxide capacitance is proportional to $V_{GS} - V_{TH}$. For $V_{GS} \geq V_{TH}$, any charge placed on the gate must be mirrored by the charge in the channel, yielding a uniform channel charge density equal to

$$Q_d = C_{OX}W(V_{GS} - V_{TH}) \tag{2.112}$$

C_{OX} is the gate/channel capacitance per unit area, where C_{OX} is multiplied by W to represent total gate capacitance per unit length. Sometimes expressions are described in terms of another capacitance known as gate/channel capacitance C_g.

$$C_g = C_{OX}WL \tag{2.113}$$

Now let the drain voltage taken be greater than zero as shown in Figure 2.37(b). Because the channel potential varies from zero at the source end to V_D at the drain, the potential difference between the gate and the channel varies from V_G to $V_G - V_D$. Thus the charge density at a point x along the channel can be written as

$$Q_d(x) = C_{OX}W[(V_{GS} - V(x) - V_{TH}] \tag{2.114}$$

where $V(x)$ is the channel potential at x. The drain current is given by

$$I_D = -C_{OX}W[(V_{GS} - V(x) - V_{TH}]v \tag{2.115}$$

In this expression the negative sign is inserted because the charge carriers are negative and v denotes the velocity of the electrons in the channel. For semiconductors $v = \mu E$, where μ is the mobility of charge carriers and $E = -dV/dx$ is the electric field.

Putting the value of v

$$I_D = C_{OX}W[(V_{GS} - V(x) - V_{TH}]\mu_n \frac{dV(x)}{dx} \tag{2.116}$$

The boundary conditions are $V(0) = 0$ and $V(L) = V_{DS}$. Although $V(x)$ can easily be found from this equation, the quantity of interest is I_D. Integrating (2.116) we get

$$\int_{x=0}^{L} I_D \, dx = \int_{V=0}^{V_{DS}} WC_{OX}\mu_n[V_{GS} - V(x) - V_{TH}]dV \tag{2.117}$$

The current equation of MOSFET in the triode region is given by

$$I_D = \mu_n C_{OX} \frac{W}{L}[(V_{GS} - V_{TH})V_{DS} - \frac{1}{2}V_{DS}^2] \tag{2.118}$$

where L is the effective channel length.

Figure 2.38 plots the parabolas given by Equation (2.118) for different values of V_{GS}. Calculating $\partial I_D / \partial V_{GS}$, one can show that the peak of each parabola occurs at $V_{DS} = V_{GS} - V_{TH}$, and the peak current is

$$I_{D,max} = \frac{1}{2}\mu_n C_{OX} \frac{W}{L}(V_{GS} - V_{TH})^2 \tag{2.119}$$

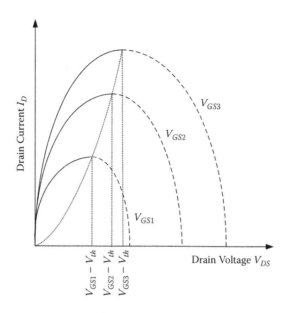

FIGURE 2.38
Drain current versus drain-source voltage in triode region.

Here, $(V_{GS} - V_{TH})$ is the *overdrive* or *effective* voltage and W/L is the *aspect ratio*. If $V_{DS} \leq V_{GS} - V_{TH}$, we say the device is operating in the *triode* or *linear* region.

$k'_n = \mu_n C_{OX}$ is known as the process transconductance parameter.

$k_n = \dfrac{W}{L} k'_n = \dfrac{W}{L} \mu_n C_{OX}$ is known as the gain factor.

Equations (2.118) and (2.119) serve as the foundation for analog and digital CMOS VLSI design.

If in (2.118) $V_{DS} < 2(V_{GS} - V_{TH})$, we have

$$I_D \approx \mu_n C_{OX} \frac{W}{L} [(V_{GS} - V_{TH})V_{DS}] \tag{2.120}$$

The drain current is a linear function of V_{DS}. This is also evident from the characteristics of Figure 2.38. For small V_{DS}, each parabola can be approximated by a straight line. The linear relationship implies that the path from the source to the drain can be approximated by a linear resistor equal to

$$R_{on} = \frac{1}{\mu_n C_{OX} \frac{W}{L}(V_{GS} - V_{TH})} \tag{2.121}$$

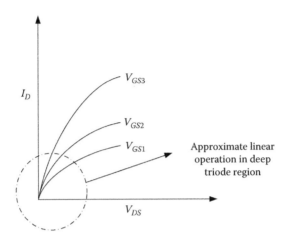

FIGURE 2.39
Linear operation in deep triode region.

A MOSFET can therefore operate as a resistor whose value is controlled by the overdrive voltage (as long as $V_{DS} < 2(V_{GS} - V_{TH})$). This is shown in Figure 2.39. Thus in triode region, MOSFET operates as a voltage controlled resistor as in Figure 2.40.

However, if in Figure 2.38 the drain-source voltage exceeds $V_{GS} - V_{TH}$, the drain current does not follow the parabolic behavior for $V_{DS} > V_{GS} - V_{TH}$. In fact, as shown in Figure 2.41, I_D becomes relatively constant, and we say that the device operates in the *saturation region* as in [38,41].

If the drain voltage remains greater than the source voltage, then the voltage at each point along the channel with respect to ground increases as we move from the source end toward the drain end. From Figure 2.42(a), this effect arises from the gradual voltage drop along the channel resistance. Because the gate voltage is constant (as the gate is conductive but carries no current in any direction) and the potential at the oxide-silicon interface rises from the source to the drain end, the potential difference between the gate

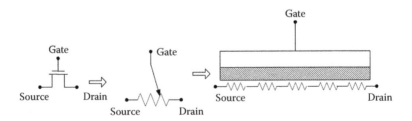

FIGURE 2.40
MOSFET as a voltage-dependent resistor.

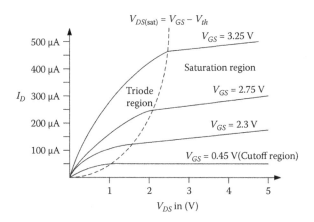

FIGURE 2.41
Plots showing saturation of drain current.

and the oxide-silicon interface decreases along the *x*-axis as in Figure 2.42(b). The density of electrons in the channel follows the same trend, falling to a minimum value at $x = L$.

The local density of inversion layer charge is proportional to $V_{GS} - V(x) - V_{TH}$. So if $V(x)$ approaches $V_{GS} - V_{TH}$, then $Q_d(x)$ drops to zero. In other words, as in Figure 2.43(a), if V_{DS} is slightly greater than $V_{GS} - V_{TH}$, then the inversion

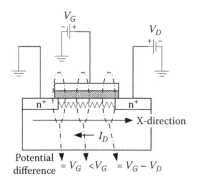

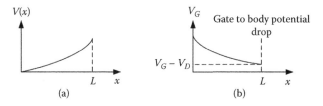

FIGURE 2.42
(a) Channel potential variation. (b) Gate-substrate voltage difference along the channel.

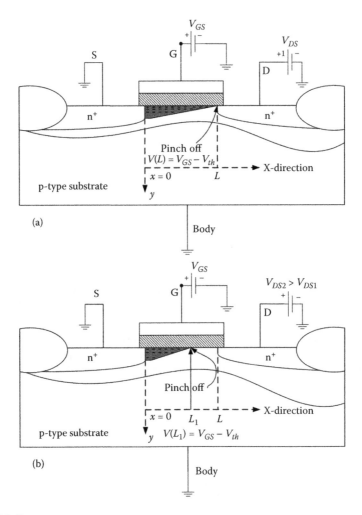

FIGURE 2.43
(a) Pinch-off condition. (b) Pinch-off point shifts to source end for increasing V_{DS}.

layer stops at $x \leq L$, and the channel is *pinched off*. As V_{DS} increases further, as in Figure 2.43(b), the point where Q_d equals zero gradually moves toward the source. So at some point along the channel, the local potential difference between the gate and the oxide-silicon interface is not sufficient to support an inversion layer.

No channel exists between L_1 and L. But the device still conducts, as illustrated in Figure 2.44. Once the electrons reach the end of the channel, they experience the high electric field in the depletion region at the drain junction and are rapidly swept to the drain terminal.

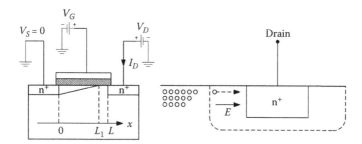

FIGURE 2.44
Detailed operation in the pinch-off region.

From the above observations, we examine (2.117) for a saturated device. The integral on the left-hand side of (2.117) must be taken from $x = 0$ to $x = L_1$, where L_1 is the point at which Q_d drops to zero, and that on the right from $V(x) = 0$ to $V(x) = V_{GS} - V_{TH}$. As a result,

$$\int_{x=0}^{L_1} I_D \, dx = \int_{V=0}^{V_{GS}-V_{TH}} W C_{OX} \mu_n [V_{GS} - V(x) - V_{TH}] dV$$

which gives

$$I_{D,\max} = \frac{1}{2} \mu_n C_{OX} \frac{W}{2L_1} (V_{GS} - V_{TH})^2 \tag{2.122}$$

Considering the approximation $L \approx L_1$, a saturated MOSFET can be used as a current source connected between the drain and the source. Current sources draw current from V_{DD} or inject current into ground as shown in Figure 2.45.

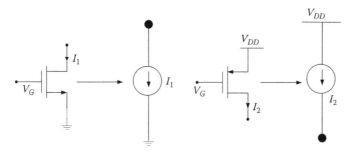

FIGURE 2.45
MOSFET acting as a current source.

2.18 Depletion MOSFET

Some devices have an already fabricated channel even at zero gate voltage. Negative gate voltage is required to turn them off [41]. Such a *normally on* device is known as a depletion-type MOSFET. The MOSFETs that are off at zero gate bias are termed *enhancement-type* MOSFET. On application of a gate voltage the device turns on.

Figure 2.46 shows an n-channel depletion-type MOSFET. A doped n-channel region exists under the oxide, meaning that an electron inversion layer already exists with zero applied gate bias. In this type of device, a negative gate voltage will induce a space charge region underneath the oxide, reducing the thickness of the n-channel region. The reduced thickness decreases the channel conductance, which in turn, reduces the drain current. A positive gate voltage will create an electron accumulation layer, increasing the drain current. The $I_D - V_{DS}$ characteristics and symbol for an n-channel depletion MOSFET are shown in Figures 2.47(a),(b). Figure 2.48 shows the $I_D - V_{GS}$ curve for depletion mode and enhancement mode MOSFET.

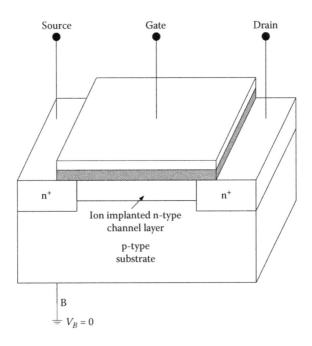

FIGURE 2.46
Structure of a depletion MOSFET.

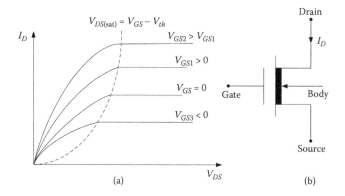

FIGURE 2.47
(a) I_D versus V_{DS} characteristics for an n-channel deletion mode MOSFET. (b) Symbol of an n-channel deletion mode MOSFET.

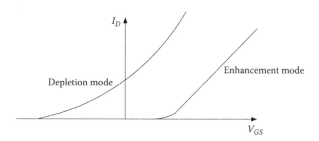

FIGURE 2.48
I_D–V_{GS} characteristics for depletion mode and enhancement mode MOSFETs.

2.19 Transconductance (g_m)

Since a MOSFET operating in saturation region produces a current in response to its gate-source overdrive voltage, we define a figure of merit that indicates how well a device converts a voltage to a current. In processing signals we deal with the changes in voltages and currents, so here we define a figure of merit as the change in the drain current divided by the changes in the gate-source voltage, called the *transconductance*, denoted by g_m, expressed as

$$g_m = \frac{\partial I_D}{\partial V_{GS}}\bigg|_{V_{DS}=cons\,tan\,t}$$

$$g_m = \mu_n C_{OX} \frac{W}{L}(V_{GS} - V_{TH}) \tag{2.123}$$

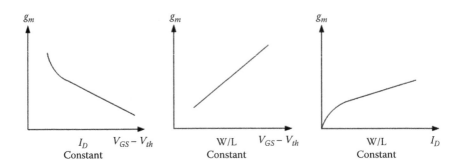

FIGURE 2.49
Variation of trans-conductance of MOSFET

g_m represents the sensitivity of the device. When g_m is high, a small change in V_{GS} results in a larger change in $I_D \cdot g_m$ in the saturation region is inverse of R_{on} in the deep triode region.

Also,

$$V_{GS} - V_{TH} = \sqrt{\frac{2I_D}{\mu_n C_{OX} \dfrac{W}{L}}} \tag{2.124}$$

Putting the value of (2.124) in (2.123), we get

$$g_m = \sqrt{2I_D \mu_n C_{OX} \frac{W}{L}} \tag{2.125}$$

$$g_m = \frac{2I_D}{V_{GS} - V_{TH}} \tag{2.126}$$

g_m of a MOS transistor can be increased by increasing its width. However, this will also increase the input capacitance and the area occupied. Equation (2.123) suggests that g_m increases with the overdrive if W/L is constant, whereas (2.126) shows that g_m decreases with the overdrive if I_D is constant. From (2.125) we see that g_m increases in square law with I_D if W/L is constant. These results are illustrated in Figure 2.49.

2.20 Channel Length Modulation

When $V_{DS} = V_{DSSat} = V_{GS} - V_{Th}$, the inversion layer charge at the drain end becomes zero. We can say that the channel is pinched off at the drain end for this bias condition. If the drain-to-source voltage is increased beyond the

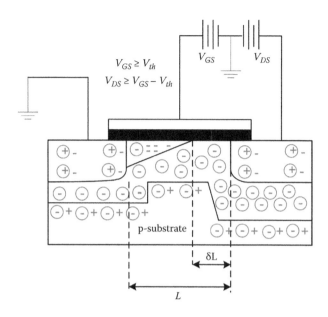

FIGURE 2.50
Reduction of channel length at saturation.

saturation voltage so that $V_{DS} > V_{DSSat}$, an even larger portion of the channel becomes pinched off. The effective channel length (i.e., the length of the inversion layer) is reduced to $L - \Delta L$, where ΔL is the length of the channel region where inversion charge is equal to zero as in Figure 2.50. The pinch-off point moves from drain end to the source end caused by increasing drain-to-source voltage. The electrons traveling from the source toward the drain traverse the inverted channel section of length L and then are injected into the depletion region of length $L - L' = \Delta L$ which separates the pinch-off point from the drain end.

The voltage remains constant at $V_{GS} - V_{TH} = V_{DSsat}$, and the additional bias applied to the drain appears as a voltage drop across the narrow depletion region between the channel end and the drain region. This voltage accelerates the electrons at the drain end of the channel and sweeps them across the depletion region into the drain. The channel length is reduced from L to $L - \Delta L$, a phenomenon known as *channel-length modulation* which is similar to base-width modulation in BJT. The shortening of the channel causes a larger current called *channel-length modulation*.

Channel-length modulation in a MOSFET is caused by the increase of the depletion layer width at the drain end with increased drain voltage. This leads to a shorter channel length and an increased drain current. An example of this is shown in Figure 2.51(a). The channel-length-modulation effect increases in small devices with low-doped substrates as in Figure 2.51(b).

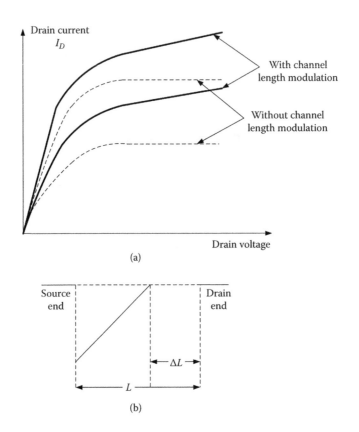

FIGURE 2.51
(a) Effect of increase in the drain current as a result of channel length modulation. (b) Channel length modulation.

From (2.122) we get

$$I_D \alpha 1 / (L - \Delta L) = 1/L \left(1 - \frac{\Delta L}{L} \right)^{-1} \tag{2.127}$$

Neglecting higher-order terms in the series, we get

$$I_D = 1/L \left(1 + \frac{\Delta L}{L} \right) \tag{2.128}$$

Let

$$\frac{\Delta L}{L} = \lambda V_{DS} \tag{2.129}$$

where λ is the channel-length modulation parameter.

Again from (2.122),

$$I_D = \mu C_{OX} \frac{W}{2L}(V_{GS} - V_t)^2 \qquad (2.130)$$

Due to channel-length modulation, the effective channel length becomes $L - \Delta L$. The drain current in saturation can be written as

$$I_D = \mu C_{OX} \frac{W}{2(L - \Delta L)}(V_{GS} - V_t)^2$$

$$I_D = \mu C_{OX} \frac{W}{2L}(V_{GS} - V_t)^2(1 + \lambda V_{DS}) \qquad (2.131)$$

The above equation is the modified current equation of a MOSFET considering channel-length modulation.

The saturation current is not dependent on drain bias if one does not consider channel-length modulation ($\lambda = 0$). However, in reality the saturation mode current increases linearly with drain bias. The slope of this current–voltage curve in the saturation region is determined by the channel-length modulation parameter λ. Thus, in the saturation region the graph is not a straight line, and there is finite slope caused by channel-length modulation as in Figure 2.51(a).

The $I_D - V_{DS}$ characteristics showing the effect of channel-length modulation are shown in Figure 2.52. The observed linear dependence of I_D on V_{DS} in the saturation region is represented in (2.131) by the factor $(1 + \lambda V_{DS})$. From Figure 2.52 it is seen that when the straight-line $I_D - V_{DS}$ characteristics are extrapolated, they intercept the V_{DS} axis at the point $V_{DS} = -V_A$, where V_A

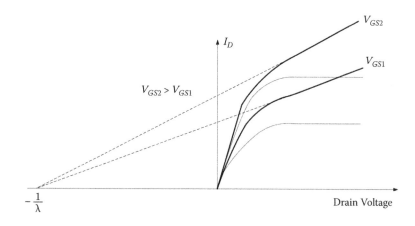

FIGURE 2.52
Effect of V_{DS} on I_D in the saturation region.

is a positive voltage. Equation (2.131) indicates that $I_D = 0$ at $V_{DS} = -1/\lambda$. It follows that $V_A = 1/\lambda$.

Therefore, V_A is a process-technology parameter with the dimensions of V. For a given process, V_A is proportional to the channel length L that the designer selects for a MOSFET. The voltage V_A is referred to as the Early voltage.

Equation (2.131) shows that when channel-length modulation is taken into account, the saturation values of I_D depend on V_{DS}. Thus, for a given V_{GS}, a change ΔV_{DS} causes a corresponding change ΔI_D in the drain current I_D. The output resistance of the current source representing I_D in saturation is no longer infinite. The *output resistance* r_o is defined as

$$r_0 = \left[\frac{\partial I_D}{\partial V_{DS}} \right]^{-1}_{V_{GS}=constant} \tag{2.132}$$

$$r_0 = \left[\lambda \frac{K_n'}{2} \frac{W}{L} (V_{GS} - V_t)^2 \right]^{-1} = \frac{1}{\lambda I_D} = \frac{V_A}{I_D} \tag{2.133}$$

Thus the output resistance is inversely proportional to the drain current [39,40].
The *output conductance* g_{DS} can be expressed as

$$g_{DS} = \left[\frac{\partial I_D}{\partial V_{DS}} \right]_{V_{GS}=constant} = \lambda I_D \tag{2.134}$$

λ is proportional to $1/L$, and I_D is proportional to $1/L$:

$$\lambda \alpha \frac{1}{L} \tag{2.135}$$

$$I_D \alpha \frac{1}{L} \tag{2.136}$$

Output conductance g_{DS} is strongly dependent on channel length, and this strong dependence is expressed by

$$g_{DS} \alpha \left(\frac{1}{L} \right)^2$$

Unlike the Early effect in bipolar devices, the amount of channel-length modulation is under the circuit designer's control. Because λ is inversely proportional to L, for a longer channel, the relative change in L for a given change in V_{DS} is smaller. We may conclude that channel length modulation is more prominent in short-channel devices as shown in Figure 2.53.

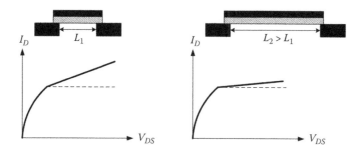

FIGURE 2.53
Channel length modulation is more prominent for short-channel devices.

2.21 Substrate Bias Effects

The derivation of linear mode and saturation mode current-voltage characteristics is done based on the assumption that the substrate potential is equal to the source potential (i.e., $V_{SB} = 0$). The threshold voltage is calculated without considering the substrate voltage. However, in many digital applications, the source potential can be larger than the substrate potential (i.e., $V_{SB} > 0$). The influence of non-zero V_{SB} must be accounted for by the threshold voltage V_T. So the modified V_T must be applied to the current characteristics. The threshold voltage is a function of the bulk-to-source voltage V_{BS} through the *substrate bias effect* or *backgate effect*. The application of V_{BS} (a negative voltage to avoid forward biasing the bulk-to source p-n junction) increases the depletion width, which increases the bulk charge as shown in Figure 2.54 and, thus, the threshold voltage. A solution to avoid the backgate effect is to make $V_{BS} = 0$ by electrically shorting the source to the bulk as in Figure 2.55.

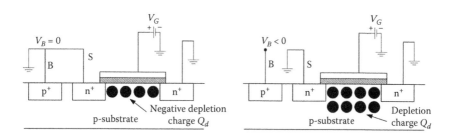

FIGURE 2.54
Variation of depletion charge with body effect.

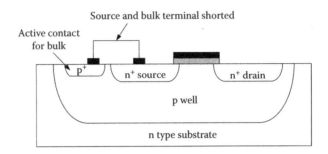

FIGURE 2.55
Source and bulk are tied together to reduce backgate effect.

2.22 MOS Transistor as a Switch

When a MOS transistor is in the linear region, the device acts as a linear resistor under gate voltage control. The transistor can be used as an on–off switch, as in Figure 2.56, if it operates in linear and cutoff regions. When the transistor is on, it operates in the linear region. The switch is turned off and the channel disappears by setting $V_{GS} = 0$. Hence, only a small amount of leakage current flows at the drain end. The switch is turned on by setting $V_{GS} = V_{DD}$, and the current path provides a resistance R_{ch}. For a p-channel MOSFET the switch is turned on if $V_{GS} = V_{DD}$. If the transistor is connected in series with a high-impedance circuit, the total current flowing through the transistor and the voltage drop $V_{DS} = I_{ds}R_{ch}$ across the channel will be very small. In this context an input voltage source signal, $V_{DS} = I_{ds}R_{ch}$, which is either zero or V_{DD} representing logical 0 or 1, respectively, is applied as shown in Figure 2.57. The output voltage V_o is either held at its previous logic value (if the switch is open) or its capacitance C is charged up to V_{in} (if the gate voltage allows the transistor path to be closed). The transistor used in this way is called a *pass transistor,* and the process of transferring the charge from the input node to the output under the influence of gate voltage is called *charge steering.*

FIGURE 2.56
MOS transistor as a switch.

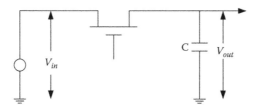

FIGURE 2.57
MOSFET as a pass transistor.

2.23 MOSFET Capacitance

The importance of capacitance is that it determines the delay or the speed of the circuit. In this section some basic facts about capacitance in a MOS device are presented. The capacitance of a MOS network depends on a number of factors, such as the physical structure of the MOS devices, the terminal voltages, and the network topology. We discuss these factors briefly.

2.23.1 Overlap Capacitance

The gate of the MOSFET is isolated from the conducting channel by the gate oxide layer that has a capacitance per unit area equal to $C_{OX} = \frac{\varepsilon_{ox}}{t_{ox}}$. From the basic drain current equation of MOSFET, it is seen that C_{OX} must be made as large as possible (i.e., the oxide thickness t_{ox} must be very thin). The total value of this capacitance is called the *gate capacitance* C_G, and it can be divided into two elements, each with a different behavior. One part of C_G contributes to the channel charge and is discussed in a subsequent section. Another part is solely due to the topological structure of the transistor. Let the transistor structure of Figure 2.58 be considered.

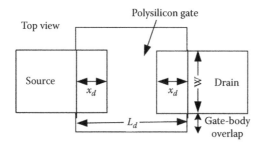

FIGURE 2.58
Overlap capacitance in MOSFET.

Ideally, the source and drain regions should end at the edge of the gate oxide. In reality, both the source and the drain tend to extend somewhat below the oxide by an amount x_d, called the *lateral diffusion length*. Hence, the effective channel of the transistor L becomes shorter than the length L_d (the length the transistor was originally designed for) by a factor $\Delta L = 2x_d$. It also gives rise to a parasitic capacitance between the gate and the source (drain) that is called the *overlap capacitance*. This capacitance is strictly linear and has a fixed value:

$$C_{GSO} = C_{GDO} = C_{OX}x_dW = C_oW \tag{2.137}$$

Because x_d is a technology-determined parameter, it is customary to combine it with the oxide capacitance to yield the overlap capacitance per unit transistor width C_o.

2.23.2 Channel Capacitance

This is the most important MOS parasitic circuit element, the *gate-to-channel capacitance* C_{GC} is divided into C_{GCS}, C_{GCD}, and C_{GCB} (being the gate-to-source, gate-to-drain, and gate-to-body capacitances, respectively), depending upon the operation region and terminal voltages. This varying distribution is best explained with the simple diagrams of Figure 2.59. When the transistor is in cutoff as in Figure 2.59(a), no channel exists, and the total capacitance C_{GC} appears between the gate and the body. In the resistive region as in Figure 2.59(b), an inversion layer is created, which acts as a conductor between the source and the drain. Consequently, $C_{GCB} = 0$ as the body electrode is shielded from the gate by the channel. From symmetry, the capacitance distributes evenly between source and drain. Finally, in the saturation mode as in Figure 2.59(c), the channel is pinched off. The capacitance between gate and drain is approximately zero, and so is the gate-body capacitance. Hence, all the capacitances are between the gate and the source.

The actual value of the total gate-channel capacitance and its distribution over the three components is best explained with the help of a number of charts. The plot in Figure 2.60(a) shows the evolution of the capacitance as a

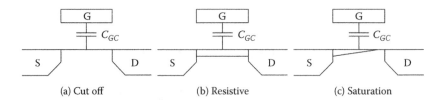

(a) Cut off   (b) Resistive   (c) Saturation

FIGURE 2.59
The gate-to-channel capacitance and their distribution over the other three terminals depending upon the operation region.

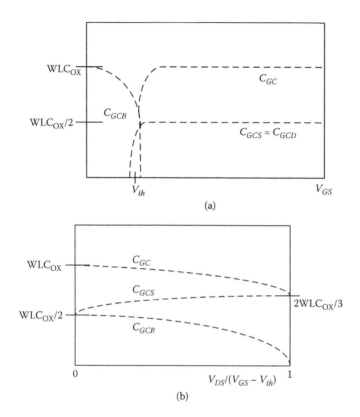

FIGURE 2.60

Distribution of the gate-channel capacitance as a function of V_{GS} and V_{DS}. (a) C_{GC} as a function of V_{GS} when $V_{DS} = 0$. (b) C_{GC} as a function of degree of saturation.

function of V_{GS} for $V_{DS} = 0$. For $V_{GS} = 0$, the transistor is off as no channel is present, and the total capacitance, equal to WLC_{ox}, appears between the gate and the body. When V_{GS} is increased, a depletion region is formed under the gate. This causes the thickness of the gate dielectric to increase, which means a reduction in capacitance. Once the transistor turns on ($V_{GS} = V_{th}$), a channel is formed and C_{GCB} drops off to 0. With $V_{DS} = 0$, the device operates in the resistive mode and the capacitance divides equally between the source and the drain, or $C_{GCS} = C_{GCD} = WLC_{ox}/2$. A designer must avoid operation in this region.

When the transistor is on, the distribution of its gate capacitance depends on the degree of saturation, measured by $V_{DS}/(V_{GS} - V_{th})$. As in Figure 2.60(b), C_{GCD} gradually drops to 0 for increasing levels of saturation, while C_{GCS} increases to 2/3 $C_{ox}WL$. This also means that the total gate capacitance decreases with an increased level of saturation.

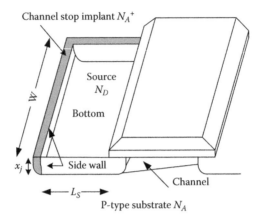

FIGURE 2.61
Schematic view of the source junction.

2.23.3 Junction Capacitances

Another capacitive component is contributed by the reverse-biased source-body and drain body p-n junctions. The depletion-region capacitance is nonlinear and decreases when the reverse bias is increased. To understand the components of the *junction capacitance* (called the *diffusion capacitance*), we must look at the source (drain) region and its surroundings. The detailed picture, shown in Figure 2.61, shows that the junction consists of two components.

The bottom-plate junction is formed by the source region (with doping N_D) and the substrate with doping N_A. The total depletion region capacitance for this component equals $C_{Bottom} = C_J W L_S$, with C_J as the junction capacitance per unit area.

The side-wall junction is formed by the source region with doping N_D and the p^+ channel-stop implant with doping level N_A^+. The doping level of this stopper is usually greater than that of the substrate, resulting in a larger capacitance per unit area. Its capacitance value equals $C_{SW} = C'_{JSW} x_j$ $(W + 2L_S)$. It is to be noted that no side-wall capacitance is counted for the fourth side of the source region, as this represents the conductive channel. Because x_j, the junction depth, is a technology parameter, it is normally combined with C'_{JSW} to give a capacitance per unit perimeter, $C_{JSW} = C'_{JSW} . x_j$. An expression for the total junction capacitance can then be derived as

$$C_{Diff} = C_{Bottom} + C_{SW} = C_J^* \text{ AREA} + C_{JSW}^* \text{ PERIMETER}$$

$$= C_J L_S W + C_{JSW} (2L_S + W) \tag{2.138}$$

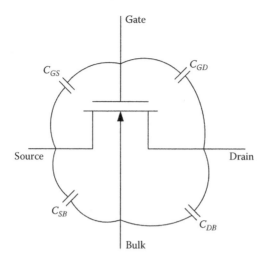

FIGURE 2.62
Different capacitances in MOSFET.

2.23.4 Different MOS Capacitors Together

All the above contributions can be combined in a single capacitive model for the MOS transistor, shown in Figure 2.62. Its components are identified based on the preceding discussions.

$$C_{GS} = C_{GCS} + C_{GSO}; \quad C_{GD} = C_{GCD} + C_{GDO} \tag{2.139}$$

$$C_{GB} = C_{GCB} \tag{2.140}$$

$$C_{SB} = C_{Sdiff}; \quad C_{DB} = C_{Ddiff} \tag{2.141}$$

2.23.5 Summary of MOS Capacitances

The structure of a MOS transistor is that of a parallel-plate capacitor. The oxide layer acts as an insulator between the two conducting plates: the gate (in polysilicon or metal) and the substrate. The parallel-plate capacitance per unit area is given as

$$C_{OX} = \frac{\varepsilon_0 \varepsilon_r}{t_{ox}} \ farad/cm^2 \tag{2.142}$$

where $\varepsilon_r = 3.9$ is the dielectric constant of the oxide, $\varepsilon_0 = 8.85 \times 10^{14}$ F/cm^2 is the free-space permittivity, and t_{ox} is the oxide thickness. The value of C_{OX} remains relatively constant but decreases slightly with increase in substrate voltage. For semiconductors, the capacitance values are very small and are expressed in units of picofarad (10^{-12} farad, denoted by pF) or femtofarad (10^{-15} farad, denoted fF). Similar parallel-plate capacitances are also formed if any overlap regions exist between the gate and the source or the gate and the drain. Even for self-aligned process, a certain amount of channel capacitance between the gate-to-source and the gate-to-drain will exist. These capacitances are denoted C_{GSO} and C_{GDO}, respectively, and must be added with C_{OX} to find the total gate capacitance C_G of the MOS transistor because they are all connected in parallel. The values of these capacitances are specified by the manufacturer with gate voltage V_G, negative or zero—that is, with no depletion region underneath the gate. With an increase in the gate voltage the phenomena of accumulation, depletion, and inversion start to take place. A depletion capacitance C_D, formed between the gate and the depletion region boundary, connected in series with C_{OX} lowers the effective gate capacitance C_G. This is shown in Figure 2.63. The value of C_D depends on the depth of the depletion region. The initial depth of the depletion region depends on the built-in contact or barrier potential (typical value 0.7 V) and then increases with increasing V_{GS}. As V_{GS} is increased so as to exceed the threshold voltage V_{th}, inversion takes place and the channel forms the conducting plate instead of the substrate, the depletion capacitances no longer exist, and the total capacitance shows a marked increase compared to its original oxide capacitance value.

Here we have assumed that the gate voltage is static or varies very slowly so that the phenomena of accumulation, depletion, and inversion take place in proper sequence. If V varies very rapidly (i.e., when the signal frequency of the gate voltage is rather high), the channel may not be formed. On average the device will appear to be in the depleted state all the time, bringing in the effects of the depletion capacitances C_D to reduce the total capacitance (Figure 2.63). This dynamic behavior of the capacitance does not really concern us because we will have to operate at considerably lower frequency to guarantee proper switching of each transistor.

It is seen that one part of the capacitances C_{GS} and C_{GD} is due to the gate overlaps at the source and the drain sides. Even if there is no overlap, it is

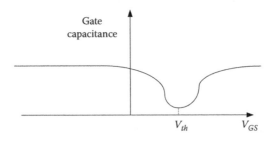

FIGURE 2.63
Variation of the gate capacitance with gate voltage.

useful to visualize C_G as consisting of a parallel connection of C_{GCS} and C_{GCD} because the channel can be viewed as a physical extension of the source and the drain regions. When the transistor is operating in the linear region, the two component capacitances have almost equal value; that is, $C_{GCS} = C_{GCD} = 0.5\,C_{GC} = 0.5\,C_{OX}WL$. When the transistor is saturated, the channel disappears near the drain end and we can assume that $C_{GCS} = C_{GC} = 2/3\,C_{OX}WL$.

2.23.6 Interconnect Capacitances

The so-called *stray or wiring capacitances* of the paths connecting the active channels could become significantly high, depending on the length of the wire, accounting for a major part of the circuit delays. This is because the wiring capacitances need to be charged up together with the gate capacitances for signals to be detected by the transistors. It is therefore convenient to be able to express the wiring capacitances in terms of the gate capacitance of a transistor of some standard size. It is also important to recognize that the diffusion wires are buried in the substrate, and their capacitance depends on two factors: the total area and the perimeter of the wire. The perimeter determines the so-called "side-wall" capacitance, which is at least 20% of the capacitance determined by the total area.

The relevant capacitances are denoted as

C_G: gate capacitance (pF/μm^2)

C_{pln}: p-channel source—drain capacitance

C_{nln}: n-channel source—drain capacitance

C_{mf}: metal-to-field oxide capacitance

C_{mp}: metal-to-polysilicon capacitance

C_{mt}: metal-to-oxide capacitance

C_{pf}: polysilicon-to-field oxide capacitance

2.24 Moore's Law

Since the invention of the first calculation machines, miniaturization has been a constant challenge to increase speed and complexity in the microelectronics industry. Linear scaling of device dimensions to a quasi-nanometer level allows building a complex system integrated on a chip which reduces the volume and power consumption per function, while increasing speed [42–45].

The steady downscaling of transistor dimensions over the past two decades has been the main stimulus to the growth of silicon integrated circuits (ICs) and the information industry. *Moore's law* [42], which states that the number of transistors on a given chip can be doubled every two years, has been the guiding principle of the continuous reduction of CMOS device dimension since Gordon Moore, co-founder of Intel, first predicted it in 1965. Over the last few decades, CMOS devices have been scaled down to the sub-100-nm regime. Although the basic device geometry has remained relatively unchanged, the gate length has been reduced from 10 mm in the 1970s to less than 0.1 μm in 2001, and the gate oxide thickness from 1000 Å to less than 20 Å [43].

2.25 Introduction to Scaling

MOS transistors are scaled primarily due to the two reasons mentioned below:

1. Increased device *packing density*: The design of high-density chips in MOS VLSI technology requires that the *packing density* of MOSFETs be as high as possible, such that the sizes of the transistors are as small as possible.

2. Improved frequency response (*transit time*) is proportional to 1/L: If the length of the device is small, then the transit time required for the charged carrier to move from the source to the drain end is small, and thus the device can be operated at higher frequencies.

Two types of scaling are common:

- Constant field scaling
- Constant voltage scaling

Constant field scaling warrants a reduction in the power supply voltage as the minimum feature size is decreased, but it yields the largest reduction in the power-delay product of a single transistor. In contrast, power supply voltage is not reduced in the *constant voltage scaling* and is therefore the preferred scaling method because it provides voltage compatibility with older circuit technologies. The disadvantage of the *constant voltage scaling* is that the electric field increases as the minimum feature length is reduced. This leads to *velocity saturation, mobility degradation*, increased leakage currents, and lower breakdown voltages.

2.26 Constant Field Scaling

The principle of constant field scaling is that the device dimension and the device voltages are to be scaled in such a way that both the horizontal and vertical electric fields remain essentially constant [46]. To ensure the reliability of the device, the electric field in the scaled device must not increase. This scaling attempts to preserve the magnitude of internal electric fields in the MOSFET, while the dimensions are scaled down. To achieve this, all voltages must be scaled down in proportion to the device dimension.

The channel length is scaled from L to αL ($\alpha > 1$). To maintain a constant horizontal electric field, drain voltage must be scaled from V_{DS} to V_{DS}/α. The gate voltage should also be scaled from V_{GS} to V_{GS}/α, so that the gate and the drain voltages remain compatible. To ensure a constant vertical electric field, gate oxide thickness must also be scaled from t_{ox} to t_{ox}/α.

Because the channel length is being reduced, the depletion width also needs to be reduced. If the substrate doping concentration is increased by the factor (α), then the depletion width is reduced approximately by the same factor α, because V_{DS} is reduced by $1/\alpha$.

2.27 Constant Voltage Scaling

In constant voltage scaling, all dimensions of the MOSFET are reduced by a factor of α (>1) as in constant field scaling. The power supply voltage and the terminal voltages, on the other hand, remain unchanged. Because the channel length is being reduced, the depletion width also needs to be reduced. If the substrate doping concentration is increased by the factor (α^2), then depletion width is reduced approximately by the factor α.

2.28 Why Constant Voltage Scaling Is More Useful than Constant Field Scaling

In constant field scaling, the scaling of voltages will be unpractical in many cases. In particular, the peripheral and interface circuitry may require certain voltage levels for all input and output voltages. To accommodate the different voltage levels, the multiple power supply arrangement is necessary, and complicated level shifters are required. To get rid of these external voltage level constraints, constant voltage scaling is preferred, knowing that it can cause serious device reliability issues. In actual technology evolution, the voltages need not be reduced with the same scaling factor.

2.29 ITRS Roadmap for Semiconductors

Due to the significant resources and investments required to develop the next generation of CMOS technologies, it is necessary to identify goals and put collective efforts toward developing new equipment and technologies.

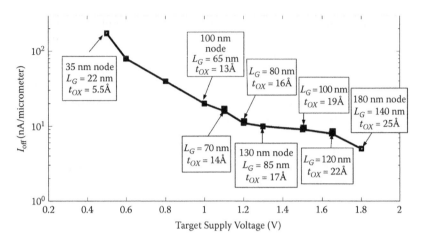

FIGURE 2.64
ITRS roadmaps for high-performance technologies for the year 1999. Plot of target I_{OFF} versus V_{DD} [47].

The semiconductor roadmap represents a consensus among industry leaders and gives projected needs based on past trends. The *International Technology Roadmap for Semiconductors (ITRS)* [47] is the standard accepted roadmap. Figure 2.64 shows the roadmap specifications for drive current and off-state leakage current for high-performance circuits, along with the associated power supply voltages and the technology nodes. The drive currents of the nMOSFET/pMOSFET are fixed at constant values of 750/350 μA/μm, while the off-state leakage current continually increases with scaling. The gate insulator needs to be aggressively scaled down to improve the drive current and to suppress short-channel effects. In the 130 nm technology node with a 70 nm physical gate length, the gate oxide thickness is only 15 Å, which is approximately six atomic layers thick. To continue past trends in CMOS scaling, a sub-10 Å effective oxide thickness will soon be required, which is about four atomic layers thick. Beyond that point, SiO_2 may lose its properties as an insulator, and we may need a different materials system. The roadmap distinguishes two different applications: high-performance and low-power circuits.

In summary, scaling improves cost, speed, and power per function with every new technology generation. All of these attributes have been improved by 10 to 100 million times in four decades—an engineering achievement unmatched in human history.

2.30 Different Groups of MOSFETs

Moving a design from an old technology to a newer one, with smaller design rules, has always been an interesting way to lower power consumption and to obtain higher speed. The overall parasitic capacitances (i.e., gates and interconnects) are decreased, the available active current per device is higher, and consequently, the same performance can be achieved with a lower supply voltage. Moving to a new technology generation, however, induces a scale down of the power supply voltage (V_{dd}), the threshold voltage (V_{th}), and the gate oxide thickness (T_{OX}). Starting from the 0.18 μm technologies, it appeared that building a transistor with a good active current (I_{on}) and a low leakage current (I_{off}) was becoming more difficult. The four main causes of limitations are as follows:

1. Voltage limits and subthreshold leakage
2. Tunneling currents
3. Statistical dispersions
4. Poly depletion and quantum effects

Two families of transistors were introduced: high-speed transistors and low-leakage transistors. The threshold voltages of the two families are tuned differently, using different channel doping. When moving to more advanced technologies, those two families are no longer sufficient regarding technological constraints. The ITRS introduces three main groups of transistors:

1. High performance (HP)
2. Low operating power (LOP)
3. Low standby power (LSTP)

A new kind of MOS device has been introduced in deep submicron technologies, starting from the 0.18 μm CMOS process generation. The new MOS, called a *low leakage* MOS device, is available as well as the original, called *high-speed MOS.*

For I/Os operating at high voltage, specific MOS devices called *high voltage MOS* are used. The high voltage MOS is built using a thick oxide, two to three times thicker than the low voltage MOS, for handling high voltages, as required by the I/O interfaces, shown in Figure 2.65(a),(b), and (c).

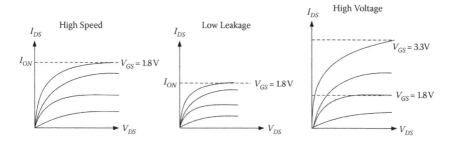

FIGURE 2.65
Three different types of MOSFETs introduced by ITRS roadmap.

2.31 Short-Channel Effects of MOSFET

A MOSFET device is considered to be short when the channel length is of the same order of magnitude as the depletion-layer widths (x_{dD}, x_{dS}) of the source and the drain junctions. Short-channel effects occur as the channel length L is reduced to increase both the operation speed and the number of components per chip.

The short-channel effects are attributed to two physical phenomena:

1. The limitation imposed on electron drift characteristics in the channel
2. The modification of the threshold voltage due to the shortening of the channel length

Reduction of channel length of MOSFETs leads to the short-channel effects such as:

- Reduction in the threshold voltage
- Drain-induced barrier lowering
- Increase in the saturation drain current with increasing V_{DS}
- Increase in the off-state leakage current
- Punch-through effect
- Mobility degradation
- Increase in the parasitic resistance and the capacitance
- Hot carrier effect

All these effects are discussed in detail in the following sections.

2.32 Reduction of the Effective Threshold Voltage

In a long-channel device, channel formation is controlled by the gate and the substrate. The gate voltage controls all the space charge induced in the channel region. As the channel length decreases, the charge in the channel region decreases. As the drain bias is increased, the reverse biased space charge region at the drain extends farther into the channel and the gate control decreases (i.e., in a short-channel device the n$^+$ type source and the drain induce a large amount of the depletion charge which cannot be neglected). For short-channel devices, the charge control of the channel is shared by the four terminals (gate, substrate, source, and drain), called *charge sharing*. The

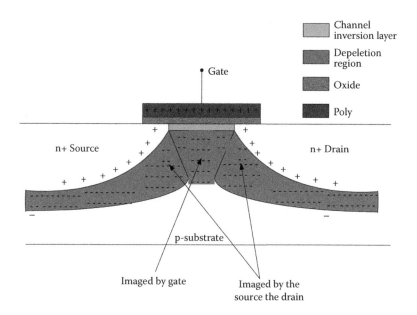

FIGURE 2.66
Charge sharing between the source/drain depletion regions and the channel depletion region.

depletion regions of the source and the drain are very close to each other. Through a charge sharing mechanism as in Figure 2.66, this phenomenon can be explained as in [48]. In case of a short-channel device, a considerable portion of the field lines emanating from the bulk charge terminate in the source and the drain regions instead of the gate. It is easier for the gate to deplete the amount of channel charge, lowering the threshold voltage of the device.

To sum up,

- *Long-channel MOS transistor*: The depletion is only due to the electric field created by the gate voltage.
- *Small-geometry transistor*: In addition to the previous contribution, the depletion charge near n+ regions contributes a significant amount of the depletion charge.

The expression of the threshold voltage in long-channel MOSFET thus overestimates the depletion charge supported by the gate voltage. Thus the amount of the gate voltage required to offset the depletion charge reduces. The estimated threshold voltage value from the threshold voltage expression of long-channel MOSFET will be larger than the actual value of the short-channel MOSFET.

The deeper depletion region is accompanied by larger surface potential, making the channel more attractive for electrons. Thus the device can conduct more current. This effect may be considered as the reduction of V_{th}, as the drain current is the function of $(V_{GS} - V_{th})$. Increase in V_{DS} and reduction of channel length will decrease the effective threshold voltage. In other words, the bulk depletion charge contributed by the gate is smaller than the expected charge, controlled solely by the gate, as a significant portion of the total depletion region charge under the gate is actually due to the source and the drain depletion junction. As the threshold voltage is a function of bulk depletion charge induced by the gate, the expression must be modified to account for this reduction in the bulk depletion charge. The threshold voltage of the short-channel MOSFET can be expressed as $V_{th\,(SC)} = V_{th} - V_{thO}$, where V_{thO} is the change in threshold voltage from the long to the short-channel MOSFET.

The depletion due to the source and the drain contacts encroaches substantially underneath the gate, decreasing the additional gate voltage required to create the strong inversion compared to the long-channel case. This is shown in Figure 2.67 where the source and the drain regions are assumed to be cylindrical with radius x_j and the depletion depth of extent x_d.

The amount of charge imaging on the gate electrode is assumed by a trapezoidal approximation to be

$$Q'_B = -qN_A x_d \left(\frac{L+L'}{2} \right) \tag{2.143}$$

In the long-channel case,

$$Q_B = -qN_A x_d L \tag{2.144}$$

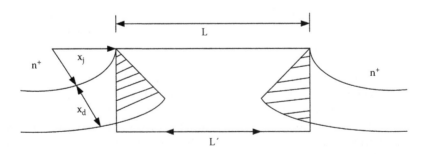

FIGURE 2.67
Schematic of the channel region of a short-channel MOSFET.

The charge in the shaded regions image is on the gate and not on the contacts. Thus the reduced bulk charge is the source of the reduced threshold voltage from

$$V_T = 2\Phi_F + V_{FB} - \frac{Q_B}{C_{OX}} \tag{2.145}$$

$$V_T' = 2\Phi_F + V_{FB} - \frac{Q_B'}{C_{OX}} \tag{2.146}$$

$$\Delta V_T = \frac{Q_B - Q_B'}{C_{OX}} = \frac{Q_B}{C_{OX}}\left(1 - \frac{Q_B'}{Q_B}\right) \tag{2.147}$$

Now as $L' \to L$, then $\frac{Q_B'}{Q_B} \Diamond 1$ (i.e., the long-channel case).

2.33 Hot Electron Effects

The longitudinal electric field in the channel increases from the source to the drain ends. For abrupt source and drain junctions, the peak field is at the drain-to-channel junction, and its value depends on V_{DS} and L. When carriers move in the electric fields that exceed the value of the onset velocity saturation, they continue to acquire kinetic energy from the electric field, but their velocity is randomized by the excessive collision such that their velocity along the electric field direction no longer increases but their random kinetic energy does. Depending on the statistics of scattering, a small fraction of the overall carrier population acquires a significant energy, and these are called *hot carriers*. The electric field heats the normal lattice electrons coming into the pinch-off region [49].

The carriers crossing from the inverted channel pinch-off point to the drain travel at their maximum saturated speed, and so gain their maximum kinetic energy in saturation. These carriers having high energy are called *hot carriers*. They travel from the source to the drain along the channel gaining kinetic energy at the expense of electrostatic potential energy in the pinch-off region, and they behave as a hot electron. Some of them obtain energy to create impact ionization with silicon lattice atoms, as a result of which new electrons and holes are created: this effect is referred to as *weak avalanche*. The new electrons created join the other channel electrons and move toward the drain. Some hot carriers in small numbers can acquire enough energy so as to surmount the Si-SiO$_2$ interface barrier and thus move into the gate oxide as in Figure 2.68. Most of the injected carriers are collected by the gate electrode resulting in the gate current I_G, reducing the input impedance. Because the barrier potential for this process is very high, the number of hot carriers injected into the gate will be much smaller compared to those

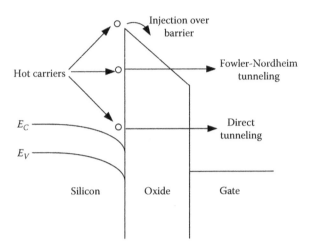

FIGURE 2.68
Three types of carrier injection into the gate causing hot carrier effects.

causing *impact ionization*. Therefore, the gate current will be smaller than the substrate current by a few orders of magnitude. It must be noted that carriers can also enter the gate oxide by tunneling. For direct tunneling, the oxide has to be very thin and the field is high. Even for the thicker oxide, the carrier with energy less than the energy barrier can tunnel through the barrier. This effect is called *Fowler-Nordheim tunneling*.

A small fraction of the high energy carriers create damage at the silicon-oxide interface which manifests itself as an increase in the interface state density, and yet another fraction becomes trapped in the oxide. The traps in the oxide significantly affect reliability. The accumulation of such traps behaves as a fixed oxide charge, causing a change in the threshold voltage of the device, and this affects the gate's control, giving rise to *oxide breakdown*.

A *lightly doped drain (LDD)* structure can reduce this hot-carrier effect. This is because, in such a case, part of the depletion region would be inside the drain, absorbing some of the potential that otherwise would exist in the pinch-off region, and lowering the maximal electric field.

2.34 Avalanche Breakdown and Parasitic Bipolar Action

An undesirable short-channel effect that occurs due to the high velocity of electrons in the presence of a large longitudinal electric field generates electron-hole pairs by *impact ionization* of the silicon atoms as in [50]. The presence of high longitudinal fields in a short-channel MOSFET can accelerate electrons that can ionize silicon atoms by impacting against them.

When the electric field in the channel is increased, due to high energetic hot electrons, *avalanche breakdown* occurs in the channel at the drain end. This increases the flow of current. The electrons are attracted by the drain, while the holes enter the substrate to form part of the parasitic substrate current.

There is also parasitic bipolar action taking place. The region between the source and the drain can act like the base of an n-p-n transistor, with the source playing the role of the emitter and the drain that of the collector. Holes generated by the avalanche breakdown move from the drain to the substrate underneath the inversion layer. The hole current forward biases the source-body p-n diode. Also if the holes coming from avalanche are collected by the source and the corresponding hole current creates a voltage drop in the substrate material of the order of 0.6V, the normally reverse biased substrate source p-n junction will conduct appreciably. The electrons are also injected as the minority carriers into the p-type substrate underneath the inversion layer from the forward biased junction, similar to the injection of electrons from the emitter to the base. They can obtain enough energy as they move toward the drain to create new e-h pairs. These electrons arrive at the drain and create further electron-hole pairs through the *avalanche multiplication*. The positive feedback between the avalanche breakdown and the parasitic bipolar action results in breakdown at lower drain voltage. The process is shown in Figure 2.69, and the steps are as follows:

Process 1: Hot carriers having sufficient energy to overcome the oxide-Si barrier are injected from the channel to the gate oxide (process 1) causing the gate current to flow. Trapping of some of this charge can change V_{th} permanently.

Process 2: Avalanching can take place producing electron-hole pairs.

Process 3: The holes produced by avalanching are collected by the substrate contact causing parasitic substrate current I_{sub}.

Process 4: Voltage drop due to I_{sub} can cause the substrate-source junction to be forward biased.

Process 5: The forward biased substrate-source junction causes the minority electrons to be injected from the source into the substrate. Some of them are collected by the reverse biased drain and cause a *parasitic bipolar action*.

2.35 DIBL (Drain-Induced Barrier Lowering)

The population of channel carriers in the long channel devices is controlled by the gate voltage that creates the vertical electric field, whereas the horizontal field controls the current between the drain and the source. The current flow in the channel depends on creating and maintaining an inversion layer on

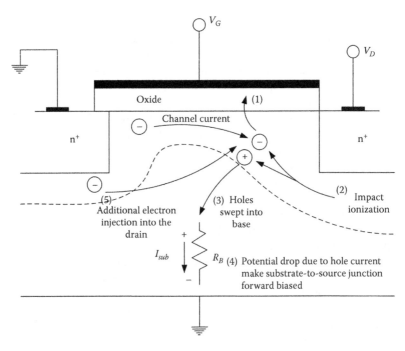

FIGURE 2.69
Impact ionization and parasitic bipolar action in a short-channel MOSFET.

the surface. If the gate voltage is not sufficient to invert the surface ($V_G < V_{th}$), the carriers (electrons) in the channel face a potential barrier that blocks the flow. Increasing the gate voltage reduces this potential barrier and eventually allows the flow of carriers under the influence of the channel electric field.

In long-channel devices, the horizontal and the vertical electric fields can be treated as having separate effects on the device characteristics. When the device is scaled down, the drain region moves closer to the source, and its electric field influences the whole channel. The drain-induced electric field also plays a role in attracting carriers to the channel without the control from the gate terminal. This effect is known as *drain-induced barrier lowering* (*DIBL*) because the drain lowers the potential barrier for the source carriers to form the channel. The threshold voltage lowers to feel the impact of this effect. DIBL attracts carriers with a loss in the gate control resulting in increased off-state leakage current.

2.36 Velocity Saturation in MOSFET

Velocity saturation due to the *mobility reduction* is important in the sub-micron devices. In the derivation of I-V relationship of the long-channel MOSFET, we explicitly assume that the mobility is a constant. However, this assumption must be modified for two reasons. Two effects are combined in the transistors to account for this mobility; reduction due to the horizontal electric field, and the mobility reduction due to the vertical electric field.

The performance of short-channeled devices is also affected by the velocity saturation, which reduces the transconductance in the saturation mode. At low electric field, the electron drift velocity V_d in the channel varies linearly with the electric field intensity [51]. However, as the electric field increases above 10^4 V/cm, the drift velocity tends to increase more slowly, and approaches a saturation value of $V_{d(sat)} = 10^7$ cm/s around the electric field $= 10^5$ v/cm at 300 K.

For a MOS device, if $V_{DS} = 5$ V and the channel length $L = 1$ µm, the average electric field is $5 * 10^4$ V/cm, and thus velocity saturation is more likely to occur in the short-channel devices for length $L < 1$ µm because drift velocity saturates around electric field $= 10^5$ V/cm.

Due to very high longitudinal electric field (drain bias) in the pinch-off region in long-channel devices, the carrier velocity saturates. This is more prominent in the short-channel devices as the corresponding horizontal electric field is generally even larger than long-channel MOSFET. For an ideal long-channel I-V relationship, the *current saturation* occurs when the inversion charge density becomes zero at the drain terminal or when $V_{DS} = V_{DS}(sat) = V_{GS} - V_{th}$.

However, velocity saturation can change this condition. Velocity saturation will yield an $I_D(sat)$ value smaller than that predicted in an ideal relation, and it will yield a smaller V_{DS} (sat) value than predicted. In a short-channel MOSFET before attaining pinch off, carrier drift velocity saturates and thus the current saturation occurs at a low value of V_{DS}. I_D will be linear with V_{GS}. The short-channel devices therefore experience an extended saturation region and tend to operate more often in saturation conditions than their long-channel counterparts, as shown in Figure 2.70.

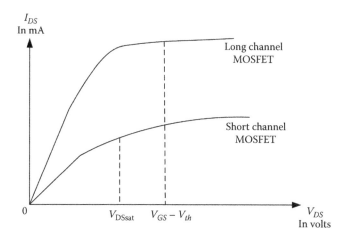

FIGURE 2.70
Current saturates in short-channel devices for small V_{DS}.

2.37 Mobility Degradation

There are two reasons for mobility reduction in MOSFET:

1. Due to the vertical electric field
2. Due to the horizontal electric field

2.37.1 Vertical Electric Field Mobility Degradation

A vertical electric field exists in MOSFET due to the applied gate voltage, which creates the conduction channel. When carriers move within the channel under the effect of the horizontal electric field, they feel the effect of the gate-induced vertical electric field, pushing carriers toward the gate oxide. This causes the carriers to collide with the oxide-channel interface. The oxide-channel interface is rough and imperfect, and carriers thus lose mobility. This effect is called *surface scattering*. It reduces mobility. If there is a positive fixed oxide charge near the oxide-semiconductor interface, the mobility will be further reduced due to additional coulomb attraction.

The effective inversion charge mobility is a strong function of temperature because of *lattice scattering*. As temperature decreases, mobility increases. The mobility used in the MOSFET model is not the mobility of electrons in the silicon crystal, called *bulk mobility*. Rather, it is a *surface mobility*. The surface mobility is lower than the bulk mobility because of increased scattering of the electrons at the silicon-oxide interface, as shown in Figure 2.71. The surface mobility depends on how much the electrons interact with the interface and, therefore, on the vertical electric field that "pushes" the electrons against the interface. The higher the electric field, the lower is the surface mobility.

2.37.2 Surface Scattering

The mobility is reduced in a small dimension compared to a larger dimension due to an increase in the average vertical field in the inversion layer. As the channel length becomes smaller due to the lateral extension of the depletion layer into the channel region, the longitudinal electric field component E_y increases, and the surface mobility becomes field dependent. Because the carrier transport in a MOSFET is confined within the narrow inversion layer, and the surface scattering (the collisions suffered by the electrons that are accelerated toward the interface by E_y) causes reduction of the mobility, the electrons move with great difficulty parallel to the interface, so that the average surface mobility, even for small values of E_y, is about half as much as that of the bulk mobility. The mobility in the inversion layer is distinctly lower than in bulk material. This is due to the fact that the electron wave-function

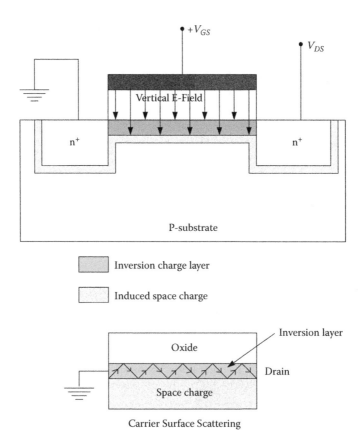

FIGURE 2.71
Vertical electric field in a short-channel MOSFET and due to that, surface scattering is seen.

extends into the oxide and the carrier mobility is lowered due to the lower mobility in the oxide.

2.37.3 Horizontal Electric Field Mobility Degradation

The mobility degradation due to the lateral field E_y (drain voltage) plays a more significant effect on the device current equations than does the normal field E_x (gate voltage). This is because an increase in the lateral field eventually causes velocity saturation of the carriers. For a given normal field, the velocity v of a carrier is proportional to E_y at low lateral fields, and the proportionality constant is the surface mobility μ_s. However, as E_y increases, the carrier velocity tends to saturate. Carriers in the short-channel devices reach the velocity saturation at lower values of V_{DS} than for the long-channel devices. This effect is due to the channel length reduction that implies higher horizontal electric fields for equivalent drain to-source voltages than the long-channel MOSFETs. The horizontal electric field within the channel is due to the voltage applied to the drain terminal. Due to this horizontal electric field, horizontal mobility also decreases.

References

1. Bart Van Zeghbroeck, *Principles of Semiconductor Devices*, Department of ECE, University of Colorado, 2007.
2. H.J. Juretschke, *Crystal Physics*, W.A. Benjamin, New York, 1974.
3. J.F. Nye, *Physical Properties of Crystals*, Oxford University Press, New York, 1979.
4. C. Kittel, *Introduction to Solid State Physics*, Wiley, New York, 1976.
5. R.H. Bube, *Electrons in Solids*, 3rd ed., Academic Press, New York, 1992.
6. B.G. Streetman, *Solid State Electronic Devices*, 4th ed., Prentice Hall, Upper Saddle River, NJ, 1995.
7. W.A. Harrison, *Solid State Theory*, Dover, New York, 1974.
8. W. VanRoosbroeck, Theory of flow of electrons and holes in germanium and other semiconductors, *Bell Syst. Techn. J.*, vol. 29, pp. 560–607, 1950.
9. R. Entner, Modeling and Simulation of Negative Bias Temperature Instability, dissertation, Technical Universität Wien, Vienna, Austria, 2007.
10. M.S. Lundstrom, *Fundamentals of Carrier Transport*, 1st ed., Addison-Wesley, Reading, MA, 1990.
11. G. Baccarani, F. Odeh, A. Gnudi, and D. Ventura, Semiconductors Part II, A critical review of the fundamental semiconductor equations, *The IMA Volumes in Mathematics and its Applications*, vol. 59, pp. 19–32, Springer-Verlag, New York, 1994.
12. Dragica Vasilesca and Stephen M. Goodnick, *Computational Electronics*, 1st ed., Morgan and Claypool, San Rafael, CA, 2006.
13. E.M. Conwell, *High Field Transport in Semiconductors*, Academic Press, New York, 1967.
14. D.L. Rode, Low field transport in semiconductors, *Semiconductors and Semimetals*, vol. 10, pp. 1–90, Academic Press, New York, 1972.
15. C. Jacoboni and L. Reggiani, The Monte Carlo method for the solution of charge transport in semiconductors with applications to covalent materials, *Rev. Mod. Phys.*, vol. 55, pp. 645–705, 1983.
16. M.V. Fischetti and S.E. Laux, Monte Carlo analysis of electron transport in small semiconductor devices including band-structure and space-charge effects, *Phys. Rev. B*, vol. 38, pp. 9721–9745, 1988.
17. T. Kunikiyo et al., A Monte Carlo simulation of anisotropic electron transport in Silicon including full band structure and anisotropic impact-ionization model, *J. Appl. Phys.*, vol. 75, pp. 297–312, 1994.
18. A. Pacelli and U. Ravaioli, Analysis of variance reduction schemes for ensemble Monte Carlo simulation of semiconductor devices, *Solid S. Electronics*, vol. 41, pp. 599–605, 1997.

19. G.E. Moore, Cramming more components onto integrated circuits, *Electronics*, vol. 38, no. 8, pp. 114–117, 1965.
20. C.H. Snowden, *Semiconductor Device Modelling*, 1st ed., Springer-Verlag, Berlin, 1989.
21. W. Shockley and W. T. Read, Statistics of the recombinations of holes and electrons, *Phys. Rev.*, vol. 87, no. 5, pp. 835–842, 1952.
22. R.N. Hall, Electron-hole recombination in germanium, *Phys. Rev.*, vol. 87, no. 2, p. 387, 1952.
23. D.J. Fitzgerald and A.S. Grove, Surface recombination in semiconductors, *IEEE Trans. Electron Devices*, vol. 15, no. 6, pp. 426–427, 1968.
24. D.K. Schroder, *Semiconductor Material and Device Characterization*, 3rd ed., Wiley Interscience, New York, 2006.
25. J.C. Maxwell, A dynamical theory of the electromagnetic field, *Royal Society Transactions*, vol. CLV, 1864.
26. T. Grasser, W. Gös, V. Sverdlov, and B. Kaczer, The universality of NBTI relaxation and its implications for modeling and characterization, in *Proc. Intl. Rel. Phys. Symp.*, pp. 1–13, 2007.
27. D.M. Caughey and R.E. Thomas, Carrier mobilities in silicon empirically related to doping and field, *Proc. IEEE*, vol. 55, pp. 2192–2193, 1967.
28. J.M. Dorkel and P.H. Leturcq, Carrier mobilities in silicon semi-empirically related to temperature, doping and injection level, *Solid State Electronics*, vol. 24, pp. 821–825, 1981.
29. N.D. Arora, J.R. Hauser, and D.J. Roulston, Electron and hole mobilities in silicon as a function of concentration and temperature, *IEEE Trans. on Electron Devices*, vol. 29, pp. 292–295, 1982.
30. D.B.M. Klaassen, A unified mobility model for device simulation-I. Model equations and concentration dependence, *Solid State Electronics*, vol. 35, pp. 953–959, 1992.
31. Silvaco Atlas User Manual, available at www.silvaco.com.
32. G. Baccarani, M. Wordeman, and R. Dennard, Generalized scaling theory and its application to 1/4 micrometer MOSFET design, *IEEE Trans. Electron Devices*, ED-31(4), p. 452, 1984.
33. R. Dennard et al., Design of ion-implanted MOSFETS with very small physical dimensions, *IEEE J. Solid-State Circuits*, SC-9, pp. 256–258, 1974.
34. N. Weste and K. Eshragian, *Principles of CMOS VLSI Design: A Systems Perspective*, Addison-Wesley, Reading, MA, 1993.
35. International Technology Roadmap for Semiconductors, available at http://www.sematech.org, 2001.
36. B. Streetman, *Solid State Electronic Devices*, Prentice Hall, Upper Saddle River, NJ, 1995.
37. C.T. Sah and H.C. Pao, The effects of bulk charge on the characteristics of metal-oxide-semiconductor transistors, *IEEE Trans. Electron Devices*, vol. ED-13, no. 4, pp. 393–409, 1966.
38. S.M. Sze, *Physics of Semiconductor Devices*, 2nd ed., Wiley, New York, 1981.
39. Y.P. Tsividis, *Operation and Modeling of the MOS Transistor*, McGraw-Hill, New York, 1999.
40. B. Razavi, *Design of Analog CMOS Integrated Circuit*, McGraw-Hill, New York, 2001.
41. N. Arora, *MOSFET Models for VLSI Circuit Simulation: Theory and Practice*, World Scientific, 2007. Reprinted from Springer-Verlag, Heidelberg, 1993.

42. Gordon Moore, *Cramming more components onto integrated circuits, Electronics,* vol.38, No. 8, April 19, 1965.
43. Claudio Fiegna, The effect of scaling on the performance of small-signal MOS amplifiers, *Proc ISCAS 2000,* pp. 733–736, May 2000.
44. R.H. Dennard, F.H. Gaensslen, L. Kuhn, and H.N. Yu, Design of micron MOS switching devices, *IEDM Dig. Techn. Pap.,* 1972.
45. R.H. Dennard, F.H. Gaensslen, H.-N. Yu, V.I. Rideout, E. Bassous, and A.R. Leblanc, Design of ion-implanted MOSFETs with very small physical dimensions, *IEEE J. Solid-State Circuits,* vol. 9, pp. 256–268, 1974.
46. R.H. Dennard, F.H. Gaensslen, E.J. Walker, and P.W. Cook, 1μm MOSFET VLSI technology: Part II— Device designs and characteristics for high-performance logic applications, *IEEE J. Solid-State Circuits,* vol. 14, pp. 247–255, 1979.
47. International Technology Roadmaps for Semiconductors (ITRS), *European Semiconductor Industry Association, JEITA, KSIA, Executive Summary,* 1999 and 2005 editions.
48. Ali Khakifirooz and Dimitri A. Antoniadis, MOSFET Performance Scaling—Part I: Historical Trend, *IEEE Trans. Electron Devices,* vol. 55, no. 6, pp. 1391–1400, 2008.
49. Ali Khakifirooz and Dimitri A. Antoniadis, MOSFET Performance Scaling— Part II: Future Directions, *IEEE Trans. Electron Devices,* vol. 55, no. 6, pp. 1401–1408, 2008.
50. S. Kang and Y. Leblebici, *CMOS Digital Integrated Circuits, Analysis and Design,* McGraw-Hill, New York 3rd ed., 1999.
51. S. Borkar, Design challenges of technology scaling, *IEEE Micro,* vol. 19, no. 4, pp. 23–29, July–August 1999.

3

Review of Numerical Methods for Technology Computer Aided Design (TCAD)

Kalyan Koley

CONTENTS

3.1 Introduction

In this chapter we discuss the numerical method used in the simulator. In an iterative method for solving, we start with a guess for the solution (often just the zero vector) and then successively renew this guess, getting closer to the solution at each stage. This iteration is usually performed until it converges to a result and a desired accuracy is achieved. The power of most iterative methods lies in their ability to achieve this convergence efficiently. However, two conflicting issues for a particular iterative method are high speed and convergence. Say, for example, to achieve a convergence for an equation a large number of iterations are needed. This will severely affect the speed and hence the time consumed. Conversely, a high-speed solver may not achieve a convergence. In the following sections the different iterative techniques implemented in technology computer aided design (TCAD) to achieve convergence efficiently are described in detail.

3.2 Numerical Solution Methods

For semiconductor devices, different solution methods are used depending upon the situation. It is also possible to use several different numerical methods to obtain solutions. And in addition, different combinations of models for a particular numerical method are also required for solving equations. There are three different types of solution techniques commonly used for obtaining solutions for semiconductor devices. These are represented by (a) de-coupled (GUMMEL), (b) fully coupled (NEWTON), and (c) BLOCK. The de-coupled technique like the Gummel method solves for each unknown in an equation while keeping the other variables constant and repeats the process until a stable solution is achieved [1]. In fully coupled techniques such as the Newton method, the total system of unknowns are solved together. Finally, in the combined or block method, the solution is obtained by solving some equations by the fully coupled method, while others are solved by the de-coupled method. Both techniques mentioned are broadly classified under the non-linear iteration method where the method converges to a solution nonlinearly and provides a more accurate result than its linear counterpart. In the next section these techniques are discussed in detail.

3.3 Non-Linear Iteration

3.3.1 Newton Iteration

For the solution of nonlinear systems, the scheme developed by Bank and Rose is applied. This scheme tries to solve the nonlinear system by using the Newton method: Each iteration of the Newton method solves a linearized version of the entire non-linear algebraic system. The size of the problem is relatively large, and each iteration takes a relatively long time. However, the iteration will normally converge quickly (in about three to eight iterations) so long as the initial guess is sufficiently close to the final solution. Strategies that use automatic bias step reduction in the event of non-convergence loosen the requirement of a good initial guess. Newton's method is the default for drift-diffusion calculations in ATLAS. There are several calculations for which ATLAS *requires* that Newton's method be used. These are DC calculations that involve lumped elements, transient calculations, curve tracing, and when frequency-domain small-signal analysis is performed. The Newton-Richardson method is a variant of the Newton iteration that calculates a new version of the coefficient matrix only when slowing convergence demonstrates that this is necessary. An automated Newton-Richardson method is available in ATLAS, and it improves performance significantly on most problems. The automated Newton-Richardson method is enabled by specifying the AUTO parameter of the METHOD statement [2]. If convergence is obtained only after many Newton iterations, the problem is almost certainly poorly defined. The grid may be very poor (i.e., it contains many obtuse or high aspect ratio triangles), or a depletion region may have extended into a region defined as an ohmic contact, or the initial guess may be very poor.

3.3.2 Gummel Iteration

Each iteration of Gummel's method solves a sequence of relatively small linear sub-problems. The sub-problems are obtained by linearizing one equation of the set with respect to its primary solution variable, while holding other variables at their most recently computed values. Solving this linear sub-system provides corrections for one solution variable. One step of Gummel iteration is completed when the procedure has been performed for each independent variable. Gummel iteration typically converges relatively slowly, but the method will often tolerate relatively poor initial guesses. The Gummel algorithm cannot be used with lumped elements or current boundary conditions. Two variants of Gummel's method can improve its performance slightly. These both limit the size of the potential correction that is applied during each Gummel loop.

The first method, called damping, truncates corrections that exceed a maximum allowable magnitude. It is used to overcome numerical ringing in the calculated potential when bias steps are large (greater than 1 V for room temperature calculations). The maximum allowable magnitude of the potential correction must be carefully specified: too small a value slows convergence, while too large a value can lead to overflow. The **DVLIMIT** parameter of the **METHOD** statement is used to specify the maximum allowable magnitude of the potential correction. By default, the value of this parameter is 0.1 V. Thus, by default Gummel iterations are damped. To specify undamped Gummel iterations, the user should specify DVLIMIT to be negative or zero.

The second method limits the number of linearized Poisson solutions per Gummel iteration, usually to one. This leads to under-relaxation of the potential update. This "single-Poisson" solution mode extends the usefulness of

Gummel's method to higher currents. It can be useful for performing low current bipolar simulations, and simulating MOS transistors in the saturation region. It is invoked by specifying the **SINGLEPOISSON** parameter of the **METHOD** statement.

3.3.3 Block Iteration

ATLAS offers several block iteration schemes that are very useful when lattice heating or energy balance equations are included. Block iterations involve solving subgroups of equations in various sequences. The subgroups of equations used in ATLAS have been established as a result of numerical experiments that established which combinations are most effective in practice.

In non-isothermal drift-diffusion simulation, specifying the BLOCK method means that Newton's method is used to update potential and carrier concentrations, after which the heat flow equation is solved in a de-coupled step. When the carrier temperature equations are solved for a constant lattice temperature, the **BLOCK** iteration algorithm uses Newton's method to update potential and concentrations [2]. The carrier temperature equation is solved simultaneously with the appropriate continuity equation to update the carrier temperature and carrier concentration.

When both the heat flow equation and the carrier temperature equations are included, the BLOCK scheme proceeds as described previously for the carrier temperature case, and then performs one de-coupled solution for lattice temperature as a third step of each iteration.

3.3.4 Combining the Iteration Methods

It is possible to start with the GUMMEL scheme and then switch to BLOCK or NEWTON if convergence is not achieved within a certain number of iterations. One circumstance where this can be very helpful is that the Gummel iteration can refine an initial guess to a point from which a Newton iteration can converge [3].

The number of initial GUMMEL iterations is limited by GUM.INIT. It may also be desirable to use BLOCK iteration and then switch to NEWTON if convergence is not achieved. This is the recommended strategy for calculations that include lattice heating or energy balance. The number of initial BLOCK iterations is limited by NBLOCKIT.

Any combination of the parameters GUMMEL, BLOCK, and NEWTON may be specified on the METHOD statement. ATLAS will start with GUMMEL if it is specified. If convergence is not achieved within the specified number of iterations, it will then switch to BLOCK if BLOCK is specified; if convergence is still not achieved the program will then switch to NEWTON.

3.4 Convergence Criteria for Non-Linear Iterations

After a few non-linear iterations, the errors will generally decrease at a characteristic rate as the iteration proceeds. Non-linear iteration techniques typically converge at a rate that is either linear or quadratic. The error decreases linearly when Gummel iteration is used (i.e., it is reduced by about the same factor at each iteration). For Newton iteration the convergence is quadratic (i.e., small errors less than one are approximately squared at each iteration). The non-linear iteration is terminated when the errors are acceptably small. The conditions required for terminations are called *convergence criteria*. Much effort has gone into developing reliable default convergence criteria for ATLAS [3]. The default parameters work well for nearly all situations, and most users will never need to change them.

3.5 Initial Guess Requirement

Non-linear iteration starts from an initial guess. The quality of the initial guess (i.e., how close it is to the final solution) affects how quickly the solution is obtained, and whether convergence is achieved. Users of ATLAS are not required to specify an initial guess strategy. If no strategy is defined, ATLAS follows certain rules that implement a sensible, although not necessarily optimum, strategy. There is some interaction between the choice of non-linear iteration scheme and the initial guess strategy. De-coupled iteration usually converges linearly, although perhaps slowly, even from a relatively poor initial guess. Newton iteration converges much faster for a good initial guess but fails to converge if started from a poor initial guess. One simple initial guess strategy is to use the most recent solution as the initial guess. Of course, there is no previous solution for the first calculation in a series of bias points. In this case, an initial solution is obtained for equilibrium conditions [3]. There is no need to solve the current continuity equations at equilibrium, and a solution of Poisson's equation is quickly obtained. It is also possible to modify the initial guess in a way that makes some allowance for the new bias conditions. Typical strategies include

- Using two previous solutions and interpolation to project a new solution at each mesh point.
- Solving a form of current continuity equation with carrier concentrations held constant. This strategy yields an improved estimate of new potential distribution.
- Modifying the majority carrier quasi-Fermi levels by the same amount as the bias changes.

- Parameters on the **SOLVE** statement can be used to specify an initial guess strategy. Five initial guess strategies are available.
- **INITIAL** starts from space charge neutrality throughout the device. This choice is normally used to calculate a solution with zero applied bias.
- **PREVIOUS** uses the currently loaded solution as the initial guess at the next bias point. The solution is modified by setting a different applied bias at the contacts.
- **PROJECTION** takes two previous solutions whose bias conditions differ at one contact and extrapolates a solution for a new applied bias at that contact. This method is often used when performing a voltage ramp.
- **LOCAL** sets the applied bias to the specified values and changes the majority carrier quasi-Fermi levels in heavily doped regions to be equal to the bias applied to that region. This choice is effective with Gummel iteration, particularly in reverse bias. It is less effective with Newton iteration.
- **MLOCAL** starts from the currently loaded solution and solves a form of the total current continuity equation that provides an improved estimate of the new potential distribution. All other quantities remain unchanged. **MLOCAL** is more effective than **LOCAL** because it provides a smooth potential distribution in the vicinity of p-n junctions. It is usually more effective than **PREVIOUS** because **MLOCAL** provides a better estimate of potential. This is especially true for highly doped contact regions and resistor-like structures.

When a re-grid is performed, the solution is interpolated from the original grid onto a finer grid. This provides an initial guess that can be used to start the solution of the same bias point on the new grid.

Although the initial guess is an interpolation of an exact solution, this type of guess does not provide particularly fast convergence.

3.6 Numerical Method Implementation

The Gummel method is effective where the system of equations is weakly coupled, with only linear convergence. The Newton method is effective where the system of equations is strongly coupled and has quadratic convergence. However, it may so happen that in a system of equations some of the quantities are weakly coupled while others are strongly coupled, or we can say a mixed coupled system. In such cases the Newton method consumes extra time solving for quantities that are essentially constant or weakly coupled. In accession to that, the Newton method requires a more accurate initial guess for the problem to obtain convergence. In order to compensate for the issues associated with the Newton method, it is better to use the block method to achieve convergence efficiently. The block method provides a comparatively faster simulation time in the mixed case over the Newton method [4]. Because Gummel can often furnish better initial guesses to problems, it is more appropriate to start a solution with a few Gummel iterations in order to generate a better initial guess and then switch to Newton to attain the final solution. The different solution methods are carried out by including the following statement for simulation:

```
METHOD GUMMEL BLOCK NEWTON
```

The exact meaning of the statement and the combination of the solution method depend upon the particular models to which it is applied. In the following sections the insights of this statement with respect to different models are described in detail.

3.7 Basic Drift Diffusion Calculations

The isothermal drift diffusion model requires the solution of three equations for potential, electron concentration, and hole concentration. Specifying GUMMEL or NEWTON alone on METHOD statement will produce simple Gummel or Newton solutions of the equations as described earlier. For almost all cases the Newton method is preferred because of its quadratic convergence to produce more accurate solution; hence, it is set as the default method. However, there are also alternative methods such as specifying:

```
METHOD GUMMEL NEWTON
```

This will cause the solver to initially start with Gummel iterations and then switch to Newton, if convergence is not achieved. This method is very robust, even though it consumes more time to obtain solutions for any device. However, this method is highly recommended for all simulations with floating regions such as silcon on insulator (SOI) transistors. A floating region is defined as an area of doping that is separated from all electrodes by a p-n junction. It may also be noted that BLOCK is equivalent to NEWTON for all isothermal drift-diffusion simulations.

3.8 Drift Diffusion Calculations with Lattice Heating

When the lattice heating model is added to drift diffusion, an extra equation is added. The BLOCK algorithm solves the three drift diffusion equations as a Newton solution and follows this with a de-coupled solution of the heat flow equation. The NEWTON algorithm solves all four equations in a coupled manner. NEWTON is preferred once the temperature is high; however, BLOCK is quicker for low temperature gradients. Typically the combination used is

```
METHOD BLOCK NEWTON
```

3.9 Energy Balance Calculations

The energy balance model requires the solution of up to five coupled equations. GUMMEL and NEWTON have the same meanings as with the drift diffusion model (i.e., GUMMEL specifies a de-coupled solution, and NEWTON specifies a fully coupled solution). However, BLOCK performs a coupled solution of potential, carrier continuity equations followed by a coupled solution of carrier energy balance, and carrier continuity equations. It is possible to switch from BLOCK to NEWTON by specifying multiple solution methods on the same line. For example,

```
METHOD BLOCK NEWTON
```

This will begin with BLOCK iterations and then switch to NEWTON if convergence is still not achieved. This is the most robust approach for many energy balance applications. The points at which the algorithms switch is predetermined but can also be changed on the METHOD statement. The default values set by TCAD tool Silvaco work well for most circumstances.

3.10 Energy Balance Calculations with Lattice Heating

When non-isothermal solutions are performed in conjunction with energy balance models, a system of up to six equations must be solved. GUMMEL or NEWTON solve the equations iteratively or fully coupled, respectively. BLOCK initially performs the same function as with energy balance calculations, and then solves the lattice heating equation in a de-coupled manner.

3.11 Setting the Number of Carriers

ATLAS can solve both electron and hole continuity equations, or only for one or none. This choice can be made using the parameter CARRIERS. For example,

```
METHOD CARRIERS = 2
```

This specifies that a solution for both carriers is required. This is the default. With one carrier the parameter ELEC or HOLE is needed. For example, for hole solutions only,

```
METHOD CARRIERS = 1 HOLE
```

To select a solution for potential only, specify

```
METHOD CARRIERS = 0
```

3.12 Important Parameters of the METHOD Statement

It is possible to alter all of the parameters relevant to the numerical solution process. This is not recommended unless you have expert knowledge of the numerical algorithms [5]. All of these parameters have been assigned optimal values for most solution conditions. It is beyond the scope of this chapter to give more details.

Two parameters, however, are worth noting at this stage:

1. **CLIMIT** or **CLIM.DD** specifies minimal values of concentrations to be resolved by the solver. Sometimes it is necessary to reduce this value to aid solutions of breakdown characteristics. A value of **CLIMIT = 1e-4** is recommended for all simulations of breakdown where the pre-breakdown current is small. **CLIM.DD** is equivalent to **CLIMIT** but uses the more convenient units of cm^{-3} for the critical concentration.

2. **DVMAX** controls the maximum update of potential per iteration of Newton's method. The default corresponds to 1 V. For power devices requiring large voltages, an increased value of **DVMAX** might be needed. **DVMAX = 1e8** can improve the speed of high-voltage bias ramps.

3. CLIM.EB controls the cut-off carrier concentration below which the program will not consider the error in the carrier temperature. This is applied in energy balance simulations to avoid excessive calculations of the carrier temperature at locations in the structure where the carrier concentration is low. However, if this parameter is set too high so that the carrier temperature errors for significant carrier concentrations are being ignored, unpredictable and mostly incorrect results will be seen.

3.12.1 Restrictions on the Choice of METHOD

The following cases all require **METHOD NEWTON CARRIERS = 2** to be set for isothermal drift-diffusion simulations. Both **BLOCK** and **NEWTON** are permitted for lattice heat and energy balance:

- Current boundary conditions
- Distributed or lumped external elements
- AC analysis
- Impact ionization

3.12.2 Pisces-II Compatibility

Previous releases of ATLAS (2.0.0.R) and other PISCES-II based programs use the **SYMBOLIC** command to define the solution method and the number of carriers to be included in the solution. In the current version of ATLAS, the solution method is specified completely on the METHOD statement. The COMB parameter that was available in earlier ATLAS versions is no longer required, as it is replaced by either the BLOCK method or the combination of GUMM EL and NEWTON.

References

1. Sentaurus TCAD Manuals, Synopsys Inc., Mountain View, CA.
2. Taurus Medici Manuals, Synopsys Inc., Mountain View, CA.
3. *ATLAS User's Manual*, Silvaco Int., Santa Clara, CA, May 26, 2006
4. M.K. Jain, S.R.K. Iyengar, and R.K. Jain, *Numerical Methods for Scientific and Engineering Computations*, New Age International, New Delhi, India, 2004.
5. M.K. Jain, *Numerical Solution of Differential Equations*, 2nd ed., New Age International, New Delhi, India, 2008.

4

Device Simulation Using ISE-TCAD

N. Mohankumar

CONTENTS

4.1 Introduction

Technology computer aided design (TCAD) is a design technique that involves computer simulation procedures to develop and characterize semiconductor processing technologies and devices. The two major components of a TCAD design process are process simulations and device simulations.

Current major suppliers of TCAD tools include Synopsys, Silvaco, and Crosslight. Synopsys provides both process and device simulation tools. It supports a broad range of applications such as complementary metal-oxide-semiconductor (CMOS), power, memory, solar cells, etc. In addition, Synopsys provides tools for interconnect modeling and parasitic extraction procedure which are required for optimizing the performance of an integrated circuit chip.

In this chapter, we discuss techniques for device simulation using the Synopsys tool suite. The tool suite includes tools for constructing device structures, meshing, simulation, visualization of electrical characteristics, plotting of various curves, and extraction of performance parameters. The working of each tool is described with the help of examples. The principles of operations involved are also discussed. The details of each tool are not provided, but readers are referred to corresponding user guides.

4.2 Design Flow

A typical design flow consists of either the creation of a device structure by a process simulation (Sentaurus Process or Taurus TSUPREME-4) tool or through computer aided design (CAD) operations and process emulation steps (using tools like Sentaurus Structure Editor). The constructed device is meshed intelligently using tools like Sentaurus Mesh/Noffset 3D. Sentaurus Device is used to simulate the electrical characteristics of the device. It simulates numerically the electrical behavior of a single semiconductor device in isolation or several physical devices combined in a circuit. Terminal currents, voltages, and charges are computed based on a set of physical device equations that describes the carrier distribution and the conduction mechanisms. Finally, Tecplot SV is used to visualize the output from the simulation in 2D and 3D, and Inspect tool is used to plot the electrical characteristics. The design flow is shown in Figure 4.1.

4.3 Sentaurus Structure Editor

Sentaurus Structure Editor is a device editor and process emulator based on CAD technology, used for constructing a semiconductor device. As shown in Figure 4.2, it has three distinct operational nodes: 2D structure editing, 3D structure editing, and 3D process emulation. Using Sentaurus Structure Editor, device structures are constructed or edited interactively using a graphical user interface (GUI) facility. Alternatively, devices can be generated

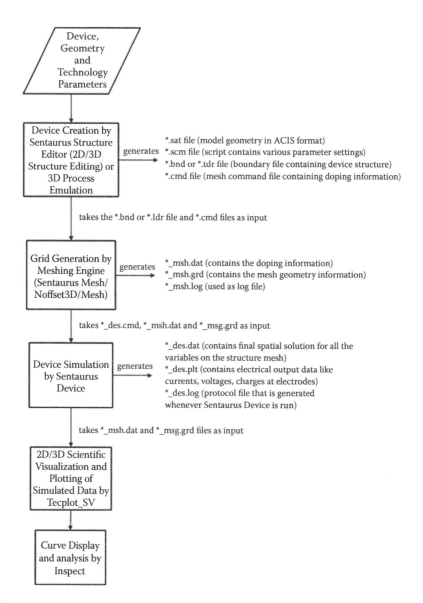

FIGURE 4.1
Design flow of device simulation process using ISE-TCAD, Synopsys.

in batch mode using Scheme scripting language. When a GUI action is performed, Sentaurus Structure Editor prints the corresponding Scheme command in the command-line window. From the GUI, 2D and 3D device models are created geometrically using 2D or 3D primitives such as rectangles, polygons, cuboids, etc. In the process emulation mode (Procem), Sentaurus Structure Editor translates processing steps such as etching and deposition,

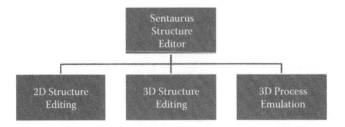

FIGURE 4.2
Operational modes of Structure Editor.

patterning, fill, and polish into geometric operations. Procem supports various options such as isotropic or anisotropic etching and deposition, rounding, and blending. The working of the Structure Editor is explained with the help of a design example, which is presented in the next sub-section.

4.3.1 Design of a 2D Bulk Metal-Oxide-Semiconductor Field-Effect Transistor (MOSFET) of Channel Length 100 nm

The 2D metal-oxide-semiconductor (MOS) structure that we attempt to create in Sentaurus Device Editor (SDE) is shown in Figure 4.3. The step-by-step design process using commands is illustrated pictorially in sequence in Figures 4.4 to 4.12.

Here we illustrate the steps required to create a rectangle that will be used as bulk silicon material. This involves the creation of a rectangle whose diagonally opposite corner coordinates are required to be specified, and commands are as follows. SDE will consider rectangular material as "Silicon" and this will be named "region_1". As this is a 2D MOS Device, the Z coordinate is considered to be 0.

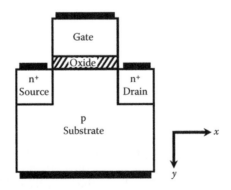

FIGURE 4.3
Schematic of 2D MOS structure. Positive X and Y axes are right-hand side and downward direction, respectively. For 2D MOS device, Z coordinate is always considered to be 0.

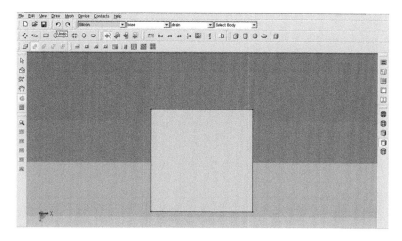

FIGURE 4.4
Schematic of bulk silicon creation in SCE by drawing a rectangle using a command whose corner coordinates are (−0.1 0.0 0.0) and (0.1 0.2 0.0).

To start Sentaurus Structure Editor, on the command line, enter:

```
sde
(sdegeo:create-rectangle
(position -0.1 0.0 0.0) (position 0.1 0.2 0.0)
"Silicon" "region_1")

(sdegeo:create-rectangle
(position -0.05 -0.002 0.0) (position 0.05 0 0.0)
"Oxide" "region_2")
```

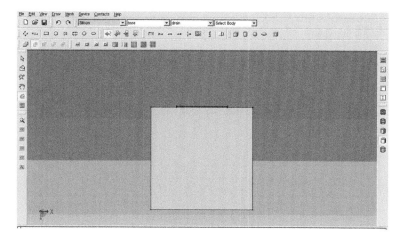

FIGURE 4.5
The creation of oxide on the bulk silicon by using sequential command.

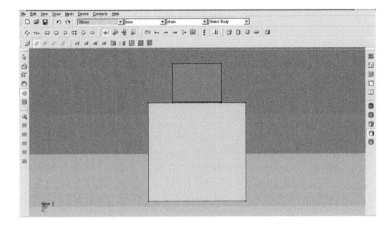

FIGURE 4.6
Molybdenum gate on oxide material. This is the actual MOS structure. Rectangular molybdenum material corner coordinates are (–0.05, –0.002, 0.0) and (0.05, –0.08, 0.0).

One needs to run the sequential command to get the entire structure [1–4]. Sequential command will now create another rectangle on the bulk silicon material which will be used as oxide or gate oxide. The material has been selected by command "Oxide" whose diagonally opposite corner coordinates are (–0.05 –0.002 0.0) and (0.05 0 0.0), and this oxide region is named "region_2".

```
(sdegeo:create-rectangle
(position -0.05 -0.002 0.0) (position 0.05 -0.08 0.0)
"Molybdenum" "region_3")

(sdegeo:define-contact-set "gate" 4 (color:rgb 1 0 0) "##")
(sdegeo:set-current-contact-set "gate")
```

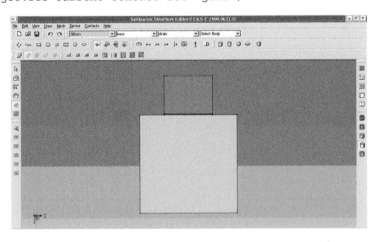

FIGURE 4.7
MOS structure with gate contact on top.

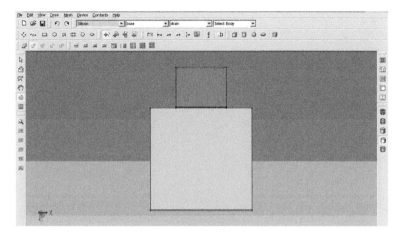

FIGURE 4.8
Body contact in which contact edge point coordinate (0.0 0.2 0) has been selected by the program.

```
(sdegeo:define-2d-contact (list (car (find-edge-id (position
0.0 -0.08 0)))) "gate");gate contact position and contact name
```

For the external connection to the 2D MOS device, a 2D gate contact has to be defined. Here contact is defined at the gate edge with proper coordinate point. Though the coordinate can be anywhere at the gate edge, for this device the middle point of the gate edge has been selected with coordinates (0.0 -0.08 0) and contact name "gate".

```
(sdegeo:define-contact-set "body" 4 (color:rgb 1 0 0) "##")
(sdegeo:set-current-contact-set "body")
```

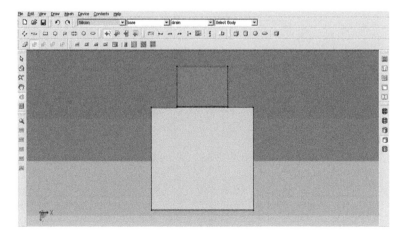

FIGURE 4.9
Source/drain 2D contact at position (–0.075 0 0) "source".

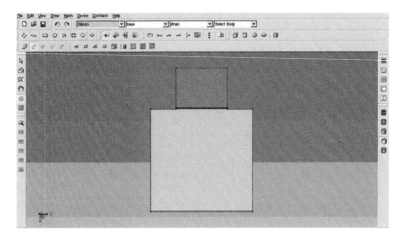

FIGURE 4.10
SDE shows the final MOS structure with all the contacts. Here another source/drain contact point is created here whose coordinate is (0.075 0 0) and name of "drain."

```
(sdegeo:define-2d-contact (list (car (find-edge-id (position
0.0 0.2 0)))) "body")

(sdegeo:define-contact-set "source" 4 (color:rgb 1 0 0) "##")
(sdegeo:set-current-contact-set "source")
(sdegeo:define-2d-contact (list (car (find-edge-id (position
-0.075 0 0)))) "source")

(sdegeo:define-contact-set "drain" 4 (color:rgb 1 0 0) "##")
(sdegeo:set-current-contact-set "drain")
```

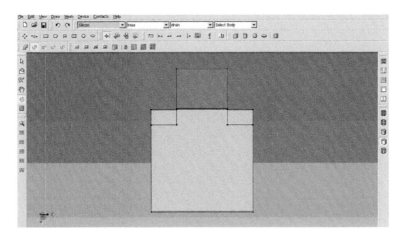

FIGURE 4.11
The SDE diagram after refinement window selection, source/drain and bulk material doping.

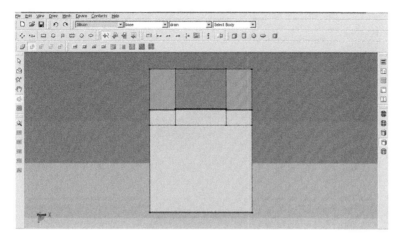

FIGURE 4.12
Complete structure of the 2D MOSFET in Sentaurus Editor after proper meshing at different places.

```
(sdegeo:define-2d-contact (list (car (find-edge-id (position
0.075 0 0))))) "drain")
```

Now the device requires proper doping at a different place to work as a MOSFET. For that purpose a different place has to be selected for doping. The techplots showing electrostatic potential and conduction band energy

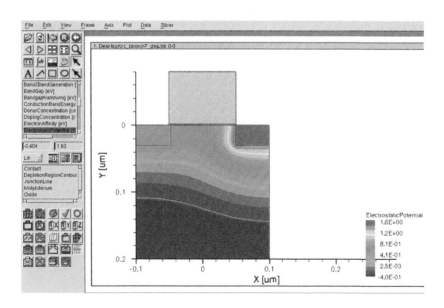

FIGURE 4.13 (See color insert)
Tecplot_sv showing electrostatic potential across the device at $V_{GS} = 2.0$ V, $V_{DS} = 2.0$ V.

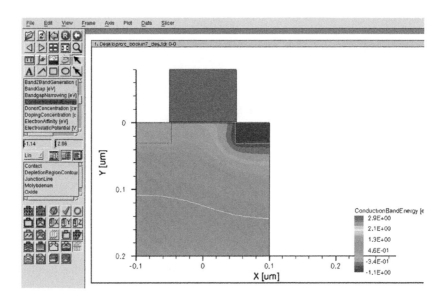

FIGURE 4.14 (See color insert)
Tecplot_sv showing conduction band energy (eV) across the device at V_{GS} = 2.0 V, V_{DS} = 2.0 V.

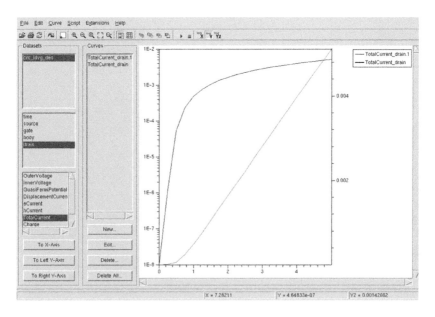

FIGURE 4.15
Inspect showing V_{GS} versus I_D both in normal and logarithmic plots at V_{GS} = 2.0 V, V_{DS} = 2.0 V.

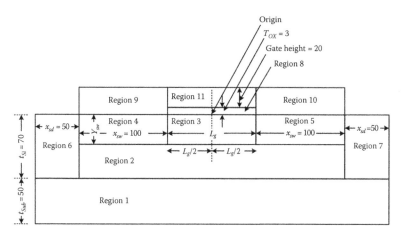

FIGURE 4.16
Construction of a proposed parameterized MOSFET.

FIGURE 4.17
Creation of Region 1 shown in Figure 4.16.

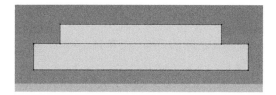

FIGURE 4.18
Creation of Region 2 shown in Figure 4.16.

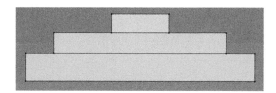

FIGURE 4.19
Creation of Region 3 shown in Figure 4.16.

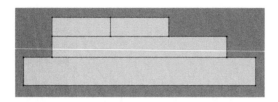

FIGURE 4.20
Creation of Region 4 shown in Figure 4.16.

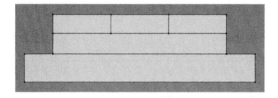

FIGURE 4.21
Creation of Region 5 shown in Figure 4.16.

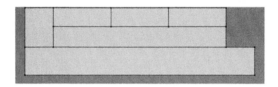

FIGURE 4.22
Creation of Region 6 shown in Figure 4.16.

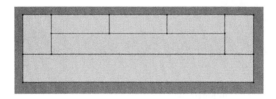

FIGURE 4.23
Creation of Region 7 shown in Figure 4.16.

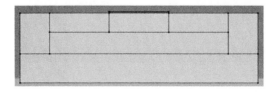

FIGURE 4.24
Creation of Region 8 shown in Figure 4.16.

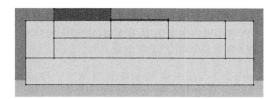

FIGURE 4.25
Creation of Region 9 shown in Figure 4.16.

across the device are shown in Figures 4.13 and 4.14. The characteristic plots in normal and logarithmic scale are shown in Figure 4.15. The diagram of a proposed parameterized MOSFET is shown in Figure 4.16. Figures 4.17 to 4.27 show the construction steps of the 11 Regions of the parameterized MOSFET structure shown in Figure 4.16. The following command [1] will select a different place as a rectangular window, and doping in that window is possible according to the requirement and command:

```
(sdedr:define-constant-profile "ConstantProfileDefinition_1"
"BoronActiveConcentration" 0.4e18)
(sdedr:define-constant-profile-material
"ConstantProfilePlacement_1"
"ConstantProfileDefinition_1" "Silicon")
```

Here the profile name is "ConstantProfileDefinition_1" and the dopant is "BoronActiveConcentration" as boron is needed to be implanted in the bulk material whose doping concentration is 0.4e18 (0.4×10^{18}) per cm^3 and whose name is mentioned as "BoronActiveConcentration". This

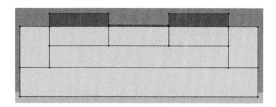

FIGURE 4.26
Creation of Region 10 shown in Figure 4.16.

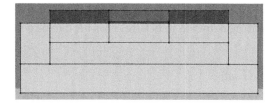

FIGURE 4.27
Creation of Region 11 shown in Figure 4.16.

"ConstantProfileDefinition_1" will be placed wherever it will find "Silicon" material [1]. In this way, the entire bulk material will be doped by boron with a doping concentration of 0.4×10^{18} per cm^3.

```
(sdedr:define-refeval-window "RefEvalWin_1" "Rectangle"
(position -0.1 0 0) (position -0.05 0.03 0))
(sdedr:define-constant-profile "ConstantProfileDefinition_2"
"ArsenicActiveConcentration" 1e+20)
position(sdedr:define-constant-profile-placement
"ConstantProfilePlacement_2" "ConstantProfileDefinition_2"
"RefEvalWin_1" 0 "Replace")
```

Now source/drain doping is needed, and for this purpose two rectangular areas must be selected. For source doping, a rectangular window has been selected by the command line whose corner coordinates are (−0.1 0 0) and (−0.05 0.03 0). This is named "RefEvalWin_1" and a doping profile has been selected "ArsenicActiveConcentration" as Arsenic is required as a dopant whose doping concentration is 10^{20} per cm^3, and finally this profile definition name is "ConstantProfileDefinition_2". Similarly, using the command below, drain doping has been done. Here the "Replace" command [1] will replace previous boron doping by new arsenic doping with the mentioned doping profile.

```
(sdedr:define-refeval-window "RefEvalWin_2" "Rectangle"
(position 0.05 0 0) (position 0.1 0.03 0)) position
(sdedr:define-constant-profile "ConstantProfileDefinition_3"
"ArsenicActiveConcentration" 1e+20)
(sdedr:define-constant-profile-placement
"ConstantProfilePlacement_3" "ConstantProfileDefinition_3"
"RefEvalWin_2" 0 "Replace")
```

4.3.2 Meshing

The input and output files of Sentaurus Structure Editor are:

- Scheme script file (.scm)

  This is a user-defined script file that contains scheme script commands describing the steps to be executed by Sentaurus Structure Editor in creating a device structure. This file can be edited to change its contents.

- ACIS SAT file (.sat)

  This file contains the model geometry in native ACIS format and cannot be edited directly.

- DF-ISE boundary file (.bnd)

    This is a boundary representation file written in the DF-ISE format. It can be directly loaded into Sentaurus Structure Editor and then to mesh engines.

- DF-ISE doping and refinement file (.cmd)

    This is a DF-ISE format file containing doping and mesh refinement information that, in conjunction with the corresponding boundary file, uniquely defines the geometry of the model.

4.4 Mesh

Mesh Generation Tools is a suite of tools that produce finite-element meshes for use which are required in semiconductor device simulation or process simulation. Once the device structure is created, meshing is usually required before the device can be numerically solved for its electrical properties. The Mesh Generation Tools are composed of three mesh generation engines: Sentaurus Mesh, Noffset3D, and Mesh. The choice of which mesh generator to use in an application depends largely on the geometry of the device. These mesh generators generate high-quality spatial discretizations for 1D, 2D, and 3D devices using a variety of mesh generation algorithms and procedures. Meshing basically involves defining a meshing strategy where the maximum and minimum sizes of the meshes are defined. These definitions are then placed in a specific region that may be a material or a device region or a user-defined refinement/evaluation (Ref/Eval) window. Ref/Eval windows are areas in which a certain mesh refinement or doping profile is to be applied. In some cases, the mesher can be instructed to refine the mesh in areas of steep doping gradients or near interfaces. For example, in the channel of a MOS transistor, a dense meshing is suitable near the silicon-oxide interface. The tightness of the grid spacing may be relaxed toward the bulk. This keeps the problem at a minimum of central processing unit (CPU) time.

Files in the Mesh Generator tool:

Input files: *.bnd or *.tdr and *.cmd files from SSE

Output Files: *_msh.dat (contains the doping information), *_msh.grd (contains the mesh geometry information), and *_msh.log (used as log file)

To summarize, after meshing, the device is divided into numbers of criss-cross points (which depend on the size of the mesh). Every point contains the following information:

1. Location of the point
2. Material of the point
3. Doping concentration of the point

This information may also be stored in a single *_msh.tdr file instead of *_msh.grd and *_msh.dat to enable the user to input a single file instead of two files to the next stage (i.e., Simulation).

4.4.1 Design Continuation

The following commands are used to mesh the generated structure:

```
(sdedr:define-refeval-window "RefEvalWin_4" "Rectangle"
(position -0.1 -0.08 0) (position 0.1 0.2 0)) ;global meshing

(sdedr:define-refinement-size "RefinementDefinition_1" 0.005
0.005 0 0.005 0.005 0)
(sdedr:define-refinement-placement "RefinementPlacement_1"
"RefinementDefinition_1" "RefEvalWin_4")

(sdedr:define-refeval-window "RefEvalWin_3" "Rectangle"
(position -0.05 -0.002 0) (position 0.05 0.03 0))

(sdedr:define-refinement-size "RefinementDefinition_2" 0.001
0.001 0 0.001 0.001 0)
(sdedr:define-refinement-placement "RefinementPlacement_2"
"RefinementDefinition_2" "RefEvalWin_3")

(sde:build-mesh "mesh" "-P -discontinuousData -f -t -d -F tdr "
"crc");
```

This last command will generate a file name crc_mesh.tdr which is required for further simulation of the device.

4.5 Sentaurus Device

Sentaurus Device is a comprehensive device simulator framework for simulating the electrical, thermal, and optical characteristics of silicon-based and compound semiconductor devices. The device simulation tool simulates the characteristics of the devices as a response to external electrical, thermal, or optical signals and boundary conditions.

4.5.1 Input-Output Files of the Tool

The input command file of Sentaurus Device consists of several command sections, each of which executes a relatively independent function. The default extension of this file is _des.cmd. The input command file typically consists of the following sections:

1. File
2. Electrode
3. Physics
4. Plot
5. Math
6. Solve

The File section describes the various input files and output files. The essential input file consists of the information regarding the device geometry and field values (_msh.tdr), for example, the doping on the structure. In addition, an optional parameter file can be specified. The device geometry information consists of the regions and materials of the device, location of the contacts, and the mesh points including the location of nodes and vertices. The optional parameter file (.par) consists of the user-defined model parameters. The Sentaurus Device simulation tool produces several output files. The Current file contains the electrical output data, such as currents, voltages, and charges at each of the contacts. The default extension of this file is _des.plt. In addition, a log file is generated that contains all the informative texts that the tool has downloaded during a run, including the error messages. The default extension of this file is _des.log.

The Electrode section consists of the definitions of the various contacts of the device, together with their initial bias conditions. Any special boundary condition for a contact can also be defined here. It may be noted with care that each electrode defined here must match (case sensitive) an existing contact name in the structure file, and only those contacts that are named in the Electrode section are included in the simulation process.

The Physics section consists of a declaration of the physical models that are to be used in the simulation procedure. Typically it consists of the carrier mobility model, the band-gap narrowing model, the carrier generation and recombination model, etc. With the use of a qualifier in the Physics section, it can be specified in which material or regions the models are to be activated. For example, Material = "[material name]", Region = "[region name]".

In this section, the models are declared on activation only. The model parameters, if different from the default, are defined and loaded using the optional Parameter file specified in the File section. The Physics section typical of a simple NMOSFET simulation is given here.

```
Physics {
Mobility (DopingDep HighFieldSat Enormal)
EffectiveIntrinsicDensity(OldSlotboom)
}
```

The Plot section is used to specify the solution variables that are to be saved in the Plot file after the simulation process is completed. The solution of these variables can later be visualized using tools like Tecplot SV.

The Math section is used to control the numeric solver involved in the simulation process. A typical Math section is shown below which may be used as a guideline.

```
Math {
      Extrapolate
      RelErrControl
      NotDamped=50
      Iterations=20
}
```

- *Extrapolate*: In quasi-stationary bias ramps, the initial guess for a given step is obtained by extrapolation from the solutions of the previous two steps (if they exist).
- *RelErrControl*: Switches error control during iterations from using internal error parameters to more physically meaningful parameters.
- *NotDamped = 50*: Specifies the number of Newton iterations over which the right-hand side (RHS) norm is allowed to increase. With the default of 1, the error is allowed to increase for one step only. It is recommended that NotDamped > Iterations be set to allow a simulation to continue despite the RHS-norm increasing.
- *Iterations = 20*: Specifies the maximum number of Newton iterations allowed per bias step (Default = 50). If convergence is not achieved within this number of steps, for a quasi-stationary or transient simulation, the step size is reduced by the factor *decrement* and simulation continues.

The Solve section defines a sequence of solutions to be obtained by the solver. To simulate the I_d–V_g characteristic, it is necessary to ramp the gate bias and obtain solutions at a number of points. By default, the size of the step between solutions points is determined by Sentaurus Device. As the simulation proceeds, output data for each of the electrodes (currents, voltages, and charges) are saved to the current file after each step and therefore the electrical characteristic is obtained. This can be plotted using Inspect. The Solve section is shown below.

```
Solve {
Poisson
Coupled {Poisson Electron}
Quasistationary (Goal {Name="gate" Voltage=2})
{Coupled {Poisson Electron}}
}
```

- Poisson: This specifies that the initial solution is of the nonlinear Poisson equation only. Electrodes have initial electrical bias conditions as defined in the Electrode section.
- Coupled {Poisson Electron}: The second step introduces the continuity equation for electrons, with the initial bias conditions applied. In this case, the electron current continuity equation is solved fully coupled to the Poisson equation, taking the solution from the previous step as the initial guess. The fully coupled or "Newton" method is fast and converges in most cases.
- Quasistationary (Goal {Name = "gate" Voltage = 2})
- {Coupled {Poisson Electron}}

The Quasistationary statement specifies that quasi-static or steady-state "equilibrium" solutions are to be obtained. A set of Goals for one or more electrodes is defined in parentheses. In this case, a sequence of solutions is obtained for increasing gate bias up to and including the goal of 2 V. A fully coupled (Newton) method for the self-consistent solution of the Poisson and electron continuity equations is specified in braces. Each bias step is solved by taking the solution from the previous step as its initial guess. If Extrapolate is specified in the Math section, the initial guess for each bias step is calculated by extrapolation from the previous two solutions.

4.5.2 Physical Models

The Physics section allows a selection of the physical models to be applied in the device simulation [2]. The physical phenomena that actually occur within semiconductor devices are very complicated and are generally described using differential equations (partial and full) of different levels of complexity. The coefficients and the boundary conditions required for solving the equations depend on the structure of the device, the principle of action, and the applied bias. Sentaurus Device allows for arbitrary combinations of transport equations and physical models.

4.5.2.1 Transport Equations

Depending on the device required to be simulated and the level of modeling accuracy required, four different simulation models can be selected:

- *Drift-diffusion isothermal simulation:* Described by basic semiconductor equations and is suitable for low-power density devices with long active regions.
- *Thermodynamic:* Accounts for self-heating and is suitable for devices with low thermal exchange, particularly, high-power density devices with long active regions.
- *Hydrodynamic:* Accounts for energy transport of the carriers. Suitable for devices with small active regions.
- *Monte Carlo:* Allows for full band Monte Carlo device simulation in the selected window of the device.

4.5.2.2 Poisson Equation and Continuity Equations

The three fundamental equations that dictate the charge transport in semiconductor devices are the Poisson equation and the electron and hole continuity equations. The Poisson equation is given as:

$$\nabla \cdot \varepsilon \nabla \varphi = -q(p - n + N_D - N_A) - \rho_{trap} \tag{4.1}$$

where ε is the electrical permittivity, q is the elementary electronic charge, n and p are the electron and hole densities, N_D is the concentration of ionized donors, N_A is the concentration of ionized acceptors, and ρ_{trap} is the charge density contributed by traps and fixed charges. The keyword for the Poisson equation is *Poisson*. The keywords for the electron and hole continuity equations are *electron* and *hole*, respectively. They are written as:

$$\nabla \cdot \vec{J_n} = qR_{net} + q\frac{\partial n}{\partial t} \qquad -\nabla \cdot \vec{J_p} = qR_{net} + q\frac{\partial p}{\partial t} \tag{4.2}$$

where R_{net} is the net electron–hole recombination rate, $\vec{J_n}$ is the electron current density, and $\vec{J_p}$ is the hole current density.

4.5.2.3 Drift-Diffusion Model

The drift-diffusion model is widely used for the simulation of carrier transport in semiconductors and is defined by the Poisson and continuity equations, see (4.1) and (4.2), where current densities for electrons and holes are given by:

$$\vec{J_n} = -nq\mu_n \nabla \Phi_n$$

$$\vec{J_p} = -pq\mu_p \nabla \Phi_p \tag{4.3}$$

where μ_n and μ_p are the electron and hole mobilities, and Φ_n and Φ_p are the electron and hole quasi-Fermi potentials, respectively.

The thermodynamic or non-isothermal model extends the drift-diffusion model to account for electrothermal effects. It assumes that the charge carriers are in thermal equilibrium with the lattice. In this model the electron and hole temperatures are assumed to be equal to the lattice temperature. The thermodynamic model is described by (4.1), (4.2), and lattice heat flow equations. Because the size of power devices is extremely large compared to that of CMOS devices, the drift-diffusion model including thermodynamic effects is usually sufficient in terms of accuracy. The drift-diffusion transport model, however, fails to describe the internal and external characteristics of deep submicron semiconductor devices. In particular, the drift-diffusion approach cannot reproduce velocity overshoot and often overestimates the impact ionization generation rates. The Monte Carlo method for the solution of the Boltzmann kinetic equation is the most general approach. However, it suffers from high computational requirements. Hence, it cannot be used for the routine simulation of devices in an industrial setting. The hydrodynamic (or energy balance) model is a good compromise. In the hydrodynamic transport model, carrier temperatures are allowed to be different from the lattice temperature.

4.5.2.4 Quantization Models

Some features of current MOSFET devices such as oxide thickness, channel width, etc., have reached their quantum-mechanical length scales. Therefore, the wave nature of electrons and holes can no longer be neglected. Quantization effects in a classical device simulation are included by incorporating potential, like quantity Λ_n in the classical carrier density formula as follows:

$$n = N_C F_{1/2}\left(\frac{E_{F,n} - E_C - \Lambda_n}{kT_n}\right) \tag{4.4}$$

An analogous quantity Λ_p is used for holes. Sentaurus Device implements four quantization models—that is, four different models for Λ_n and Λ_p. They differ in accuracy, computational expense, and robustness:

- The *van Dort model* is a numerically robust, fast, and proven model. However, it is only suited to bulk MOSFET simulations. The important terminal characteristics are well described by this model, but it does not give the correct density distribution in the channel.

- The *1D Schrödinger equations* make up the most accurate quantization model. It can be used for MOSFET simulation, and quantum well and ultrathin silicon-on-insulator (SOI) simulation. However, the simulation procedure is slow and often leads to convergence problems that restrict its use to situations with small current flow. It is used mainly for the validation and calibration of other quantization models.

- The *density gradient model* is numerically robust but significantly slower than the van Dort model. It can be applied to MOSFETs, quantum wells, and SOI structures. It gives reasonable description of terminal characteristics and charge distribution inside a device. Compared to the other quantization models, it can describe 2D and 3D quantization effects.
- The *modified local-density approximation* (MLDA) is a numerically robust and fast model. It can be used for bulk MOSFET simulations and thin SOI simulations. Although it sometimes fails to calculate accurate carrier distribution in the saturation regions because of its one-dimensional characteristic, it is suitable for three-dimensional device simulations because of its numeric efficiency.

Due to the wide usage of the density gradient model, we briefly discuss this below.

The density gradient model for Λ_n in (4.4) is given by a partial differential equation:

$$\Lambda_n = -\frac{\gamma \hbar^2}{12 m_n} \left\{ \nabla^2 \ln n + \frac{1}{2} (\nabla \ln n)^2 \right\} = -\frac{\gamma \hbar^2}{6 m_n} \frac{\nabla^2 \sqrt{n}}{\sqrt{n}} \tag{4.5}$$

where γ is a fitting parameter. The density gradient equation for electrons and holes is activated by the eQuantumPotential and hQuantumPotential switches in the Physics section. These switches can also be used in region wise or material wise in the Physics section. In metal regions, the equations are never solved. Apart from activating the equations in the Physics section, the equations for the quantum corrections must be solved by using eQuantumPotential or hQuantumPotential, or both in the Solve section. For example:

```
Physics {
eQuantumPotential
}
Plot {
eQuantumPotential
}
Solve {
Coupled {Poisson eQuantumPotential}
Quasistationary (
Do Zero InitialStep=0.01 MaxStep=0.1 MinStep=1e-5
Goal {Name="gate" Voltage=2}
){
Coupled {Poisson Electron eQuantumPotential}
}
}
```

4.5.2.5 Mobility Models

Sentaurus Device provides several options for the description of carrier mobilities. The various causes of mobility degradation can be individually modeled. In the simplest case, the mobility is considered to be a function of the lattice temperature. This is referred to as the constant mobility model and accounts only for phonon scattering. This should only be used for undoped materials. In doped semiconductors, scattering of the carriers by charged impurity ions leads to degradation of the carrier mobility. This is modeled in Sentaurus Device [4]. In the channel region of a MOSFET, the high transverse electric field forces carriers to interact strongly with the semiconductor–insulator interface. Carriers are subjected to scattering by acoustic surface phonons and surface roughness. This phenomenon of mobility degradation at interface is also modeled in Sentaurus Device. The carrier-carrier scattering effect can also be modeled. The carrier-carrier contribution to overall mobility degradation is combined with mobility contributions from other degradation models following Matthiessen's rule. The Philips unified mobility model unifies the description of majority and minority carrier bulk mobilities. In addition to describing the temperature dependence of the mobility, the model takes into account electron–hole scattering, screening of ionized impurities by charge carriers, and clustering of impurities. The mobility degradation due to high electric field can also be modeled in Sentaurus Device.

4.5.2.5.1 Doping-Dependent Mobility Degradation

The models for the mobility degradation due to impurity scattering are activated by specifying the DopingDependence flag to Mobility:

```
Physics {Mobility (DopingDependence...)...}
```

Different models are available and are selected by options to DopingDependence: Physics {Mobility (DopingDependence ([Masetti | Arora |UniBo])...)...}

If DopingDependence is specified without options, Sentaurus Device uses a material-dependent default. The default model used by Sentaurus Device to simulate doping-dependent mobility in silicon was proposed by Masetti et al. This is as follows:

$$\mu_{dop} = \mu_{\min 1} \exp\left(-\frac{P_c}{N_{tot}}\right) + \frac{\mu_{const} - \mu_{\min 2}}{1 + (N_{tot}/C_r)^\alpha} - \frac{\mu_1}{1 + (C_s/N_{tot})^\beta} \tag{4.6}$$

The reference mobilities $\mu_{\min 1}$, $\mu_{\min 2}$, and μ_1; the reference doping concentrations P_c, C_r C_s; and the exponents α and β are accessible in the parameter set DopingDependence. The corresponding values for silicon are given in Table 4.1.

TABLE 4.1

Masetti Model: Default Coefficients

Symbol	Parameter Name	Electrons	Holes	Unit
μ_{min1}	Mumin1	52.2	44.9	Cm^2/Vs
μ_{min1}	Mumin2	52.2	0	Cm^2/Vs
μ_1	Mu1	43.4	29.0	Cm^2/Vs
P_c	Pc	0	9.23×10^{16}	cm^{-3}
C_r	Cr	9.68×10^{16}	2.23×10^{17}	cm^{-3}
C_s	Cs	3.34×10^{20}	6.10×10^{20}	cm^{-3}
α	Alpha	0.680	0.719	1
β	Beta	2.0	2.0	1

4.5.2.5.2 Mobility Degradation at Interfaces

In the channel region of a MOSFET, the high transverse electric field forces carriers to interact strongly with the semiconductor–insulator interface. The carriers are thus subjected to scattering by acoustic surface phonons and surface roughness. The models in this section describe mobility degradation caused by these effects. To select the calculation of field perpendicular to the semiconductor–insulator interface, specify the Enormal option to Mobility:

```
Physics {Mobility (Enormal...)...}
```

The surface contribution due to acoustic phonon scattering has the form:

$$\mu_{ac} = \frac{B}{F_n} + \frac{C(N_{tot}/N_0)^\lambda}{F_n^{1/3}(T/300K)^k} \tag{4.7}$$

And the contribution attributed to surface roughness scattering is given by:

$$\mu_{sr} = \left(\frac{(F_n/F_{ref})^{A^*}}{\delta} + \frac{F_n^3}{\eta} \right)^{-1} \tag{4.8}$$

These surface contributions to the mobility are then combined with the bulk mobility according to Mathiessen's rule:

$$\frac{1}{\mu} = \frac{1}{\mu_b} + \frac{D}{\mu_{ac}} + \frac{D}{\mu_{sr}} \tag{4.9}$$

where $F_{ref} = 1$ V/cm, F_n = Normal electric field, $D = \exp(-x/l_{crit})$ (where x is the distance from the interface and l_{crit} is a fit parameter). In the Lombardi model, the exponent in (4.8), A^* is equal to 2. According to another study, an improved fit to measured data is achieved if A^* is given by:

$$A^* = A + \frac{\alpha_\perp (n+p) N_{ref}^v}{(N_{tot} + N_1)^v}$$ (4.10)

The respective default parameters that are appropriate for silicon are given in Table 4.2.

4.5.2.5.3 High-Field Saturation

In high electric fields, the carrier drift velocity is no longer proportional to the electric field; instead, the velocity saturates to a finite speed v_{sat}. The high-field saturation models include three sub-models: the actual mobility model, the velocity saturation model, and the driving force model. With some restrictions, these models can be freely combined. The actual mobility model is selected by flags eHighFieldSaturation or hHighFieldSaturation. The default is the Canali model whose parameter values for Silicon are given in Table 4.3. The Canali model originates from the Caughey–Thomas formula but has temperature-dependent parameters, which were fitted up to 430 K by Canali et al. [5]:

$$\mu(F) = \frac{(\alpha + 1)\mu_{low}}{\alpha + \left[1 + \left(\frac{(\alpha+1)\mu_{low} F_{hfs}}{v_{sat}}\right)^\beta\right]^{1/\beta}}$$ (4.11)

TABLE 4.2

Lombardi Model: Default Coefficients for Silicon

Symbol	Parameter Name	Electrons	Holes	Unit
B	B	4.75×10^7	9.925×10^6	Cm/s
C	C	5.80×10^2	2.947×10^3	$Cm^{5/3}/V^{-2/3}s^{-1}$
N_O	N_O	1	1	Cm^{-3}
λ	Lambda	0.1250	0.0317	1
K	K	1	1	1
δ	Delta	5.82×10^{14}	2.0546×10^{14}	Cm^2/Vs
A	A	2	2	1
$\alpha_\perp$	Alpha	0	0	cm^{-3}
N_1	N1	1	1	cm^{-3}
V	Nu	1	1	1
η	Eta	5.82×10^{30}	2.0546×10^{30}	$V^2cm^{-1}s^{-1}$
l_{crit}	L-crit	1×10^{-6}	1×10^{-6}	cm

TABLE 4.3

Canali Model Parameters (Default Values for Silicon)

Symbol	Parameter Name	Electrons	Holes	Unit
β_0	Beta0	1.109	1.213	1
β_{exp}	Betaexp	0.66	0.17	1
α	alpha	0	0	1

where μ_{low} denotes the low-field mobility, and F_{hfs} is the driving field. The exponent β is temperature dependent according to:

$$\beta = \beta_0 \left(\frac{T}{300K} \right)^{\beta_{exp}}$$

(4.12)

4.5.3 Design Continuation

This file "crc_des.cmd" is required to apply voltage at its different terminals. This file script is given here.

```
* Quantum

File{
    Grid      = " crc_mesh.tdr "
    Plot      = "crc_des.tdr"
    Current   = "crc_des.plt"
    Output    = "crc_des.log"
}

Electrode{
    {Name="source" Voltage=0.0}
    {Name="drain"   Voltage=0.0}
    {Name="gate"    Voltage=0.0}
    {Name="body" Voltage=0.0}
}

Physics{
    * DriftDiffusion
    eQuantumPotential
    EffectiveIntrinsicDensity(OldSlotboom)
    Mobility(
        DopingDep
        eHighFieldsaturation(GradQuasiFermi)
        hHighFieldsaturation(GradQuasiFermi)
```

```
      Enormal
   )
   Recombination(
      SRH(DopingDep)
   )
}
Plot{
*- Density and Currents, etc
   eDensity hDensity
   TotalCurrent/Vector eCurrent/Vector hCurrent/Vector
   eMobility hMobility
   eVelocity hVelocity
   eQuasiFermi hQuasiFermi

*- Temperature
   eTemperature Temperature * hTemperature

*- Fields and charges
   ElectricField/Vector Potential SpaceCharge

*- Doping Profiles
   Doping DonorConcentration AcceptorConcentration

*- Generation/Recombination
   SRH Band2Band * Auger
   AvalancheGeneration eAvalancheGeneration
hAvalancheGeneration

*- Driving forces
   eGradQuasiFermi/Vector hGradQuasiFermi/Vector
   eEparallel hEparallel eENormal hENormal

*- Band structure/Composition
   BandGap
   BandGapNarrowing
   Affinity
   ConductionBand ValenceBand
   eQuantumPotential
}

Math {
   Extrapolate
   Iterations=20
   Notdamped=100
   RelErrControl
   ErRef(Electron)=1.e10
   ErRef(Hole)=1.e10
}
```

```
Solve {
   *- Build-up of initial solution:
   NewCurrentFile="init"
   Coupled(Iterations=100){Poisson eQuantumPotential}
   Coupled{Poisson Electron Hole eQuantumPotential}

   *- Bias drain to target bias
   Quasistationary(
      InitialStep=0.01 Increment=1.35
      MinStep=1e-5 MaxStep=0.2
      Goal{Name="gate" Voltage=2.0 }
){Coupled{Poisson Electron Hole eQuantumPotential}
}

 *- gate voltage sweep
 NewCurrentFile=""

Quasistationary(
   InitialStep=1e-3 Increment=1.35
   MinStep=1e-5 MaxStep=1.1
   Goal{Name="drain" Voltage=2.0}
){Coupled{Poisson Electron Hole eQuantumPotential}
   CurrentPlot(Time=(Range=(0 1) Intervals=20))
}
* none
}
```

This will generate output files "crc_des.tdr", "crc_des.plt", and "crc_des.log".

4.6 Tecplot

Tecplot is a plotting software with extensive 2D and 3D capabilities for visualizing data from simulations and experiments. Tecplot can be started at the command prompt without loading any data file:

```
> tecplot_sv
```

4.6.1 Input Files

Two types of files can be loaded into Tecplot SV. The first type is the .tdr file. This file is used to describe a device structure, corresponding meshing, and the values of the field quantities existing in the corresponding device. The other type is the .plt file. Datasets included in this file are used by Tecplot SV to generate X-Y plots. Loading can be performed initially when Tecplot SV is started from the command line or interactively after Tecplot SV has started.

4.7 Inspect

Inspect tool is used for efficient viewing of *X-Y* plots such as doping profiles and I-V curves. An Inspect curve is a sequence of points defined by an array of *x* coordinates and *y* coordinates. Inspect extracts parameters such as junction depth, threshold voltage, and saturation currents from the respective *X-Y* plot. It is possible to manipulate curves interactively by using scripts. Inspect features a large set of mathematical functions for curve manipulation such as differentiation, integration, and to find min/max. The inspect script language is open to tool command language (TCL) and therefore inherits all the power and flexibility of TCL. To start inspect, at the command line type: inspect.

4.8 Parameterized Scripting

This section describes the use of script language for constructing parameterized devices. This technique enables the user to generate a device of various geometries without explicitly specifying the coordinates while constructing. This considerably reduces the device construction time.

```
; Model for Halo doped/Conventional DG Tunnel FET device
; Model for Asymmetric Heavily/Lightly Doped DG MOSFET
; All Dimensions in Nanometer initially

(sde:clear)
(define Lg 100)        ;Gate length (considered the X direction)
(define tsi 70)        ;Channel Thickness (considered the Y
                        direction)
(define Toxf 3)        ;Oxide thickness
(define xsw 100)       ;shallow source drain length
(define ysw 33)        ;shallow source drain depth
(define sub 50)        ;substrate depth
(define GateHght 20)   ;Gate Height thickness
(define Xsd (/Lg 2))   ;S/D width along X direction
(define Nsd 3.7e20)    ;Constant Source drain doping (p++ type)
(define Nssd 2e20)     ;Constant shallow source Drain doping
                        (n++ type)
(define Ns 1e16)       ;Constant reto Channel doping (n type)
(define Na 1e18)       ;Constant sub Channel doping (n type)
(define Nd 1e16)       ;Constant delta Channel doping (n++ type)

; TCAD Sentaurus by default assumes all dimensions in
micrometer. Thus, all the definitions stated above are
```

considered in micrometer. However, to convert the dimension from micrometer to nanometer, the following steps are performed.

```
(set! Lg (/Lg 1e3))
```
Let us consider the first command i.e., (set! Lg (/Lg 1e3)). Here, the originally defined value of Lg=100 micrometer is divided by 1e3 (1×103) to finally set the value of Lg to 100 nanometer.

```
(set! tsi (/tsi 1e3))
(set! Toxf (/Toxf 1e3))
(set! Xsd (/Xsd 1e3))
(set! Xsp (/Xsp 1e3))
(set! xsw (/xsw 1e3))
(set! ysw (/ysw 1e3))
(set! sub (/sub 1e3))
(set! GateHght (/GateHght 1e3))

(define d_Xmax (+ (+ Xsd (+ Xsd Lg)) (* xsw 2)))
(define d_Ymax (+ (+ (+ Toxf GateHght) tsi) sub)) ;
```

Let us illustrate the steps required to create region 1. This involves the creation of a rectangle, the coordinates of whose diatonically opposite corners A and B are to be specified.

```
;create substrate
(sdegeo:create-rectangle
  (position (* (+ (+ (/Lg 2) xsw) Xsd) -1) tsi 0)
  (position (+ (+ (/Lg 2) xsw) Xsd) (+ sub tsi) 0)
  "Silicon" "Body_3")

;create channel_2
(sdegeo:create-rectangle
  (position (+ (/Lg 2) xsw) tsi 0)
  (position (* (+ (/Lg 2) xsw) -1) ysw 0)
  "Silicon" "Body_2")

;create channel_1
(sdegeo:create-rectangle
  (position (/Lg -2) 0 0)
  (position (/Lg 2) ysw 0)
  "Silicon" "Body_1")

;create shallow source
(sdegeo:create-rectangle
  (position (/Lg -2) 0 0)
  (position (- (/Lg -2) xsw) ysw 0)
  "Silicon" "source_sh")
```

```
;create shallow drain
(sdegeo:create-rectangle
  (position (+ (/Lg 2) xsw) 0 0)
  (position (/Lg 2) ysw 0)
  "Silicon" "drain_sh")

;create source
(sdegeo:create-rectangle
  (position  (* (+ (/Lg 2) xsw) -1)           0       0)
  (position  (* (+ (+ (/Lg 2) xsw) Xsd) -1)   tsi     0)
  "Silicon" "Source")

;create drain
(sdegeo:create-rectangle
  (position  (+ (/Lg 2) xsw)              0      0)
  (position  (+ (+ (/Lg 2) xsw) Xsd)    tsi     0)
  "Silicon" "Drain")

;create oxide
(sdegeo:create-rectangle
  (position (/Lg -2) 0 0)
  (position (/Lg 2) (* Toxf -1) 0)
  "SiO2" "oxide")

;create oxide spacer
(sdegeo:create-rectangle
  (position  (* (+ (/Lg 2) xsw) -1)                    0       0)
  (position  (* (/Lg 2) -1)(* (+ Toxf GateHght) -1)   0)
  "SiO2" "oxide_2")

;create oxide spacer
(sdegeo:create-rectangle
  (position  (+ (/Lg 2) xsw)              0       0)
  (position  (/Lg 2)          (* (+ Toxf GateHght) -1)   0)
  "SiO2" "oxide_1")

;create gate electrode
(sdegeo:create-rectangle
(position (/Lg -2) (* Toxf -1) 0)
(position (/Lg 2) (* (+ Toxf GateHght) -1) 0)
"PolySi" "gate")
```

4.9 Sentaurus Workbench

Sentaurus Workbench is the complete graphical environment that integrates various Sentaurus tools. It is used throughout the semiconductor industry to design, organize, and run simulations. Simulations are comprehensively organized into projects. Sentaurus Workbench automatically manages the information flow, which includes preprocessing of user input files, parameterizing projects, setting up and executing tool instances, and visualizing results. Sentaurus Workbench allows users to define parameters and variables in order to run comprehensive parametric analyses. The resulting data can be used with statistical and spreadsheet software.

4.10 Summary

TCAD refers to the use of computer simulations to model semiconductor processing and devices. The two major functionalities of TCAD are device simulation and process simulation. The device simulation process starts from construction of the devices based on geometry and process parameters. Subsequently, the device is meshed intelligently and simulated. The various electrical characteristics can be visualized and plotted. This chapter demonstrates the device simulation procedure through examples using Sentaurus TCAD tool of Synopsys.

References

1. Integrated Systems Engineering (ISE) TCAD Manuals, 2006, Release 10.0.
2. Saha, Samar. Extraction of substrate current model parameters from device simulation, *Solid-State Electronics*, Volume 37, Issue 10, October 1994, Pages 1786–1788.
3. Saha, Samar. Design considerations for 25 nm MOSFET devices, *Solid-State Electronics*, Volume 45, Issue 10, October 2001, Pages 1851–1857.
4. MOSFET test structures for two-dimensional device simulation, original research article, *Solid-State Electronics*, Volume 38, Issue 1, January 1995, Pages 69–73.
5. Canali, C., G. Majni, R. Minder, and G. Ottaviani, Electron and hole drift velocity measurements in silicon and their empirical relation to electric field and temperature, *IEEE Trans. on Electron Devices*, vol. ED-22, pp. 1045–1047, 1975.

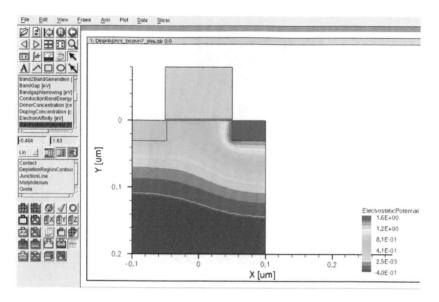

COLOR FIGURE 4.13
Tecplot_sv showing electrostatic potential across the device at $V_{GS} = 2.0$ V, $V_{DS} = 2.0$ V.

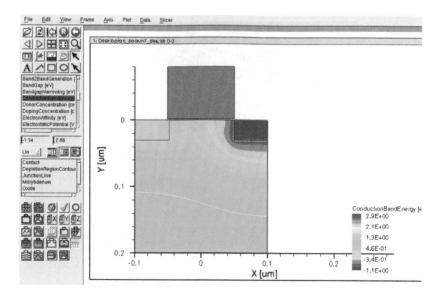

COLOR FIGURE 4.14
Tecplot_sv showing conduction band energy (eV) across the device at $V_{GS} = 2.0$ V, $V_{DS} = 2.0$ V.

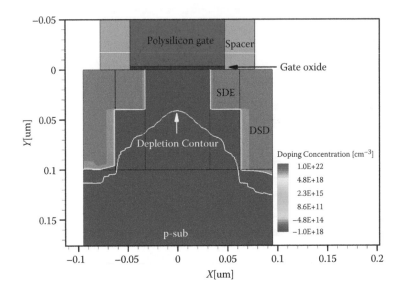

COLOR FIGURE 6.3
Bulk MOSFET structure created using TCAD.

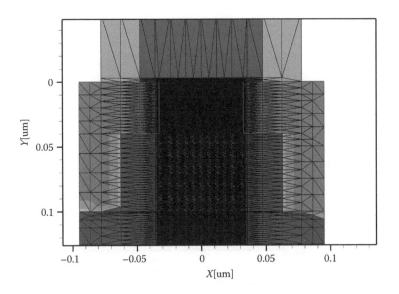

COLOR FIGURE 6.4
A zoomed-in view of the active region of the meshed structure.

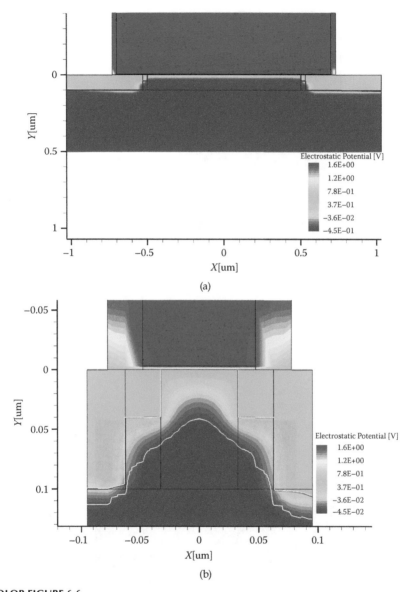

COLOR FIGURE 6.6

Electrostatic potential contours in (a) long channel $L_{eff} = 1\,\mu m$ and (b) short channel $L_{eff} = 65\,nm$ $(V_{gs} = 1.0V, V_{ds} = 0.05V)$.

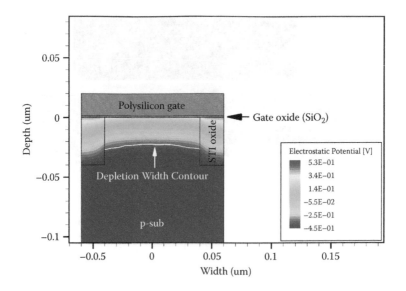

COLOR FIGURE 6.17
Cross-section along the width of a trench-isolated MOSFET.

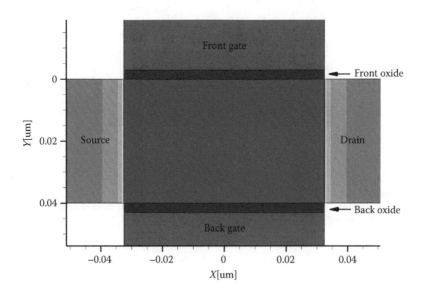

COLOR FIGURE 6.26
A typical DG-MOSFET structure as simulated in TCAD.

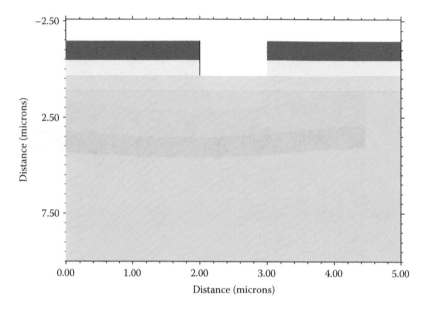

COLOR FIGURE 8.2
Deposition of negative photoresist of thickness 1 μ*m* on grown oxide; a portion (2 to 3 μ*m*) of the photoresist is being etched out by using mask NBL.

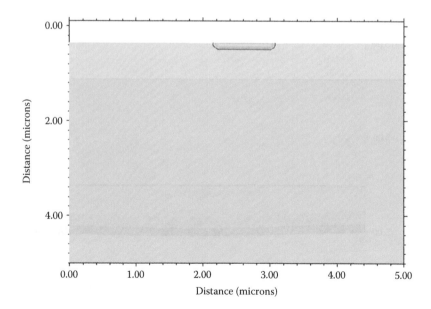

COLOR FIGURE 8.3
Implantation of antimony of pearson tilt = 7, dose = 1.0 e^{15}, and energy = 100, which will be used as NBL.

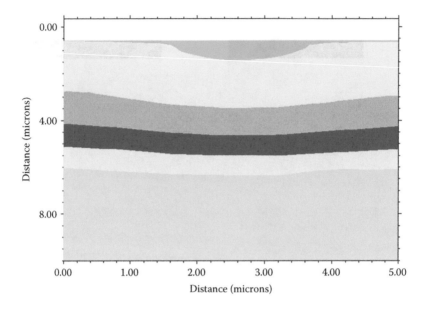

COLOR FIGURE 8.4
Placement of antimony dopant in the wafer after the application of drive-in voltage on it.

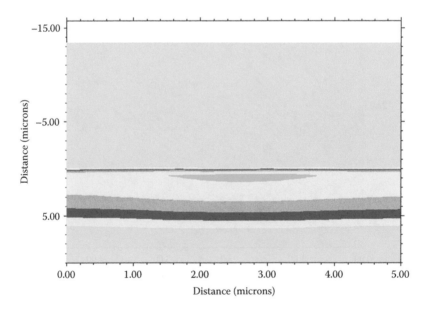

COLOR FIGURE 8.5
Placement of NBL and epitaxial growth on the initial wafer.

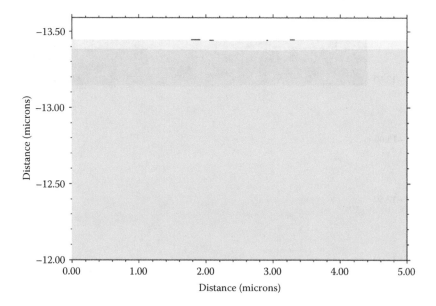

COLOR FIGURE 8.6
Pad oxide on the wafer is shown; here the y-axis has been chosen up to –12 μ*m*.

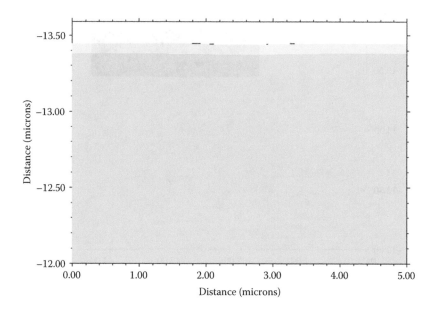

COLOR FIGURE 8.7
Structure of the wafer after formation of the gate oxide.

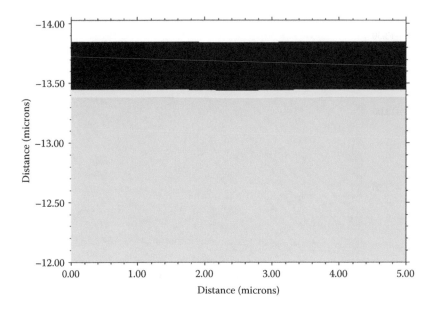

COLOR FIGURE 8.8
Polysilicon material deposition on the grown gate oxide material.

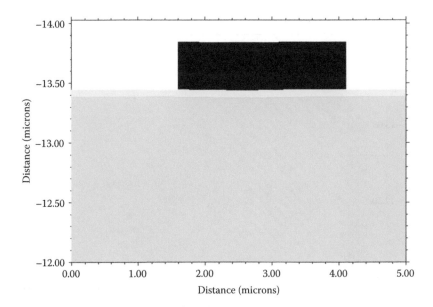

COLOR FIGURE 8.9
Polysilicon gate formation on the gate oxide after selective polysilicon material by etching from the wafer; polysilicon gate length is 2.5 μm.

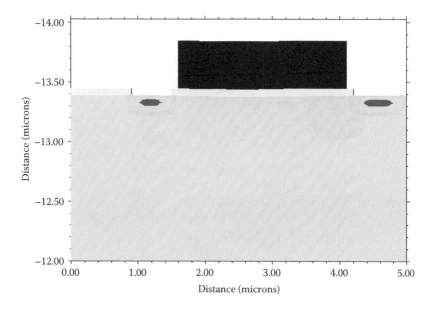

COLOR FIGURE 8.10
Device structure after source and drain formation.

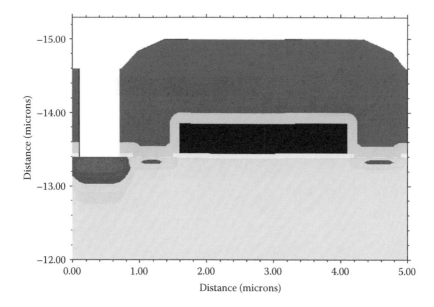

COLOR FIGURE 8.11
Formation of the p region in the wafer of length 0.1 to 0.7 μm.

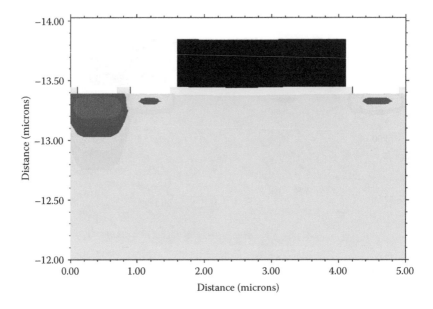

COLOR FIGURE 8.12
The structure achieved by executions of etch photoresist and etch nitride all statements.

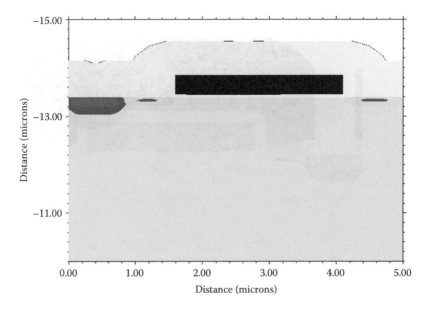

COLOR FIGURE 8.13
Borophosphosilicate glass deposition before the first layer of metal contact to the device.

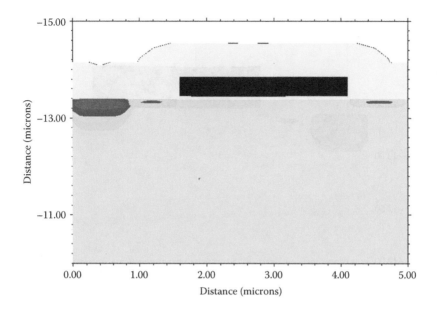

COLOR FIGURE 8.14
Structure of the device after application of annealing step for BPSG.

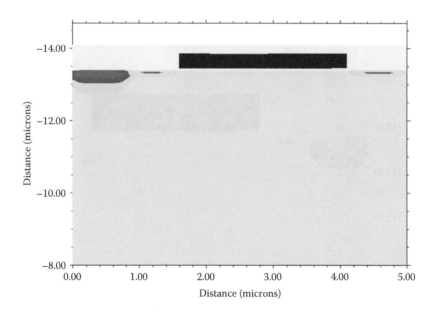

COLOR FIGURE 8.15
Polishing of the device top surface by removal of excess oxide.

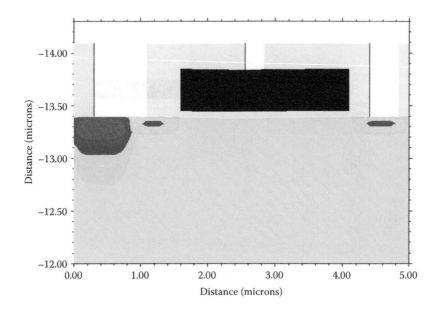

COLOR FIGURE 8.16
Selective etching has been performed to deposit aluminum through it for the first layer of metal contacts.

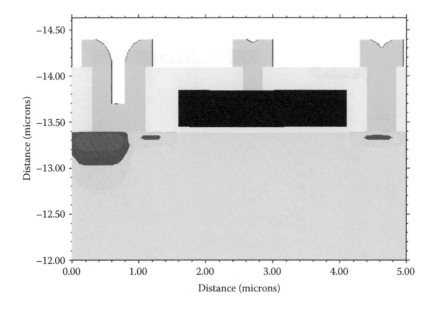

COLOR FIGURE 8.17
First layer of metal deposition through the opening of BPSG for the direct contacts from the device.

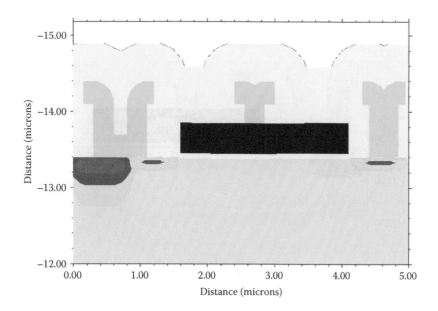

COLOR FIGURE 8.18
Deposition of IMD after formation of the first layer of metal contacts.

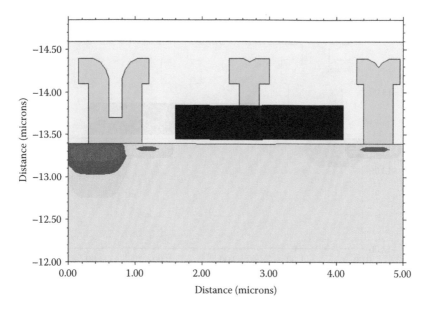

COLOR FIGURE 8.19
Structure achieved by smoothening the top surface by the polishing operation again.

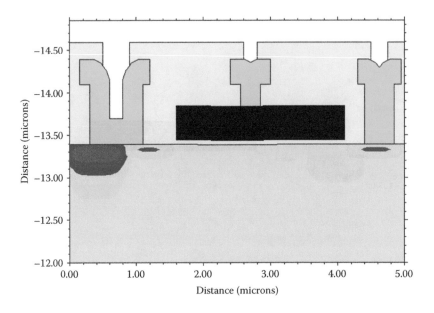

COLOR FIGURE 8.20
Opening region has been created by etching through which the second layer of metal will be deposited.

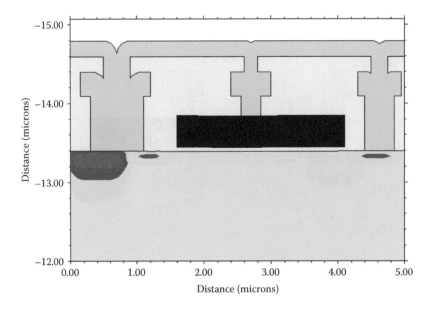

COLOR FIGURE 8.21
Second layer of metal is being deposited and connected to the first layer of the metal through the openings.

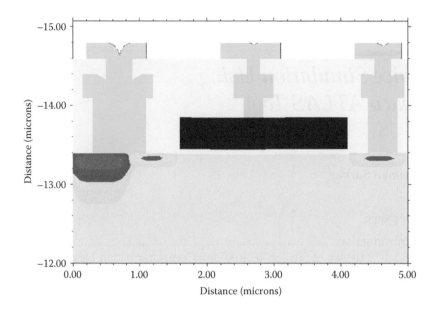

COLOR FIGURE 8.22
Final device structure, y coordinate is taken up to –12.0 μm.

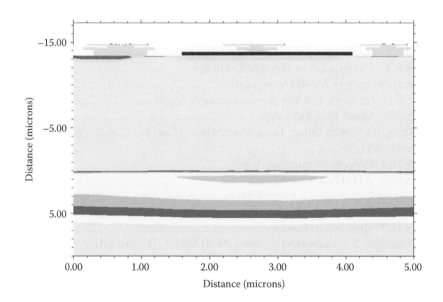

COLOR FIGURE 8.23
Final simulated structure of the 5 μm MOSFET.

5

Device Simulation Using Silvaco ATLAS Tool

Angsuman Sarkar

CONTENTS

5.1 Introduction

One of the critical issues for the fabrication of integrated circuits (ICs) is precise design of the operation of the circuits containing huge numbers of transistors. It is quite natural to predict the device operation by computer calculations using the simulators and device models. Devices scaled down to deca-nanometer range, operating at their physical limits, put stringent requirements on the modeling and simulation of device characteristics [1]. Computer aided modeling and simulation plays a crucial role in the development and prediction of the properties of modern technologies. Because of trial manufacturing and circuit redesign, the cost of modern, highly dense ICs containing deep sub-micron devices is very high. Simulation allows visualization and better understanding of the microscopic physical phenomenon and effects taking place over very small lengths or over small periods in macroscopic dimensions. To achieve these goals, over the past twenty years, two/three-dimensional numerical technology computer aided design (TCAD) device simulation tools have evolved into a well-accepted and extremely important branch of electronic design and automation (EDA) tools. It is suitable for the analysis and characterization of semiconductor structures and devices standing alone and/or coupled in integrated circuits [2]. TCAD has already been considered an invaluable tool in the research and development of new technology at the level of semiconductor process and device design.

The goal of this chapter is to introduce the device simulation of metal-oxide-semiconductor field-effect transistor (MOSFET) using Silvaco (Silicon Valley Corporation) TCAD device simulation tools using selected examples. The chapter illustrates the key aspects of Silvaco TCAD tools, showing their capability to understand the physical behavior and potential of a device structure. Silvaco TCAD device simulators provide unique insight into the internal operation of the analyzed device structure using a variety of complex physics-based models and advanced numerical solvers securing stable calculations.

Any semiconductor device is represented by a structure whose electrical and physical properties are discretized onto a mesh of nodes. The two/three-dimensional device structure may be the output of the process simulator or can be supplied from an input file containing the mesh information, types of materials, doping profiles in specific regions, names of the terminals, and properly defined boundary conditions with applied external electrical, optical, mechanical, magnetic, and thermal fields. An extensive set of electro-physical models is also supplied to characterize the behaviors of various physical effects present in a semiconductor. A device simulator calculates the output characteristics by solving a set of partial differential equations through iterative numerical techniques.

Two-dimensional device simulation with properly selected calibrated models and a well-defined appropriate mesh structure are very useful for predictive parametric analysis of novel device structures.

5.1.1 History of Silvaco Technology Computer Aided Design (TCAD)

Examples of early numeric simulation of semiconductors can be found in [3,4]. The journey of modern commercial TCAD started with the development of two famous general-purpose simulation software programs. SUPREM (Stanford University Process Engineering Models) and PISCES (Poisson and Continuity Equation Solver) came as an outgrowth of the research done at Stanford University. SUPREM3 is a one-dimensional process simulator, while SUPREM4 is a two-dimensional process simulator. PISCES is the corresponding two-dimensional device simulator. TSUPREM4 and MEDICI are the versions of these programs of Technology Modeling Associates (TMA) formed in 1979. ATHENA and ATLAS are commercial equivalent alternatives of these programs, as Silvaco later licensed these programs from Stanford University. The other major TCAD vendor is Integrated Systems Engineering (ISE). Their equivalent alternative products are DIOS and DESSIS from ISE.

Many other specific MOSFET device simulators have been developed, such as MINIMOS from the Technical University of Vienna, SEQUOIA Device Designer, FLOOPS/FLOODS from the University of Florida, IBM's FEDSS/FIELDAY, Agere's PROPHET/PADRE, APSYS, C-SUPREM and PROCOM from Crosslight, and Visual TCAD from Cogenda.

5.1.2 Device Simulation Challenges

There was always skepticism within the industry about the ability of TCAD device simulation procedure to correctly predict the experimental results. The primary reason for these discrepancies includes the lack of adequate models to describe the physical behavior of the actual device, the selection of appropriate electro-physical models, proper optimization of mesh structure, and the constant evolution of technology. As the technology scales down,

MOS transistors with channel lengths as small as 10 nm are now being actively studied both theoretically and experimentally [5]. Advances in scaling of MOSFET resulted in new physical effects like quantum mechanical effects, stress effects, introduction of new materials, etc., which are becoming pertinent to deep sub-micron devices. Therefore, the current challenge in device simulation is to address those new physical effects [5,6] in order to obtain accurate results.

5.1.3 Application of Device Simulation

The goal of the device simulation procedure is to use the output of the simulation process for predictive analysis of the properties and behavior of the simulated device structure with a unique insight into the internal process and structure operation, along with the possibility of further optimization and development. Two- and three-dimensional modeling and the simulation process contribute to a better understanding of the properties and behavior of the new devices by identifying the inevitable parasitic devices attributing to standard malfunction behaviors and degraded performances. Based on the interpretation of experimentally obtained data along with the result of device simulation, new structures and devices with modified layouts and concentration profiles can be designed and verified. More on TCAD applications can be found in [7,8].

5.2 How the Device Simulator ATLAS Works

ATLAS is a device simulation tool. The framework of ATLAS combines several one-, two-, and three-dimensional simulation tools into one comprehensive device simulation package. This allows for the simulation of a wide variety of modern semiconductor devices.

ATLAS is a physically based predictive device simulator that predicts the electrical characteristics associated with specified physical structures and bias conditions to provide insight within device operation and behavior.

To simulate a device in ATLAS, a description of the device is required. Descriptions of the device meshed with a two- or three-dimensional grid are provided via ASCII command line instructions supplied to ATLAS. The two- or three-dimensional grid approximating the device structure consists of a number of grid points known as nodes. The maximum number of grid points is limited to 20,000, a constraint set by ATLAS.

Figure 5.1 illustrates the main components of semiconductor device simulation. There are two main strongly coupled sets of equations that must be solved consistently. They are the transport equations governing the flow of the charges within the semiconductor, and the fields driving such charge

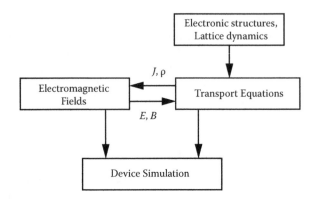

FIGURE 5.1
The ATLAS device simulation approach.

flow. Such fields are obtained from the solution of Maxwell's equations. These quasi-static fields are calculated from Poisson's equations with the aid of available boundary conditions. These fields are the driving forces for the charge transport.

In ATLAS, the transport of carriers is calculated at every node of the grid by applying a set of differential equations, derived from Maxwell's laws along with physical models (i.e., with appropriate numerical solvers invoked by the user). A set of differential equations on the grid is applied. With a bias point specified, the properties of the carriers in the device are solved through an iterative procedure. This facilitates the user to analyze electrical, thermal, and optical characteristics of the devices through simulation without having to manufacture the actual device and also to determine static and transient terminal currents and voltages in DC, AC, or transient modes of operation.

In order to complete a simulation run, ATLAS solves six equations for every point on the mesh structure defined. They are the Poisson's equation, two carrier continuity equations, two energy balance equations, and the lattice heat flow equation. The choice of techniques in solving these numerical equations can strongly affect the convergence time of a complete simulation run. However, in some circumstances it is sufficient to solve only one (either hole or electron) carrier continuity equation. It is possible to supply parameters to the METHOD statement to specify which carrier (electron, hole, or both) continuity equation is to be solved. The choice can be made using the parameters CARRIERS. For example, to include both electron and hole, the user must specify "CARRIERS = 2" and for only holes "CARRIERS = 1 HOLE" in the METHOD statement.

The user has to provide the structural information of the device to be simulated (appropriate mesh structure) to invoke the appropriate physical models and their associated numerical solvers and to set the desired bias profile for ATLAS to predict the electrical behavior of a particular device.

5.3 ATLAS Inputs and Outputs

To set up a device simulation, the first step required is to define the mesh structure of the device. Mesh structure can be created by entering text-based command line instructions in DECKBUILD or graphically by DEVEDIT. The text-based commands like Mesh, region, electrode, doping, etc., require definition of a structure given in a particular order, shown in Figure 5.2. Examples in Sections 5.11, 5.12, and 5.14 demonstrate the creation of device structure using command line instructions. It is possible to create the same structures using DEVEDIT (GUI-based structure and mesh editor) which work in conjunction with ATLAS. DEVEDIT is normally preferred to solve the problems of inadequate or excessive triangles to generate non-uniform mesh structure. DEVEDIT is also useful for re-meshing a device structure suffering with unsatisfactory mesh density during the process and the device simulation, or in an intermediate stage between the process and the device simulation. But it is worth mentioning that numerical process simulator ATHENA should not be replaced by DEVEDIT when it is important to obtain a device structure close to the real-world fabricated device. DEVEDIT produces two types of output: (1) standard Silvaco structure file format (*.str) containing mesh, region, impurity information, etc., and (2) command format (*.de) containing list of DEVEDIT instructions to specify the current state of mesh development. The Meshbuild command creates a mesh using the mesh parameters available. DEVEDIT file should be saved into two formats: (1) standard Silvaco structure (*.str) to be used in subsequent device simulation using tools like ATLAS and TONYPLOT or (2) command file (*.de) with the possibility to change the structure and mesh at a later time. The output of the process simulator ATHENA also produces structure file to be simulated using ATLAS (demonstrated via example in Section 5.13).

In summary, simulations generally use two inputs: a text file that has commands for ATLAS to execute and a structure file that defines the structure of the device. The text file with simulator-specific commands is created with DECKBUILD. The syntax of the input file statement is:

```
<statement> <parameter>=<value>.
```

The parameters can be real, integer, character, or logical.

The designer uses ATLAS to solve the input files to generate and visualize output data. ATLAS provides three types of output:

1. *Run-time output*: The progress of the simulation and error or warning messages are given by the run-time output. The various parameters displayed during the SOLVE statement are (a) "proj" denoting the initial guess methodology used (previous, local, or init); (b) "i,j,m" indicates the iteration number of outer loop, inner loop for decoupled

---

(a) Structure Specification

- Structure MESH specification

- Structure REGION specification

- ELECTRODE placement

- DOPING profile specification

(b) Model Specification

- Defining MATERIAL type for each region (SiO2 or Al, etc.)

- Invoke appropriate physical MODELS

- Setting boundary conditions and CONTACT information

- INTERFACE specification

(d) Solution Specification

- LOG file declaration to contain I-V characteristics

- Solution of device using SOLVE with intervals for bias voltage (or drive current) specification

- Specifying data types to be solved (e.g., recombination parameter, bad energy, etc.).

(e) Result Analysis

- Parameter extraction via EXTRACT command

- Analyzing device structure with different bias condition using graphical post-processing tool TONYPLOT

(c) Numerical Method Selection

- Invoke suitable numerical solvers such as Gummel, Newton, or Bulk METHOD for different operating conditions with appropriate parameters such as number of iteration, error limit, etc.

FIGURE 5.2
The order of each command group to be specified in ATLAS from (a) to (e).

method and numerical method used; m = G, B or N denoting G = Gummel, B = Block, and N = Newton, respectively. (c) x,rhs denotes norms of the equation being solved, and (d) (*) indicates the error measure.

2. *Log files*: These contain the electrical information, or more precisely I-V data, generated with the currents and terminal voltages applied during device analysis.

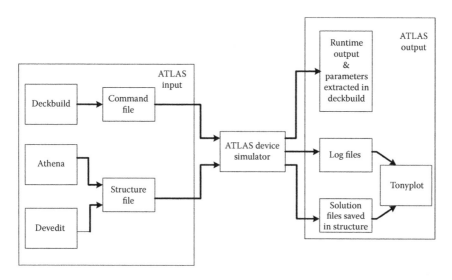

FIGURE 5.3
Input and output in ATLAS.

3. *Silvaco general structure file*: The structure files store the two- and three-dimensional data relating to the values of solution variables such as electric field, electrostatic potentials, etc., within the device for a single bias point.

The log files and the solution structure files are visualized using TONYPLOT. Figure 5.3 summarizes the various input and output techniques available for device simulation using ATLAS.

5.4 Simulation Setup

Once a structure file is defined in ATLAS, the remaining part of the setup (except for result analysis) consists of providing command line instructions that must be used to instruct ATLAS in order to complete a simulation run. The order in which these instructions are given is important and can be divided into five groups of statements outlined in Figure 5.2. Failing to adhere to this order may result in premature termination of a simulation run, or in an error message being produced. The order of statements within each group such as structural definition, model specification, and solution groups is also important. Failing to place these statements in proper order may result in aforementioned complications.

5.5 Brief Review of Electro-Physical Models Employed in ATLAS

The accuracy of the results obtained by device simulation process using ATLAS depends on the models used in the simulation process. Normally physics-based models are used to account for the complex dependencies of the device properties on dimensions and other process variables. Generally, the model parameters are derived from measurements and characteristics of the devices.

The drift-diffusion (DD) model is the simplest current transport model and is derived from Boltzmann's transport equation (BTE). It is famous for its simplicity and most efficient approach. Over a long period of time, this model has been at the core of semiconductor device simulation. The robust discretization of the DD equations proposed by Scharfetter and Gummel [9] is still in use today. In isothermal simulation, this simplest DD model includes three partial differential equations (PDEs): Poisson's equation and current continuity equations for electrons and holes. The total current density at any point (sum of electron and hole current) of the structure is found after solving the free electron and hole concentrations using potential and variable parameters like mobility, electric field and generation-recombination rate, etc.

As technology moves forward, devices scale down to sub-micron regime, without scaling the supply voltages proportionately. This results in a large electric field inside the device. As this large electric field changes quickly over small lengths, it brings about non-local and hot-carrier effects that tend to dominate device performance. Therefore, non-stationary carrier transport phenomenon such as velocity overshoot, which substantially influences the on-current (mainly the drift-current) of the MOSFET should be taken into account in the simulation. This makes the suitability of the DD model highly questionable for the simulation of nanometer MOSFET, as it neglects non-stationary effects.

As a result, transport models have been refined and extended to capture the practical transport phenomena more accurately. To address this issue, extensions of the DD model have been proposed. The extensions add a balance equation for the average carrier energy. They also add a driving term to the current relation, proportional to the gradient of the carrier temperature.

A non-isothermal thermodynamic model is chosen where self-heating effects are involved. Therefore, additional PDEs corresponding to non-isothermal temperature distribution must be coupled [10].

A more complex hydrodynamic (HD) or energy balance model is chosen for deep sub-micron devices, which involves solution of six coupled partial differential equations: three basic semiconductor equations and three energy balance equations. The individual electron and hole temperatures are calculated from the energy balance equation.

In the HD model the equation set of the DD model is extended by the energy balance equation to allow for non-stationary carrier transport. The disadvantages of the HD model are that it is less stable than DD, it is less efficient in terms of computation time than DD, and it sometimes overestimates the current in MOSFET [11].

The typical hierarchy of the MOSFET simulation approach consists of DD, HD, and Monte Carlo (MC) methods. MC is most reliable for calculation of on-current. Unfortunately, the added accuracy comes at the expense of higher computational complexity as compared to HD and DD. Thus MC is not suitable for investigations where large numbers of different transistors are to be simulated. Even 3D simulations necessary for FinFETs are difficult with MC. Furthermore, because of its statistical nature, the MC method has serious problems in calculating the very low subthreshold currents of MOSFETs accurately.

In a recent study, it was shown that the DD on-current is 10% less than MC-current calculated for a 100-nm MOSFET [12]. For shorter length MOSFET such as 40 nm, the difference of on-current is about 40% [13]. However, the on-current simulated by DD can be improved by adjusting the mobility model used. For the calculation of subthreshold current, the DD model is best suited for both long and short channel MOSFETs [14]. Granzner et al. [15] suggested that modifications of the velocity-field characteristics in the DD simulations are suggested to improve the accuracy of the DD model, and for the simulation of subthreshold current the standard DD model is best suited.

To incorporate quantum-mechanical effects (QMEs) that are significant for highly scaled deca-nanometer devices, the Schrödinger equation should be solved self-consistently in order to obtain the most accurate simulation result.

Therefore, to achieve maximum predictive capability one needs to move toward the quantum transport regime of the hierarchical structure shown in Figure 5.4. The green function's approach [16] shown in Figure 5.4 is most difficult in terms of complexity of equations and numerical efficiency.

The details of various advanced electro-physical models to characterize the properties and behavior of analyzed semiconductor device structures can be found in the user manual of simulator ATLAS [17].

Physical models are specified in the "MODELS" statement depending on the material used and physical nature of the device. For example, the statement "MODELS CVT SRH FERMIDIRAC" enables the Lombardi mobility model (constant voltage and temperature, CVT), Shockley-Read-Hall recombination with fixed carrier lifetimes (SRH) and Fermi-Dirac statistics (FERMIDIRAC) for common long-channel Si-MOSFET simulation.

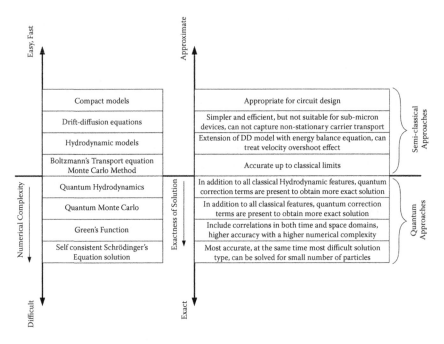

FIGURE 5.4
Hierarchy of transport model in Silvaco ATLAS.

5.6 Choice of METHOD in ATLAS

The equations can be solved either by a fully coupled (Newton), de-coupled (Gummel), or combined (Block) manner. The coupled solutions with all equations solved at once are best when the interactions between the equations are strong (e.g., high current producing sufficient local heating). Newton's method is a fully coupled procedure that solves the equations simultaneously, through a generalization of the Newton-Raphson's method for solving the roots of an equation. However, they need a good initial guess as to the solution variables for reliable convergence. Unless special techniques such as projection are used for calculating the initial guess, the voltage step size during a bias ramp in a fully coupled solution might be small in order to obtain reliable convergence. Also, the Newton method may spend extra time solving the quantities that are weakly coupled or almost constant. Newton's method is the default method or drift-diffusion calculation in ATLAS. Other than that, ATLAS requires Newton's method for DC calculations using lumped elements, transient calculations, and frequency-domain small-signal analysis.

On the other hand, de-coupled solutions where a subset of the equation is solved while others are held constant has shown an advantage when the interaction between the equations is small (typically low voltage and current levels). They tolerate a relatively poor initial guess for convergence. They tend to either diverge or take excessive central processing unit (CPU) time once the interaction among the equations becomes stronger. Gummel's method cannot be used with lumped elements or current boundary conditions. In a typical fully de-coupled procedure, if we choose quasi-Fermi level formulation, at first non-linear Poisson's equation is solved. The potentials obtained are substituted into continuity equations that are now linear and solved directly. The results in terms of quasi-Fermi levels are substituted back to Poisson's equation until the convergence is reached, as shown in Figure 5.5.

In general, Gummel's method is useful where the system of equations is weakly coupled, but it has only linear convergence. The Newton method is useful when the system of equations is strongly coupled and has quadratic convergence.

Gummel's method can provide better initial guess to the problems. Therefore, it is possible to start a solution using few Gummel's iterations to generate a good initial guess and then switch to Newton to complete the simulation [18], as shown in Figure 5.6.

The combined method will solve some equations fully coupled and the remaining others using a de-coupled method. The Block method can provide faster simulation than Newton's method for solving quantities that are weakly coupled. They involve solving a subgroup of equations in various sequences. In non-isothermal drift-diffusion simulation using Block method,

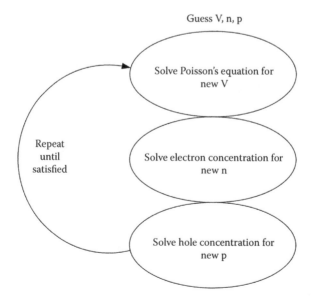

FIGURE 5.5
Uncoupled numerical solution.

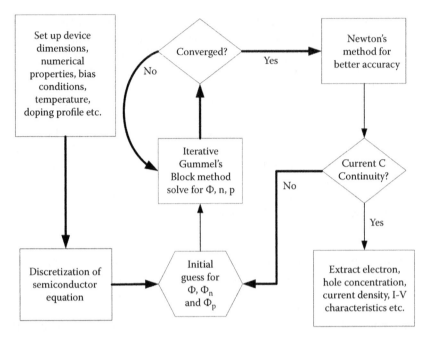

FIGURE 5.6
Steps for numerical solutions.

Newton's method is solved for a constant lattice temperature to update carrier concentration and potential, after which a heat flow equation with the appropriate continuity equation is solved using a de-coupled method to update carrier temperature and carrier concentration.

5.7 Mobility Models in ATLAS

The choice of carrier mobility models is one of the most important decisions of TCAD device simulation. This is governed by physical parameters, ambience, and operating conditions.

Electrons and holes accelerate by electric fields but lose momentum due to various scattering processes like lattice vibrations (phonons), impurity ions or other carriers, surfaces, and other material imperfections. Various microscopic phenomena affect the macroscopic mobility specified in the transport equation. Hence these mobilities are functions of the local electrical field, doping concentration, lattice temperature, and many other parameters. Based on the operating region, mobility models can generally be classified into four types: (1) low field, (2) high field, (3) bulk semiconductor region, and (4) inversion region.

From an alternative perspective, mobility models can be categorized into three types: (1) physically based, (2) semi-empirical, and (3) empirical.

Physical-based mobility models are obtained from fundamental calculations using a lot of approximations, and they rarely match the experimental data. In a semi-empirical model, to obtain an agreement between the model and the experimental result, the coefficients of the physical-based models are allowed to vary, keeping the power law dependencies preserved. In contrast, it is possible to vary the power-law dependencies in a purely empirical model, exhibiting the lowest physical content and a narrower range of validity. The most popular mobility models used in ATLAS are listed in Table 5.1.

Low field mobility degrades upon phonon and impurity scattering and is valid for carriers present in the bulk under low electric field. The high

TABLE 5.1

Summary of the Popular Mobility Model Used in ATLAS

Model	Behavior	Syntax	Features
Constant low-field mobility model	Low-field behavior	MUN, MUP, TMUN, and TMUP are specified in MOBILITY statement	Constant low-field mobility with only temperature variation
Concentration-dependent low-field model	Low-field behavior	CONMOB in MODELS statement and parameters in MOBILITY statement	Look up table based, valid for Si, GaAs at 300 K
Concentration, temperature-dependent analytical model	Low-field behavior	ANALYTIC in MODELS and parameters in MOBILITY statement	Based on Caughey-Thomas formulae, valid between 77 and 450 K
Arora's analytical model	Low-field behavior	ARORA in MODELS and parameters in MOBILITY statement	Alternatives to analytic model, doping and temperature dependent
Masetti analytical model for low-field mobility	Low-field behavior	MASETTI in MODELS and parameters in MOBILITY statement	Optimized for room temperature, concentration dependent
Carrier-carrier scattering model	Low-field behavior with carrier-carrier scattering	CCSMOB parameter in MODELS and parameters in MOBILITY statement	Temperature, doping, and carrier-carrier scattering dependent, useful when carrier concentration is very high such as forward bias power devices
Klassen unified low-field mobility model	Low-field behavior	KLA parameter in MODELS and KLA.N and KLA.P parameters in MOBILITY statement	Temperature dependent, doping dependent, includes the effect of lattice scattering, impurity scattering, carrier-carrier scattering, and impurity clustering. Uses separate mobility for majority and minority carriers, useful for bipolar devices

TABLE 5.1 (CONTINUED)

Summary of the Popular Mobility Model Used in ATLAS

Model	Behavior	Syntax	Features
Lombardi (CVT) inversion layer mobility model	Inversion layer	CVT on the MODELS statement and parameters in MOBILITY statement	Transverse field, doping, and temperature-dependent parts of mobility are combined. Overrides any other mobility model used in MODELS statement
Yamaguchi inversion layer mobility model	Inversion layer	YAMAGUCHI in MODELS and parameters in MOBILITY statement	Low-field, doping-dependent mobility with surface-degradation dependent on parallel filed included
Tasch model	Inversion layer	TASCH in MODELS statement and parameters in MOBILITY statement	Explicitly for MOSFETs, includes transverse field dependence for planar devices with very fine mesh structure
Shirahata model	Inversion layer	SHI in MODELS statement and SHI.N and SHI.P parameters in MOBILITY statement	General-purpose MOSFET mobility model, an alternative surface mobility model that can be combined with Klassen model
Watt surface mobility model	Perpendicular electric field dependent	SURFMOB parameter on the MODELS statement and parameters in MOBILITY statement	Includes phonon scattering, surface roughness scattering, and charged impurity scattering
Saturation velocity model	Parallel electric field dependent mobility	FLDMOB parameter on the MODELS statement and BETAN and BETAP parameters in MOBILITY statement	Caughey and Thomas expression is used to calculate field-dependent mobility, getting reduced at high field due to velocity saturation effect, model parallel field dependence for Si and GaAs

electric field dependent parameter decreases with increasing high electric field, making the velocity of the carrier constant known as saturation velocity.

Bulk mobility modeling generally requires:

1. Identification of low field mobility as a function of doping and lattice temperature
2. Identification of saturation velocity as a function of temperature
3. Description of transition between high and low field region

In ATLAS, low field mobility for the bulk region includes:

1. Constant mobility model
2. Caughey and Thomas model (doping and temperature dependent) [19]
3. Arora model (doping, temperature and carrier-carrier scattering dependent) [20]
4. Klaassen unified low-field mobility model (unified minority and majority carrier mobility, includes lattice scattering, carrier-carrier scattering, impurity-clustering effects at high concentrations) [21,22]

It is important to model mobility dependent on the transverse electric field associated with inversion layers in order to obtain accurate results. ATLAS supports five major transverse field-dependent mobility models. They are the CVT (Lombardi) MODEL [23], Watt model [25], Shirahata model [26], Yamaguchi model [27], and Tasch model [28].

In ATLAS, the CVT [23] model is selected by specifying CVT on the model statement. This model overrides any other mobility model specification. In the CVT model, three different components based on transverse field and doping concentrations are combined using Mathiessen's rule [24]. The components are $\mu_{AC}, \mu_{sr},$ and μ_b. μ_{AC} is the surface mobility limited by scattering with acoustic phonons. μ_{sr} is a mobility component limited by surface roughness. μ_b is the mobility component limited by scattering with optical intervalley phonons. According to Mathiessen's rule the total mobility in a CVT model is given by $\mu_T^{-1} = \mu_{AC}^{-1} + \mu_{sr}^{-1} + \mu_b^{-1}$.

The Watt model (SURFMOB) [25] is a surface mobility model and is activated by specifying the parameter WATT in the MODEL statement. The Watt model takes into consideration the phonon scattering, surface scattering, and charge impurity scattering mechanisms in the inversion layer.

The Shirahata model [26] is enabled by specifying the SH parameter of the model statement. It is a general-purpose MOS mobility model that takes into account the screening effect in inversion layers and perpendicular field dependence for thin gate oxides. When the user enables Shirahata model, ATLAS automatically enables Klaassen's model [21,22].

The Yamaguchi model [27] and the Tasch model [28] are selected by setting YAMAGUCHI and TASCH, respectively, on the MODEL statement. The model overrides all mobility models other than CVT.

ATLAS invokes filed dependent mobility models if FLDMOB is specified in the MODEL statement. FLDMOB should always be specified unless one of the inversion layer mobility models (exhibits parallel field dependency itself) is specified.

In ATLAS, it is possible to use more than one mobility model. The default mobility model is the constant low field mobility μ_{no} and μ_{po} for electron and hole, respectively. Values of μ_{no} and μ_{po} may be specified in the MATERIAL statement using the parameters "MUN" and "MUP". This particular constant mobility is of no practical use because it leads to unrealistic high carrier velocity at high electric fields. Therefore, in order to achieve a successful prediction of a MOSFET device, it is necessary to use multiple non-conflicting mobility models simultaneously. It is also necessary to know which models are overriding others when conflicting mobility models are defined.

In ATLAS the mobility model to be used is specified in the MODEL statement. Detailed parameters associated with the chosen mobility models are specified on a separate MOBILITY statement.

For example, the ATLAS model statement shown below uses the constant voltage and temperature (CVT) Mobility Model for MOSFET in an ATLAS program. It is beyond the scope of this chapter to discuss in detail the parameters of the mobility model used.

```
models srh conmob fldnob b.electrons=2 b.holes=1 evsatmod=0
hvsatmod=0 cvt boltzmann print numcarr=1 electrons temperature=300

mobility material=CVT parameter bn.cvt=4.75e+07 bp.cvt=9.92
5e+06 cn.cvt=174000 cp.cvt=884200 taun.cvt=0.125
taup.cvt=0.0317 gamn.cvt=2.5 gamp.cvt=2.2 mu0n.cvt=52.2
mu0p.cvt=44.9 mu1n.cvt=43.4 mu1p.cvt=29 mumaxn.cvt=1417
mumaxp.cvt=470.5 crn.cvt=9.16e+16 crp.cvt=2.23e+17 csn.cvt=3
43e+20 csp.cvt=6.1e+20 alphn.cvt=0.68 alphp.cvt=0.71
betan.cvt=2 bctap.cvt=2 pcn.cvt=0 pcp.cvt=2.3e+15
deln.cvt=5.82e+14 delp.cvt=2.0546e+14
```

5.8 Benchmarking of MOSFET Simulations

Computer simulations allow modified devices to be tested within a few hours, although this can stretch out to days, and simulations also do not consume valuable raw materials or production time. However, the accuracy of computer models is always subject to question. This has led to the development of many different models for carrier mobility as well as different carrier transport models.

5.8.1 Method of Simulator Calibration

Typical TCAD tools, including both the process and the device simulators, are accurate, or predictive, but only for a sufficiently stable and mature technology, and after a lengthy calibration procedure [29]. This poses the issue of the relevance of device simulation for the development of a new technology node, in which new processes and materials are introduced which are not under complete control until the technology is finally released, and for which doping profiles and geometry are not known with sufficient accuracy. However, although not fully predictive, the positive aspects of TCAD that make them useful are that they still provide optimization guidelines, explanations of the characterization results, and insights into the transport mechanisms. Therefore, it is extremely important to calibrate the process and device simulator tools not just to reproduce qualitative behaviors but also to obtain accurate device characterization.

5.8.2 Calibration of Process Simulator

Selection of the appropriate process coefficients and the process models of processes like diffusion, ion-implantation, etc., is the most critical part of the entire calibration procedure of a process simulator creating a virtual two-dimensional device simulator. Thereafter the lengthiest task is to match the profiles of the actual device with the virtual device, for example, doping profile extraction from a physical device. The matching procedure effectively calibrates the simulator for a given process. It is a tedious task to tune the large number of undefined coefficients present in the empirical-based models in the process simulator considering the length of the subsequent process simulation runs.

The matching process is iterative. An exact match is rarely obtained. Generally there will be slight discrepancies between the structure generated by the process simulator and the physical fabricated device.

5.8.3 Calibration of Device Simulator

To calibrate a device simulator (ATLAS), the experimental I-V characteristics of the device obtained are set to best match the output of the DECKBUILD ATLAS program based on the same device dimension and structure. In order to obtain a match between the electrical characteristics of a real fabricated device and that with a simulated one, different advanced mobility models taking care of different scattering along with different doping profiles and different carrier transport models are analyzed iteratively, as shown in Figure 5.7.

5.9 Importance of Mesh Optimization

A mesh refers to a collection of volumetric elements whose union defines the interior and exterior of the device. The large number of small surface mesh elements allows finer geometrical resolution resulting in more accurate simulation.

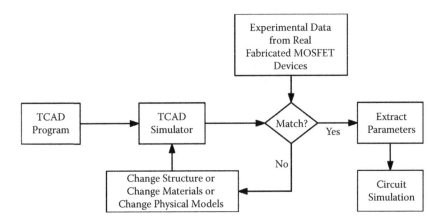

FIGURE 5.7
Method for calibration.

In ATLAS, advanced nonlinear iterative solvers are employed for numerical solution of PDEs. Higher accuracy is correlated with the large number of nodes obtained from a dense mesh, but the tradeoff is between a longer elapsed time and memory. In order to obtain the desired accuracy and efficiency of solution, the mesh structure should be optimized. Very dense mesh is used in the region of interest of the device structure where the gradient of impurity and potential and current density is high. For example, in a MOSFET, very dense mesh is used in the channel region under the gate-oxide interface and in the drain region where electric field is the highest. The tradeoff between the requirement of accuracy and computational time (complexity) dominated by the specification of the mesh structure causes problems for users. This problem always causes a dilemma for users. The time needed to complete a simulation run is roughly proportional to N^a, where N is the number of nodes (grid points) and a is a constant that varies from 2 to 3 depending upon the complexity of the problem. Unfortunately, the size of the device pushes the number of nodes used close to 20,000—the maximum number of nodes permitted by ATLAS. The meshing procedure is thus extremely relevant to the way nodes are distributed throughout the device. ATLAS uses triangular meshes. It is shown that although optimal mesh generation is not exactly deterministic, guidelines and heuristics for defining satisfactory meshes with good triangulation scheme yield better results [30].

5.9.1 Strategy to Obtain a Satisfactory Mesh

1. Generate enough points to provide the required accuracy.
2. Do not generate many unnecessary points that impair accuracy.
3. Minimize or avoid generation of obtuse triangles and long, thin triangles which tends to impair accuracy, convergence, and robustness.

4. Transition from a region with larger-sized triangles to a region with small triangles must be smooth (i.e., avoid abrupt discontinuity).

5. Generate dense mesh in critical areas where the movements of electrons and holes are rapid in order to prevent information loss and accuracy.

6. Use coarse mesh density for non-critical areas.

7. For a symmetrical structure, simulators allow simulation of one-half of the structure and then reflection of the results on the other half to save computational time and memory.

5.9.2 Mesh Re-Gridding

Re-gridding is used for refinement of regions of the initial/base mesh according to some specified criterion. If the value of the specified solution variable (such as carrier concentration, doping concentration, electric field, etc.) exceeds a certain value or when the change in the value within a mesh triangle exceeds a certain value, then mesh refinement will take place. ATLAS supports mesh re-gridding based upon doping or a wide range of suitable solution variables as the basis for mesh refinement. Regrid algorithms will search for the triangle that satisfies the criteria specified for refinement. Once they are identified, those triangles will be divided into four congruent sub-triangles. Grid solution quantities (electric field potential, carrier concentration, etc.) are interpolated into the new nodes using linear or logarithmic interpolation. The initial or base mesh is referred to as "level 0," and new triangles are referred to as "level 1." After all level 0 triangles are examined, level 1 triangles will be examined by the same procedure. Therefore, sub-triangles of level 1 become level 2 triangles. This process continues until no more triangles meet the refinement criterion. It is possible to specify the "maximum level" which is the limiting factor for amount of refinement and size of the grid after refinement. Re-gridding can produce obtuse triangles causing inaccurate results. Therefore, smoothing should be performed on both the initial grids and subsequent re-grids.

5.10 Introduction to Other Tools from Silvaco Used in Conjunction with ATLAS

Virtual Wafer Fab (VWF) is a suite of software programs used to create a multifunctional environment for the simulation of semiconductor technology. VWF of Silvaco allows cost and yield estimation as well as comprehensive parametric analysis of semiconductor processing by integrating process simulation, device simulation, and parameter extraction within an interactive graphical user-friendly interface. ATLAS can be used as a stand-alone or as a core tool in a VWF environment.

DECKBUILD is the software to run the code and provides the general input and output interface for Silvaco (for all TCAD modules) for changing and altering the code. ATLAS works in conjunction with DECKBUILD. The top half of the DECKBUILD window is the command input listed in the form of a program created by a text editor. The bottom half of the DECKBUILD window is where program execution and "extract" information is listed as the program runs and the execution/results are displayed.

DEVEDIT is a program that allows for structure editing, structure specification, and grid generation graphically by drawing on the screen. All of Silvaco's programs use a mesh or grid. Mesh or grid is used to determine the level of detailing the simulation will generate in a specific area of the device. Therefore, it allows users to cut down the simulation time by removing detailing from areas with less interest containing uniform or no reaction to change/alter simulation results. The creation of these meshes is the main function of DEVEDIT; however, it is also used for the editing and specification of two- and three-dimensional devices created with the VWF tools.

TonyPlot is the graphical plotting program used to plot the data extracted from the simulation using ATLAS. Simulation results do not automatically load into TonyPlot when a simulation is complete. Users have to save results into a file that can be opened directly from TonyPlot. The data can be plotted as desired by the user either in 1D *x-y* data, 2D contour data, Smith charts, or polar charts. Measured data can also be imported and plotted in the above-mentioned types. The overlay feature helps in comparing the multiple simulation runs. It annotates plots to create meaningful figures for reports and presentations. It enables 2D structure plots to be cut by multiple, independently controlled 1D slices. It supports plotting of user-defined equations with the variables being either electrical data (e.g., drain current) or physical parameters (e.g., electric field).

S-Pisces and Blaze are two primary simulators used for silicon device and advanced heterojunction devices, respectively. In addition, there are other simulators like Giga, MixedMode, ESD, TFT, Luminous, and LASER to supplement available process and device simulators with specialized capabilities. Giga supports non-isothermal calculations. ESD provides the simulation of electrostatic discharge phenomenon. TFT provides the support for simulation of devices with amorphous and polycrystalline materials. MixedMode provides the ability to simulate circuits using a combination of SPICE models and ATLAS devices. Luminous and Laser support general optoelectronic and semiconductor laser devices, respectively.

Mixed-mode simulations with mesh-based device structures within a circuit defined by SPICE models are also supported in order to enhance the capability. For the transient mode of simulation, the device properties are re-solved at any increment of time. In addition to the device design, the mixed device/circuit simulation environment allows users to evaluate the device performance in a real circuit, and there is no limitation as to whether a compact model ever exists or not for the device under test. By examining the simulation results for the new structure, designers can make tradeoffs among design parameters to achieve optimal device characteristics. MIXEDMODE is a circuit simulator of Silvaco to provide mixed-mode circuit simulation of multiple device structures simulated using device simulation and compact circuit models.

Device 3D, INTERCONNECT3D, and THERMAL3D provide support for three-dimensional device capabilities for three-dimensional simulation, parasitic extraction, and three-dimensional thermal analysis.

5.10.1 Process Simulation Tools

Semiconductor process simulation involves the numerical solution of equations describing the physics of dopant diffusion, silicon oxidation, lithography, ion implantation, etching, and deposition steps resulting in geometry and doping profiles that define a device. A number of programs each specialized to solve a specific set of equations are used to simulate the entire process flow. However, the various programs use different solution strategies.

ATHENA is a framework program that integrates several smaller programs into a more complete process simulation tool. This program focuses upon the simulation of fabrication processes. In ATHENA, devices are created through simulation of the fabrication process. To optimize the device characteristics, changes in process parameters supplied to the ATHENA process simulator environment are required, as certain process parameters change the device characteristics. ATHENA consists of four primary and several secondary tools. The primary tools are SSuprem4 for simulating ion implantation, diffusion, oxidation, and silicidation process for silicon; Flash for simulating implantation and diffusion for advanced materials; Elite for topography simulation; and Optolith for lithography simulation. ATHENA also provides options for modeling silicides, Monte Carlo modeling of ion implantation, etc.

5.10.2 ATHENA and ATLAS

ATLAS is very often used in conjunction with the ATHENA process simulator to take advantage of the automatic interface between them. ATHENA predicts the physical structures that result from the processing steps. The resulting physical structures are used as input by ATLAS, which then predicts the electrical characteristics for a particular bias. Therefore, it is possible to determine the effect of process parameters on device characteristics by the combination of ATHENA and ATLAS. However, it is much more difficult to control the actual device parameters and its operation in the ATHENA process simulator environment in comparison to the ATLAS device simulator environment. It is possible to precisely control the device structure, materials, and doping concentrations in specific regions through the given code syntax in ATLAS. However, a change in the individual process parameter affects the entire structure of the device in ATHENA, which makes a device constructed in ATHENA more difficult to characterize. However, a device realized in ATHENA is much closer to a true fabricated transistor.

5.11 Example 1: Bulk *n*-Channel MOSFET Simulation

Figure 5.8 shows the schematic cross-sectional diagram of an n-channel bulk MOSFET with channel length $L = 80$ nm, oxide thickness $t_{OX} = 2$ nm, junction

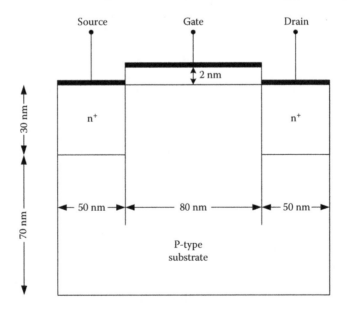

FIGURE 5.8
An n-channel bulk MOSFET.

depth X_j = 30 nm, n+ source/drain having uniform doping concentration 10^{20} cm^{-3}, and with a p-type substrate doping of 10^{18} cm^{-3}. To simulate this device using ATLAS, the following steps are required. The ATLAS program code provided in the text demonstrates the use of these steps to simulate the device.

Step 1: Generate the device structure file using ATHENA/Atlas/ DEVEDIT.

```
1.1: simulator specification
1.2: mesh definition
1.3: region definition
1.4: electrode specification
1.5: doping specification
1.6: contact specification
```

Step 2: Set the material model.

Step 3: Set the method used to do the calculation.

Step 4: Obtain the initial solution.

Step 5: Run the simulator to obtain a solution for a different bias condition.

Step 6: Display the results.

5.11.1 Program for Bulk n-Channel MOSFET Simulation

```
# Program to simulate n-channel bulk MOSFET

# In DECKBUILD # indicates a comment line, not a part of the program.

# Step 1: Generate the device structure

# 1.1 simulator specification
go atlas

# 1.2 mesh definition
mesh space.mult=1.0

# mesh definition in x direction
# loc stands for location, specifying the location of the grid
line
x.mesh loc=0.00 spac=0.01
# spac stands for spacing, specifying mesh spacing at a given
location
x.mesh loc=0.05 spac=0.001
x.mesh loc=0.09 spac=0.004
x.mesh loc=0.13 spac=0.001
x.mesh loc=0.18 spac=0.01

# mesh definition in y direction
y.mesh loc=-0.002 spac=0.0005
```

```
y.mesh loc=0 spac=0.0004
y.mesh loc=0.03 spac=0.008
y.mesh loc=0.10 spac=0.01

# 1.3 region definition
region num=1 y.min=0 silicon
region num=2 y.max=0 oxide

# 1.4 electrode declaration
electrode name=gate number=1 x.min=0.05 x.max=0.13 top
electrode name=source number=2 left length=0.05 y.min=0 y.max=0
electrode name=drain number=3 right length=0.05 y.min=0 y.max=0
electrode name=substrate number=4 bottom

# 1.5 doping specification of distribution, type
doping uniform conc=2e18 p.type region=1
doping uniform conc=1e20 n.type x.left=0 x.right=0.05 y.min=0
y.max=0.03
doping uniform conc=1e20 n.type x.left=0.13 x.right=0.18
y.min=0 y.max=0.03

# 1.6 contact specification
# n.poly sets n+ doped polysilicon as contact material with
workfuction=4.17eV
contact name=gate n.poly
contact name=source neutral
contact name=drain neutral
contact name=substrate neutral

#Step 2: Set the material model
models mos print

#Step 3: Set the specific method used to do the calculation
method newton trap

#Step 4: generate initial solution at zero bias
solve init

#Step 5: Run the simulator to obtain solution for different
bias condition
#solution for different drain bias
solve vdrain=0.1 outf=solve_vdrain1
solve vdrain=0.2 outf=solve_vdrain2
solve vdrain=0.3 outf=solve_vdrain3
solve vdrain=0.4 outf=solve_vdrain4

# ramp gate bias with a specific drain bias solution
load infile=solve_vdrain1
```

```
#output the result in a specific log file
log outf=gate1.log
solve name=gate vgate=0 vfinal=1.2 vstep=0.1

load infile=solve_vdrain2
log outf=gate2.log
solve name=gate vgate=0 vfinal=1.2 vstep=0.1

load infile=solve_vdrain3
log outf=gate3.log
solve name=gate vgate=0 vfinal=1.2 vstep=0.1

load infile=solve_vdrain4
log outf=gate4.log
solve name=gate vgate=0 vfinal=1.2 vstep=0.1

#Step 6: Display the results
# display all the log files overlaid together
tonyplot -overlay gate1.log gate2.log gate3.log gate4.log
quit
```

5.11.2 Simulation Results

Figure 5.9 shows the simulated $I_d - V_{gs}$ characteristics for the MOSFET device structure shown in Figure 5.8 with drain bias 0.1, 0.2, 0.3, and 0.4 V.

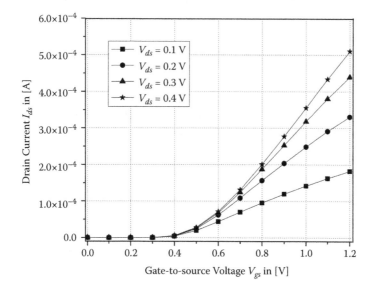

FIGURE 5.9

$I_d - V_{gs}$ characteristics of a bulk n-channel MOSFET with channel length $L = 80$ nm, $t_{OX} = 2$ nm.

5.12 Example 2: Silicon-on-Insulator (SOI) MOSFET Simulation

A fully depleted silicon-on-insulator (SOI) device is a silicon-based device built upon a thick SiO_2 layer as an insulating substrate that starts at or within the depletion layer. SOI technology appears as an interesting alternative to standard planar bulk devices [31,32]. This thick silicon dioxide layer is also known as the buried oxide (BOX) layer. The silicon layer above the BOX is where the device is fabricated. A fully depleted SOI device offers many advantages over bulk silicon MOSFETs, like reduction in junction capacitances between drain (source) and body, decrease in leakage currents, improvement in cross-talk immunity [33], low level of dielectric loss with high resistivity SOI substrates important for radio frequency (RF) applications [34], and higher immunity to radiation effects. The SOI structure not only kills the latch-up and improves digital error immunity, but it also allows for better control of the channel, leading to an improved subthreshold slope and lower short-channel effects. However, the SOI circuits suffer from several dynamic floating body effects [35]. In this example, a simplified version of the SOI MOSFET device is created using ATLAS with a facility to vary parameters like gate length, doping, etc., easily. To perform this simulation, the main Silvaco tools used are DECKBUILD, ATLAS, and TONYPLOT. The following text illustrates the use of the commands with required explanations to complete the device simulation.

5.12.1 Program Description for SOI MOSFET Simulation

Step 1: Declare and initialize variables used.

Variables are useful for ease of changing certain parameters when a code is being run multiple times with different values. For a new design, only the values declared with a set command need to be altered.

```
# Program for SOI Device Simulation
# To open ATLAS in DECKBUILD
go atlas

#location of the midpoint of the silicon thickness i.e. sMp=t_si/2
set sMp=0.0125

#location of the bottom of the silicon thickness i.e. sTsi=t_si
set sTsi=0.025

#location of the end of the bottom oxide layer thickness
set sToxb=0.1
```

```
# location of the start of the front (gate) oxide layer thickness
set sToxf=-0.002

# minimum x location of gate electrode
set gmin=1

# maximum x location of gate electrode
set gmax=2
```

Step 2: Define the mesh to construct the structure.

The first step in defining the structure of a SOI device is to set the mesh with a larger number of points in the areas of interest. The first mesh command must be the mesh space multiplier command. This command will tell ATLAS the scaling factor of the mesh. In this case, the device is rectangular, not cylindrical. The default mesh symmetry is rectangular. To specify cylindrical symmetry, a cylindrical parameter must be appended in the first mesh statement.

The mesh will be less dense for a large number and more dense for a smaller number. The value for this is normally set to equal 1. Mesh statements are entered in as vertical and horizontal lines in microns and as distance from the center line. ATLAS divides the grid using a triangle format. The user can use a dense mesh density in the region of interest.

```
mesh space.mult=1.0

# mesh definition in x direction
x.mesh loc=0.00 spac=0.50
x.mesh loc=1.15 spac=0.02
x.mesh loc=1.5 spac=0.1
x.mesh loc=1.85 spac=0.02
x.mesh loc=3 spac=0.5

# mesh definition in y direction
y.mesh loc=$sToxf spac=0.02
y.mesh loc=0.00 spac=0.005
y.mesh loc=$sMp spac=0.02
y.mesh loc=$sTsi spac=0.01
y.mesh loc=$sToxb spac=0.25
```

Step 3: Define the regions with different numbers and corresponding locations.

After defining the mesh structure, it is necessary to define the regions. For this example of SOI MOSFET, three regions are to be defined: (1) gate oxide layer for insulating gate contact, (2) silicon region with thickness t_{Si}, and (3) buried oxide (BOX).

The regions will be used to assign materials and properties to the device. The regions must be defined along the mesh lines, and the statements will be similar to those used for the mesh states. ATLAS allows the user to define up to 200 different regions for one device. If the designer overlaps any of the two regions, ATLAS will assign the material type to the last region that was defined. The entire two-dimensional mesh area must be defined into regions or ATLAS will not run successfully.

```
# region definition
region number=1 x.min=0 x.max=3 y.min=$sToxf y.max=0
material=Oxide
region number=2 x.min=0 x.max=3 y.min=0 y.max=$sTsi
material=Silicon
region number=3 x.min=0 x.max=3 y.min=$sTsi y.max=$sToxb
material=Oxide
```

Step 4: Define the names of each electrode with corresponding location.

The electrodes are to be defined in order to tell the program about the location of the metal contacts like gate, source, and drain. One can observe the dissimilarities between the fabrication process and ATLAS program because the top oxide layer and the electrodes are defined before the silicon is doped. ATLAS requires the structure to be defined in this order in order to achieve easier calculation. Because the gate electrode is above the top oxide layer, both the *y.min* and *y.max* values can be set to *sToxf* (–0.002), which will put it above the oxide layer, but with no thickness. Similarly, the substrate electrode is located below the bottom oxide layer.

```
# electrode definition
electrode name=gate number=1 x.min=$gmin x.max=$gmax
y.min=$sToxf y.max=$sToxf
electrode name=source number=2 x.min=0 x.max=0.5 y.min=0 y.max=0
electrode name=drain number=3 x.min=2.5 x.max=3 y.min=0 y.max=0
electrode name=substrate number=4 x.min=0 x.max=3 y.min=$sToxb
y.max=$sToxb
```

Step 5: Set the contact and interface properties.

The command "contact" is used to tell ATLAS how to treat the electrode. In the default condition, an electrode in contact is assumed to be ohmic. If the designer wants the electrode to be treated like a Schottky contact, the design must use the *workfunction*. The gate contact material is set to n^+ doped polysilicon by the *n.poly* parameter. In another approach, instead of using the material name (*n.poly*) required, *workfunction* can be mentioned by setting the *workfunction* parameter to the required value. For example, the statement "contact name=gate workfunction=4.17" is equivalent to the statement "contact name=gate n.poly" because the workfunction of n^+ polysilicon equals 4.17 eV.

The fabrication process, no matter how well controlled, introduces interface states at the SiO_2-Si interface, which critically affects the electrical characteristics of the device. Interface states, whether inherent, process related, or operationally generated, were found to cause degradation in device parameters such as transconductance, carrier mobility, and threshold voltage and to generally reduce device reliability and lifetime [36,37]. The "interface" statement is used to define the interface charge density interfaces between semiconductors and insulators. It indicates that all interfaces between semiconductors and oxide have a fixed charge of 3 × 10^{10} C/cm^2 [37].

```
contact name=gate n.poly
interface qf=3e10
contact name=source neutral
contact name=drain neutral
contact name=substrate neutral
```

Step 6: Define doping in the MOSFET structure.

The next action, doping, is one of the most important actions a designer does to affect the electrical properties of the structure being designed. Silvaco allows the designer to specify the type of dopant and the concentration. It also allows the designer to specify the distribution of the doping material. ATLAS has the ability to distribute the dopants in a uniform, analytical Gaussian, or other supported profiles. For this device, all the doping will be defined in region 2, where the silicon is located. Initially a p-type uniformly distributed doping is specified in the whole region 2. It is followed by a Gaussian distribution n-type doping specification targeting the source and the drain region of the structure. The concentration listed in the command for Gaussian doping in the source and the drain will have the peak concentration = 1 × 10^{20} cm^{-3} at $y = 0$ and the characteristics = 0.05 which is the principal characteristic length of the implant (standard deviation) for which the doping level will drop off in a vertical direction. The lateral fall-off outside the x-coordinates mentioned in the doping statement is defined by the *lat.char* parameter.

```
# p-type doping with a uniform concentration throughout all of
the silicon doping uniform conc=2e17 p.type direction=y regions=2

# Gaussain doping profile in the source
doping gaussian characteristic =.05 conc=1e20 n.type x.left=0
x.right=$gmin y.top=0 lat.char=0.05 direction=y
```

```
# Gaussain doping profile in the drain
doping gaussian characteristic =.05 conc=1e20 n.type
x.left=$gmax x.right=3 y.top=0 lat.char =.05 direction=y
```

Step 7: Save and display the structure file.

The "master" after the name of the out file specifies that the output file needs to be written as a standard structure file instead of binary format. The generated MOSFET structure can be visualized with the help of TonyPlot.

```
struct outf=SOI.str master
tonyplot SOI.str
```

Step 8: Select the models used in this simulation.

In this simulation, for mobility standard concentration dependent (conmob) and parallel field dependent mobility (fldmob) to model, the velocity saturation effect is chosen. For carrier statistics, Fermi-Dirac and band-gap narrowing (bgn) are chosen. The parameters *evsatmod* and *hvsatmod* with *b.electrons* and *b.holes* supplied are used to select a particular field-dependent equation to be used. Carrier generation and recombination are the processes by which the semiconductor material is moved away from thermal equilibrium and returned to equilibrium after being disturbed from it. For recombination Shockley-Read-Hall (SRH) and Auger recombination models are chosen.

In the presence of heavy doping (greater than 10^{18}cm^{-3}), the p-n product in silicon becomes doping dependent [38]. As doping increases, the band gap decreases, where the conduction band decreases the same amount as the valence band is raised. This feature is considered by using the model "BGN" (band-gap narrowing).

```
# model specification
models auger srh conmob fldmob b.electrons=2 b.holes=1 evsat-
mod=0 hvsatmod=0 bgn temperature=300
```

Step 9: Define the numerical methods.

The Newton method is chosen with a maximum number of iteration equal to 25. To overcome the problem with diverging solution with a poor initial guess, a "trap" statement is used, where *maxtrap* is the maximum allowed number of trials (default = 4) with the bias step reduced by the factor supplied in the parameter known as *atrap*. To enhance the performance for slow convergence, a variant of Newton's method known as the Newton-Richardson method is enabled by the *autonr* parameter.

```
# method specification
method newton itlimit=25 trap atrap=0.5 maxtrap=4 autonr
```

Step 10: Solve for specified bias.

The initial guesses for potential and carrier concentrations must be made from the doping profile at zero bias by using the "solve init" statement, when no previous solutions are available. If the user omits this statement, ATLAS automatically evaluates this before executing the first "solve" statement.

It is most difficult to obtain good convergence for the first two non-zero solve statements because they use solution at zero bias provided by the "solve init" statement as a poor initial guess. This is why the first two non-zero solve statements should use small voltage steps. However, once their solutions are obtained, by the use of the projection algorithm, the remaining solve statements obtain a good initial guess, resulting in good convergence.

```
# Set the initial values and solve
# solve the device for zero bias
solve init
# solve for drain bias with 0.01 then 0.05 to finally to 0.1
# first two non-zero solve statements are given small bias step
solve vdrain=0.01
solve vdrain=0.02
solve vdrain=0.1

# Extraction of Id-Vds characteristics or different Vgs
solve vgate=1.0 outf=solve_vgate1
solve vgate=1.5 outf=solve_vgate2
solve vgate=2.0 outf=solve_vgate3
solve vgate=2.5 outf=solve_vgate4

load infile=solve_vgate1
log outf=SOI11.log
solve name=drain vdrain=0 vfinal=2.0 vstep=0.1

load infile=solve_vgate2
log outf=SOI21.log
solve name=drain vdrain=0 vfinal=2.0 vstep=0.1

load infile=solve_vgate3
log outf=SOI31.log
solve name=drain vdrain=0 vfinal=2.0 vstep=0.1

load infile=solve_vgate4
log outf=SOI41.log
solve name=drain vdrain=0 vfinal=2.0 vstep=0.1
```

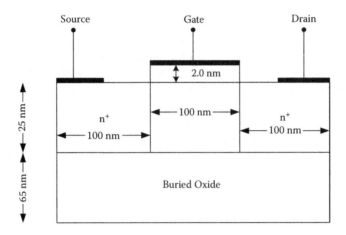

FIGURE 5.10
Cross-sectional diagram of a fully depleted n-channel SOI MOSFET.

Step 11: Display the output characteristics.

```
# plot of the log files overlaid together
tonyplot -overlay SOI11.log SOI21.log SOI31.log SOI41.log
# denotes end of the program
quit
```

5.12.2 Simulation Results

Figure 5.11 shows the simulated I_d-V_{ds} characteristics for the n-channel fully depleted SOI MOSFET device structure displayed in Figure 5.10 for gate bias 1.0, 1.5, 2.0, and 2.5 V, respectively.

5.13 Example 3: 0.18 µm Bulk nMOS Transistor with Halo Implant

The device used for this simulation is a LDD 0.18 µm [39] n-channel MOSFET. The device is fabricated using silicon with a <100> orientation. The p-type substrate is formed by doping with 3×10^{13} atoms/cm^3 of boron. The n$^-$ region is formed by implanting a dose of 10^{15} atoms/cm^3, and the n$^+$ region is formed by implanting 5×10^{15} atoms/cm^3 of arsenic. The program starts with a mesh definition. The mesh is defined for one-half of the symmetric MOSFET structure. After all the process fabrication steps, the structure will be mirrored

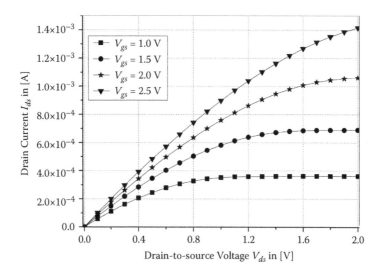

FIGURE 5.11

Plot of drain current I_{ds} as a function of drain-to-source voltage V_{ds} for gate-to-source voltage $V_{gs} = 1.0$ V, 1.5 V, 2.0 V, and 2.5 V, respectively, obtained by simulation for an n-channel fully depleted SOI MOSFET.

to generate full MOSFET structure. The different steps performed to simulate the fabrication of n-channel MOSFET followed by extraction of different MOSFET parameters of it are described below. The status of the MOSFET structure and the effect of every successive process step can be visualized by inserting a command for plotting the structure (tonyplot structure_name. str) after every step described below.

5.13.1 Program for 0.18 μm Bulk nMOS

Step 1: Provide mesh definition and structure declaration.

```
# start process simulator ATHENA
go Athena

# Define mesh for x-plane
line x loc=0.15 spac=0.1
line x loc=0.2 spac=0.006
line x loc=0.4 spac=0.006
line x loc=0.6 spac=0.01

# Define mesh for y-plane
line y loc=0.0 spac=0.002
line y loc=0.2 spac=0.005
```

```
line y loc=0.5 spac=0.05
line y loc=0.8 spac=0.15

# structure declaration
struct outfile=nmos_bulk.str
```

Step 2: Initialize silicon substrate of crystal orientation 100 with 1.0 × 10^{15}/cm^3 boron dopant added. A space multiplier (Space.mult = 2) is used to speed up the process. Decrease the space multiplier parameter to obtain a denser mesh with more accuracy.

```
init orientation=100 c.boron=1.0e15 space.mul=2
```

Step 3: Grow a smooth thin layer of SiO$_2$ of depth 10 to 15 nm on the substrate to reduce the channeling effect and to prevent contamination of the substrate. The simplest deposit method in ATHENA is conformal deposition. It is used when the exact shape of the deposited layer is not critical. Dry oxidation is performed for 30 minutes at 1000°C. Then etch to obtain a uniform blanket of oxide of 0.02 μm thick.

```
diffus time=30 temp=1000 dryo2 press=1.00 hcl=3
etch oxide thick=0.02
```

Step 4: Use the default dual Pearson model to choose a boron implant with a dose of 8.0 × 10^{12} ions/cm^2 with energy 100 keV. An n-channel MOS transistor must be developed on p-type silicon as this material under the gate must be inverted. Therefore, the next step is implantation of boron to create a p-well in the substrate. The nMOS fabrication could have started with an initial p-type substrate, but p-well implantation reviewed here is common in industry. ATHENA offers three different models for ion implantation [40]: (1) dual Pearson (default), (2) single Pearson, and (3) Monte Carlo.

```
implant boron dose=3.0e13 energy=200 pearson
```

Step 5: Move and settle the boron atoms. As a result of the ion implantation step, the net doping peaks at an average penetration depth with a specific doping concentration. In order to enhance the doping uniformity, the substrate is heated to high temperatures so that the boron atoms are given enough energy to move and settle more uniformly in the substrate. As a result of this heating in the presence of oxygen, an oxide is formed. Wet oxidation is used.

```
diffus temp=950 time=100 weto2 hcl=3
```

Step 6: Further propel the p-well into the substrate and increase the doping uniformity by performing more diffusion steps with varying temperatures, temperature change rates, and processing environments known as *welldrive*.

```
diffus time=50 temp=1000 t.rate=4.000 dryo2 press=0.10 hcl=3
diffus time=220 temp=1200 nitro press=1
diffus time=90 temp=1200 t.rate=-4.444 nitro press=1
```

Step 7: Etch all present oxide layers in order to obtain a surface on which to begin the process of defining physical MOSFET parameters.

```
etch oxide all
```

Step 8: Perform sacrificial cleaning in order to ensure that the surface is free from damage due to previous fabrication steps. In this process, a thin layer of silicon is sacrificed by first oxidation and then removal of the oxide produced.

```
diffus time=20 temp=1000 dryo2 press=1 hcl=3
etch oxide all
```

Step 9: Perform deposition of gate oxide by dry oxidation, with the thickness of the gate oxide playing a major role in determining the MOSFET characteristics that can be altered by controlling time and temperature.

```
diffus time=3 temp=895 dryo2 press=1.00 hcl=1
```

Step 10: Define the threshold voltage by implanting boron through the gate oxide. A higher dose of boron implant will lead to higher threshold voltage because it will be more difficult to invert the p-channel.

```
implant boron dose=1.5e13 energy=45 pearson
```

Step 11: Use conformal deposition of polysilicon to create the gate.

```
depo poly thick=0.2 divi=10
```

Step 12: Begin patterning of the polysilicon gate by etching 0.35 μm from the left side.

```
etch poly left p1.x=0.51
```

Step 13: Perform implantation through the deposited gate oxide layer *t* from the light drain/source.

```
method fermi compress
diffuse time=5 temp=900 weto press=0.8
implant arsenic dose=1.0e15 energy=30 pearson
```

Step 14: Implant p-doped halo (NMOS).

```
implant boron dose=3.0e13 energy=15 tilt=30 fullrotat
```

Step 15: Form an oxide spacer to provide a barrier of isolation and to aide in patterning for the next implantation.

```
depo oxide thick=0.10 divisions=8
etch oxide dry thick=0.10
```

Step 16: Form the heavy drain/source region by implantation with arsenic instead of phosphorus as in the case of the light drain/source.

```
implant arsenic dose=5e15 energy=60 pearson
```

Step 17: Perform annealing in the presence of nitrogen to diffuse the drain/source created.

```
method fermi compress
diffuse time=1 temp=1000 nitro press=1.0
```

Step 18: Etch the oxide layer above the drain/source region to pattern the source/drain contact metal.

```
etch oxide left p1.x=0.35
```

Step 19: Deposit aluminum to create electrodes with a low ohmic contact.

```
deposit alumin thick=0.03 divi=2
```

Step 20: Etch away unwanted materials.

```
etch alumin right p1.x=0.33
```

Step 21: Mirror the structure to obtain the full symmetrical device.

```
structure mirror right
```

Step 22: Define the electrodes and save the obtained structure.

```
electrode name=gate x=0.59 y=0.1
electrode name=source x=0.2
```

```
electrode name=drain x=1.0
electrode name=substrate backside
save outfile=nmos_bulk.str
```

Step 23: Extract different MOSFET parameters and plot the structure.

```
# to extract gate oxide thickness
extract name="gateox" thickness oxide mat.occno=1 x.val=0.59

# extract final S/D junction depth Xj
extract name="nxj" xj silicon mat.occno=1 x.val=0.2 junc.
occno=1

# extract the N++ regions sheet resistance...
extract name="n++ sheet rho" sheet.res material="Silicon" mat.
occno=1 x.val=0.2 region.occno=1

# extract the sheet rho under the spacer, of the LDD region...
extract name="ldd sheet rho" sheet.res material="Silicon" mat.
occno=1 x.val=0.49 region.occno=1

# extract the surface conc under the channel....
extract name="chan surf conc" surf.conc impurity="Net Doping"
material="Silicon" mat.occno=1 x.val=0.6

# potting the structure
tonyplot nmos_bulk.str
```

The program to extract different parameters, threshold voltage, and I_{ds}-V_{gs} characterization using the structure generated with ATHENA code is described above. The threshold voltage is extracted by finding the intersection of the maximum slope of the I_{ds}-V_{gs} curve with the x-axis. The output of the extract command will appear in the lower DECKBUILD window after running the code.

```
go atlas
# IMPORT THE structure to use the auto-interface between
ATHENA and ATLAS
mesh inf=nmos_bulk.str

# model definition
models cvt srh numcarr=2

# method specifiatiion
method newton itlimit=25 trap atrap=0.5 maxtrap=4 autonr
nrcriterion=0.1 tol.time=0.005 dt.min=1e-25

#solve for id-vgs by ramping gate voltage and store the answer
in the log file
```

```
solve init
solve vdrain=0.1
log outf=bulkHalo.log master
solve name=gate vgate=0.1 vfinal=1.5 vstep=0.1

# plot the log file
tonyplot bulkHalo.log

# extraction of threshold voltage
extract name="vt" (xintercept(maxslope(curve(abs(v."gate"),abs
(i."drain"))))) - abs(ave(v."drain"))/2.0)
quit
```

5.13.2 Simulation Results

Figure 5.12 shows the I_{ds}-V_{gs} characteristics of an n-channel MOSFET simulated using program 5.13.1 for two different devices. One of them is a pocketed device that considers the halo implantation process step, and the other is a non-pocketed device that omits the halo implantation process step. A process called *pocket implant* or *halo implant* is introduced in which the locally high doping concentration in the channel near the source/drain junctions is created. The technology has been given prominence to tailor the short-channel performances of diminuend devices. Therefore, it is widely used to reduce threshold voltage roll-off and punch-through [41–46].

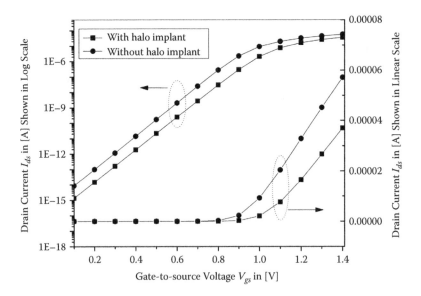

FIGURE 5.12
Simulation data of the drain–current I_{ds} versus the gate–source voltage V_{gs} for devices with and without halo implantation.

A highly doped halo surrounding either the source or drain proves to be effective in increasing switching speed by more than a factor of two over the conventional MOSFET [47]. It reduces leakage current and, hence, reduces static power dissipation [48]. However, careful tuning of pocket implant parameters in order to obtain desired electrical performance of nano-MOSFETs is always necessary. Variations of pocket implant energy, implant dosage, and tilt angle influence the characteristics of nano-n-MOSFETs [49,51]. However, careful tradeoffs between minimum channel length and other device electrical parameters are required.

Figure 5.12 demonstrates an increase of the subthreshold current due to the drain-induced barrier lowering suffered by the non-pocketed device compared to the pocketed device. At low gate voltages, the additional potential barriers created by the pockets are strong enough for controlling the current, resulting in a comparative lower current for halo-doped MOSFETs in the subthreshold region.

5.14 Example 4: Volume Inversion Double-Gate (DG) MOSFET

As CMOS scaling is approaching its limits, DG MOSFET is becoming a very promising solution for the fabrication of high-performance devices for low-power applications [35].

A DG MOSFET is considered electrostatically superior (channel is controlled by the gate voltage, not by the drain to source voltage) compared to single-gate MOSFET due to its strong coupling between the conduction channel and gate electrodes.

The use of symmetric DG MOSFET allows the suppressing of short-channel effects like drain-induced barrier lowering (DIBL) and subthreshold slope degradation, making unnecessary the conventional use of high channel doping densities and gradients compared to a bulk MOSFET [52,53]. Here symmetric means that the same voltage bias is applied to the two gates having the same work function. The use of an undoped body [54,55] also results in enhanced mobility by reducing charged-impurity scattering and in reduced device parameter variation by eliminating the statistical fluctuation of dopant concentration [56–58]. Therefore, doping of the DG MOSFET is not desired and is usually not used, but there is always a small unintentional doping density ~10^{15} cm^{-3} during fabrication [59]. Therefore, a lightly doped DG MOSFET is considered in this simulation.

In this example, ATLAS is used to simulate 50 nm DG n-MOSFET having abrupt source/drain junctions (i.e., $L_{eff} = L_{met} = 50$ nm [60]). The device parameters are as follows: channel width $W = 50$ nm, channel length $L = 50$ nm, silicon thickness $t_{Si} = 20$ nm, equivalent gate oxide thickness $t_{ox} = 2$ nm, doping concentration of the silicon channel $N_A = 10^{15}$ cm^{-3}, doping concentration of the source/drain contact regions $N_D = 10^{20}$ cm^{-3}, and mid-gap metal gate with workfunction 4.74 eV.

5.14.1 Program for Structure of DG MOSFET

```
# Program to generate the structure of DG MOSFET

go atlas
mesh space.mult=1.0

# mesh definition in x direction
x.mesh loc=0.00 spac=0.01
x.mesh loc=0.05 spac=0.001
x.mesh loc=0.075 spac=0.005
x.mesh loc=0.1 spac=0.001

x.mesh loc=0.15 spac=0.01
# mesh definition in y direction
y.mesh loc=-0.002 spac=0.001
y.mesh loc=0 spac=0.0004
y.mesh loc=0.01 spac=0.001
y.mesh loc=0.02 spac=0.0004
y.mesh loc=0.022 spac=0.001

# region definition
region num=1 y.min=0 y.max=0.02 silicon
region num=2 y.min=-0.002 y.max=0 oxide
region num=3 y.min=0.02 y.max=0.022 oxide

# electrode definition
electrode name=gate number=1 x.min=0.05 x.max=0.1 top
electrode name=gate1 number=2 x.min=0.05 x.max=0.1 bottom
electrode name=source number=3 left length=0.05 y.min=0
y.max=0
electrode name=drain number=4 right length=0.05 y.min=0
y.max=0

# doping specification
doping uniform conc=1e15 p.type region=1
doping uniform conc=1e20 n.type x.left=0 x.right=0.05 region=1
doping uniform conc=1e20 n.type x.left=0.1 x.right=0.15
region=1

# save and display the structure
save outfile=dgmos.str
tonyplot dgmos.str
quit
```

Quantum confinement effects will not be taken into account here, because silicon film thickness greater than 10 nm [61] and length greater than 10 nm are being considered. Quantum mechanical tunneling from source to drain (through the barrier) degrades device performance for extreme short length devices, but for channel lengths longer than about 10 nm, MOSFETs behave classically [62].

The surface potential Ψ_s is the most natural variable for the formulation of MOS device physics. It is defined as the difference between the electrostatic potential at the SiO_2/Si interface and the potential in the neutral bulk region due to band bending. Recently there has been wide consensus in the compact modeling community that traditional threshold-voltage-based models have reached the limit of their usefulness and need to be replaced with more advanced models based on surface potential referred to as surface-potential-based models, because an accurate physical description of MOSFET can be easily obtained by the use of surface-potential-based models.

5.14.2 Program to Obtain Potential Variation of DG MOSFET

```
# Program for obtaining variation of surface and body center
potential

go atlas

# IMPORT THE MESH structure
mesh inf=dgmos.str

# contact specification
contact name=gate n.poly workfunction=4.74
# two separate electrodes gate and gate1 are shorted by
"common" parameter
contact name=gate1 n.poly workfunction=4.74 common=gate
contact name=source neutral
contact name=drain neutral

# model declaration
models auger srh conmob fldmob bgn temperature=300

# method definition
method newton itlimit=25 trap

# solution for specific gate and drain bias
solve init
solve vdrain=0.0
solve vgate=0.0

# saving the structure file
save outf=dg.str

# extraction of surface potential where y.val=0.0
(semiconductor front gate oxide interface) along the depth and
saving the curve in a file
```

```
extract name="srp_profile1" curve(depth, potential
material="Silicon" mat.occno=1 y.val=0.0) outfile="extract1.
dat"

# extraction of channel center potential where
y.val=0.01(center of the SOI silicon semiconductor body
y=t_si/2) along the depth and saving the curve in a file
extract name="srp_profile2" curve(depth, potential
material="Silicon" mat.occno=1 y.val=0.01) outfile="extract2.
dat"

# plot the files in the same window overlaid
tonyplot -overlay extract1.dat extract2.dat
quit
```

5.14.3 Simulation Results

Figure 5.14 shows the variation of surface and channel center potential of DG MOSFET shown in Figure 5.13 using the simulation of codes listed in Sections 5.14.1 and 5.14.2. Figure 5.14 portrays the potential with gate and drain bias equal to zero. However, it is possible to obtain potentials for various bias conditions by changing the gate and the drain bias in the program listed in Section 5.14.2, subjected to convergence of the solution attained.

Figure 5.15 presents the surface potential Ψ_S and potential at the center of the channel Ψ_0, evaluated for gate voltage varying from 0 V to 1.5 V in steps of 0.1 V by changing the program for every step change. It is observed that below threshold, while the semiconductor charge is small, there is volume inversion, and the potential remains essentially flat ($\Psi_S = \Psi_0$) throughout the

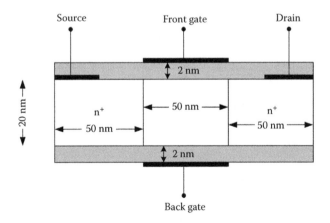

FIGURE 5.13
A fully depleted thin-film DG SOI MOSFET.

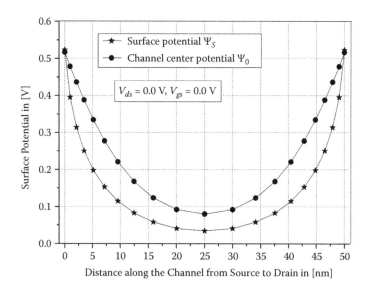

FIGURE 5.14
Plot of surface potential and channel center potential versus position along the channel from source to drain for $V_{ds} = 0$ V and $V_{gs} = 0$ V.

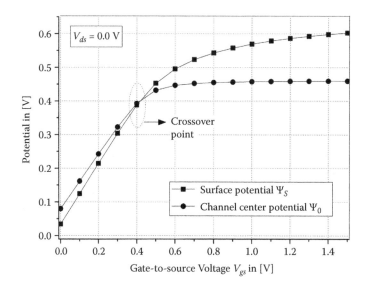

FIGURE 5.15
Variation of surface potential Ψ_S and channel center potential Ψ_0 versus gate-to-source voltage V_{gs} with x = $L/2$ for a fixed channel length $L = 50$ nm and $V_{ds} = 0.0$ V.

entire silicon film thickness, with both Ψ_S and Ψ_0 closely following the gate voltage. As the gate voltage increases toward threshold, the electron density becomes significant. As a result, Ψ_S continues to increase slowly, and Ψ_0 starts to depart from Ψ_S and later saturates after being pinned to a maximum value. The mobile charge near the silicon surfaces screens the gate field from the center of the silicon film, and Ψ_S and Ψ_0 become de-coupled (i.e., there is no volume inversion). The channel potential versus gate voltage characteristics for the devices having equal lengths but different thicknesses pass through a single common point termed the *crossover point* [63], which is also shown in Figure 5.15.

5.14.4 Program to Obtain I_{ds}-V_{gs} Characteristics of DG MOSFET

```
# Program to obtain Id-Vgs characteristics

go atlas
# IMPORT THE MESH
mesh inf=dgmos.str

# contact specification
contact name=gate n.poly workfunction=4.74
contact name=gate1 n.poly workfunction=4.74 common=gate
contact name=source neutral
contact name=drain neutral

# model declaration
models auger srh conmob fldmob bgn temperature=300

# method definition
method newton itlimit=25 trap

# initial solution and drain bias specification
solve init
solve vdrain=0.05
#solve vdrain=0.1
#solve vdrain=0.5
#solve vdrain=1.5

# output log file declaration
log outf=dg1.log

# solution to obtain solution for id-vgs characteristics by
ramping Vgs
solve name=gate vgate=0 vfinal=1.2 vstep=0.05

# plotting the structure
tonyplot dg1.log
quit
```

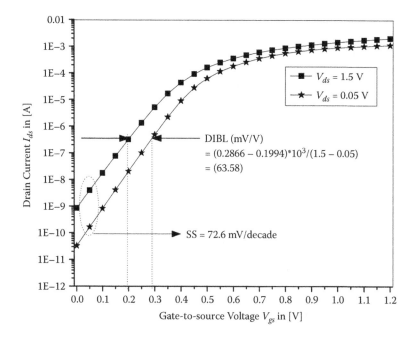

FIGURE 5.16
Drain current as a function of gate-to-source voltage for two different values of V_{ds}. Also shown is the calculation of DIBL and the subthreshold slope.

5.14.5 Simulation Results

Figure 5.16 shows the horizontally translated I_{ds}-V_{gs} characteristics for low ($V_{ds} = 0.05$ V) and high ($V_{ds} = 1.5$ V). The translation is known as DIBL and is characterized by the number of millivolts of translation per volt of change in drain voltage. In this case the DIBL is equal to 63.58 mv/V, which is satisfactory because well-designed MOSFETs typically exhibit DIBL < 100 mV/V.

Figure 5.16 also indicates subthreshold swing (SS) characterized by the slope of the I_{ds}-V_{gs} curve in the subthreshold region, which is the number of millivolts of increase in gate voltage needed to increase the drain current by a factor of 10. The theoretical lower limit of SS is 60 mV/decade at room temperature, while for well-designed MOSFETs the tolerable value for SS is less than 80 mV/decade in order to ensure a rapid transition between the off and on states (i.e., a small SS).

5.15 Summary

This chapter presents a comprehensive overview about MOSFET simulation using Silvaco TCAD tools. The strategy and methodology applied for MOSFET device simulation using Silvaco are emphasized. An overview of the software developed by Silvaco in order to meet the simulation needs for researching conventional and advanced MOSFET structures is also presented. In addition, several examples with source codes are provided for the task of simulating different types of MOSFETs that are in use today. It was shown how it is possible to obtain unique insight into the behavior of MOSFETs by performing device simulation using Silvaco TCAD tools.

References

1. Semiconductor Industry Association, International Roadmap for Semiconductors, 2011 (available at http://public.itrs.net/)
2. H. Masuda, T2CAD: Total design for sub-m process and device optimization with technology-CAD, in *Simulation of Semiconductor Devices and Processes*, H. Ryssel and P. Pichler, Eds. Vienna, Austria: Springer, 1995, vol. 6, pp. 408–415.
3. J.J. Barnes and R.J. Lomax, Finite element methods in semiconductor device simulation, *IEEE Trans. on Electron Devices*, vol. ED-24, August 1977, pp. 1082–1088.
4. P.E. Cottrell and E.M. Buturla, Steady-state analysis of field effect transistors via the finite element method, *IEDM Technical Digest*, December 1975, pp. 51–55.
5. M.S. Lundstrom, R.W. Dutton, D.K. Ferry, and K. Hess, Eds., *IEEE Trans. Electron Devices* 47(10), 2000. *Special Issue on Computational Electronics: New Challenges and Directions.*
6. J. Dabrowski, H.-J. Mussig, M. Duane, S.T. Dunham, R. Goossens, and H.-H. Vuong, Basic science and challenges in process simulation, *Advances in Solid State Physics*, vol. 38, 1999, pp. 565–582.
7. M. Duane, TCAD needs and applications from a user's perspective, *IEICE Transactions*, vol. E82-C, no. 6, 1999, pp. 976–982.
8. J. Mar, The application of TCAD in industry, *SISPAD Proceedings*, 1996, pp. 64–74.
9. D.L. Scharfetter and D.L. Gummel, Large signal analysis of a silicon read diode oscillator, *IEEE Transaction on Electron Devices*, vol. ED-16, 1969, pp. 64–77.
10. M.V. Fischetti and S.E. Laux, Monte Carlo simulation of submicron Si MOSFETs, in *Simulation of Semiconductor Devices and Processes*, G. Baccarani and M. Rudan, Eds. Bologna: Technoprint, 1988, vol. 3, pp. 349–368.
11. K. Banoo and M. S. Lundstrom, Electron transport in a model Si transistor, *Solid State Electron.*, vol. 44, no. 9, pp. 1689–1695, Sept. 2000.
12. S.E. Laux and M.V. Fischetti, Monte Carlo study of velocity overshoot in switching a 0.1-micron CMOS inverter, *Proc. IEDM*, Washington, DC, 1997, *Tech. Dig. IEDM*, pp. 877–880.
13. J.D. Bude, MOSFET modeling into the ballistic regime, *Proceedings of the SISPAD*, Seattle, WA, 2000, pp. 23–26.
14. C. Fiegna, H. Iwai, M. Saito, and E. Sangiori, Application of semiclassical device simulation to trade-off studies for sub-0:1 _m MOSFETs, *Proc. IEDM*, San Francisco, CA, 1994, *Tech. Dig. IEDM*, pp. 347–350.
15. R. Granzner, V.M. Polyakov, F. Schwierz, M. Kittler, and T. Doll, On the suitability of DD and HD models for the simulation of nanometer double-gate MOSFETs, *Solid State Electronics, Physica E*, vol. 19, 2003, pp. 33–38.

16. G.D. Mahan, *Many-Particle Physics,* New York: Kluwer Academic/Plenum, 2000.
17. Silvaco International, Santa Clara, CA, ATLAS User's Manual, 2010.
18. Kazautaka Tomizawa, Numerical Simulation of Submicron Semiconductor Devices, Norwood, MA: Artech House, 1993.
19. D.M. Caughey and R.E. Thomas, Carrier mobilities in silicon empirically related to doping and field, *Proc. IEEE,* vol. 55, 1967, pp. 2192–2193.
20. N.D. Arora, J.R. Hauser, and D.J. Roulston, Electron and hole mobilities in silicon as a function of concentration and temperature, *IEEE Trans. Electron Devices,* vol. 29, 1982, pp. 292–295.
21. D.B.M. Klaassen, A unified mobility model for device simulation—I. Model equations and concentration dependence, *Sol. St. Elec.,* vol. 35, no. 7, 1992, pp. 953–959.
22. D.B.M. Klaassen, A unified mobility model for device simulation—II. Temperature dependence of carrier mobility and lifetime, *Sol. St. Elec.,* vol. 35, no. 7, 1992, pp. 961–967.
23. C. Lombardi, S. Manzini, A. Saporito, and M. Vanzi, A physically based mobility model for numerical simulation of nonplanar devices, *IEEE Trans. Comp. Aided Design,* vol. 7, 1992, pp. 1154–1171.
24. M. Lundstrom, *Fundamentals of Carrier Transport*, Cambridge, UK: Cambridge University Press, 2000.
25. J.T. Watt, Surface Mobility Modeling, presented at Computer Aided Design of IC Fabrication Processes, Stanford University, California, August 1988.
26. M. Shirahata, H. Kusano, N. Kotani, S. Kusanoki, and Y. Akasaka, A mobility model including the screening effect in MOS inversion layer, *IEEE Trans. Computer Aided Design,* vol. 11, no. 9, September 1992, pp. 1114–1119.
27. Ken Yamaguchi, Field-dependent mobility model for two-dimensional numerical analysis of MOSFETs, *IEEE Trans. Electron Devices,* vol. 26, 1979, pp. 1068–1074.
28. G.M. Yeric, A.F. Tasch, and S.K. Banerjee, A universal MOSFET mobility degradation model for circuit simulation, *IEEE Trans. Computer Aided Design,* vol. 9, 1991, pp. 1123–1126.
29. M.E. Law, K.S. Jones, L. Radic, R. Crosby, M. Clark, K. Gable, and C. Ross, Process modeling for advanced devices, *Proc. Mater. Res. Soc. Symp.,* vol. 810, 2004, pp. C3.1.1–C.1.7.
30. Chae Soo-Won and Klaus-Jürgen Bathe, On automatic mesh construction and mesh refinement in finite element analysis, *Computers & Structures,* vol. 32, no. 3–4, 1989, pp. 911–936.
31. J.-P. Colinge, Fully-depleted SOI CMOS for analog applications, *IEEE Trans. Electron Dev.,* vol. 45, no. 5, 1998, pp. 1010–1016.
32. D. Flandre et al. Fully depleted SOI CMOS technology for heterogeneous micropower, high-temperature or RF microsystems, *Solid-State Electron.,* vol. 45, no. 4, 2001, pp. 541–549.
33. J.-P. Raskin, A. Viviani, D. Flandre, and J.-P. Colinge, Substrate crosstalk reduction using SOI technology, *IEEE Trans. Electr. Dev.,* vol. 44, no. 12, 1997, pp. 2252–2261.
34. C. Raynaud et al. Is SOICMOSa promising technology for SOCs in high frequency range? 207th Electrochemical Society Meeting, *Proc. Silicon Insulator Technol. Dev.,* Quebec City, Canada, May 2005, Electrochemical Society, Pennington, N.J., pp. 331–344.

35. J.-P. Colinge, *Silicon-on-Insulator Technology: Materials to VLSI*, Dordrecht: Kluwer Academic, 1991.
36. E.H. Poindexter, MOS interface states: Overview and physicochemical perspective, *Semicond. Sci. Technol.*, vol. 4, 1989, pp. 961–969.
37. E. Takeda, C.Y.-W. Yang, and A. Miura-Hamada, *Hot-Carrier Effects in MOS Devices*. San Diego, CA: Academic Press, 1995.
38. A. Schenk, Finite-temperature full random-phase approximation model of band gap narrowing for silicon device simulation, *J. Appl. Phys.*, vol. 84, no. 7, 1998, pp. 3684–3695.
39. A. Burenkov, K. Tietzel, and J. Lorenz, Optimization of 0.18 μm CMOS devices by coupled process and device simulation, *Solid-State Electron.*, vol. 44, no. 4, 2000, pp. 767–774.
40. J.D. Plummer, M.D. Deal, and P.B. Griffin, *Silicon VLSI Technology—Fundamentals, Practice and Modeling*, Upper Saddle River, NJ: Prentice Hall, 2000.
41. Y. Okumura, M. Shirahata, T. Okudaira, A. Hachisuka, H. Arima, T. Matsukawa, and T. Tsubouchi, A novel source-to-drain nonuniformly doped channel (NUDC) MOSFET for high-current drivability and threshold voltage controllability, *IEDM Tech. Dig.*, 1990, pp. 391–394.
42. B. Yu, H. Wang, O. Milic, Q. Xiang, W. Wang, J.X. An, and M.-R. Lin, 50 nm gate-length CMOS transistor with super-halo: Design, process, and reliability, *IEDM Tech. Dig.*, 1999, pp. 653–656.
43. C.F. Codella and S. Ogura, Halo doping effects in submicron DI_LDD device design, *IEDM Tech. Dig.*, 1985, pp. 230–231.
44. D.G. Borse, M. Rani K.N., N.K. Jha, A.N. Chandorkar, J. Vasi, V.R. Rao, B. Cheng, and J.C.S. Woo, Optimization and realization of sub-100 nm channel length single halo p-MOSFETs, *IEEE Trans. Electron Devices*, vol. 49, no. 6, June 2002, pp. 1077–1079.
45. T. Hori, A 0.1-μm CMOS technology with tilt-implanted punchthrough stopper (TIPS), *IEDM Tech. Dig.*, 1994, pp. 75–78.
46. R. Gwoziecki, T. Skotnicki, P. Bouillon, and P. Gentil, Optimization of *V*th roll-off in MOSFETs with advanced channel architecture-retrograde doping and pockets, *IEEE Trans. Electron Devices*, vol. 46, no. 7, July 1999, pp. 1551–1161.
47. A. Akturk, N. Goldsman, and G. Metze, Increased CMOS inverter switching speed with asymmetric doping, *Solid State Electron.*, vol. 47, no. 2, February 2003, pp. 185–192.
48. A. Bansal and K. Roy, Asymmetric halo CMOSFET to reduce static power dissipation with improved performance, *IEEE Trans. Electron Devices*, vol. 52, no. 3, March 2005, pp. 397–412.
49. J.-G. Su, C.-T. Huang, S.-C. Wong, C.-C. Cheng, C.-C. Wang, H.-L. Shiang, and B.-Y. Tsui, Tilt angle effect on optimizing HALO PMOS and NMOS performance, in *Proc. of IEEE IEDM*, 1997, pp. 11–14.
50. Heng-Sheng Huang, Shuang-Yuan Chen, Yu-Hsin Chang, Hai-Chun Line, and Woei-Yih Lin, TCAD simulation of using pocket implant in 50 nm n-MOSFETs, presented at International Symposium on Nano Science and Technology, Tainan, Taiwan, 20–21 November 2004.
51. B. Yu, C.H. Wann, E.D. Nowak, K. Noda, and C. Hu, Short channel effect improved by lateral channel engineering in deep-submicrometer MOSFETs, *IEEE Trans. Electron Devices*, vol. 44, April 1997, pp. 627–633.

52. H.-S.P. Wong, K.K. Chan, and Y. Taur, Self-aligned (top and bottom) double-gate MOSFET with a 25-nm thick silicon channel, *IEDM Tech. Dig.*, 1997, pp. 427–430.

53. S. Tang, L. Chang, N. Lindert, Y.-K. Choi, W.-C. Lee, X. Huang, V. Subramanian, J. Bokor, T.-J. King, and C. Hu, FinFET—A quasiplanar double-gate MOSFET, *ISSCC Tech. Dig.*, 2001, pp. 118–119.

54. Y. Taur, Analytic solutions of charge and capacitance in symmetric and asymmetric double-gate MOSFETs, *IEEE Trans. Electron Devices*, vol. 48, no. 12, December 2001, pp. 2861–2869.

55. Y. Taur, D.A. Buchanan, W. Chen, D.J. Frank, K.E. Ismail, S.-H. Lo, G.A. Sai-Halasz, R.G. Viswanathan, H.-J.C. Wann, S.J. Wind, and H.-S. Wong, CMOS scaling into the nanometer regime, *Proc. IEEE*, vol. 85, no. 4, April 1997, pp. 486–504.

56. K. Takeuchi, R. Koh, and T. Mogami, A study of the threshold voltage variation for ultra-small bulk and SOI CMOS, *IEEE Trans. Electron Devices*, vol. 48, no. 9, September 2001, pp. 1995–2001.

57. A.R. Brown, J.R. Watling, and A. Asenov, A 3-D atomistic study of archetypal double gate MOSFET structures, *J. Comput. Electron.*, vol. 1, no. 1/2, July 2002, pp. 165–169.

58. H.-S.P. Wong and Y. Taur, Three-dimensional "atomistic" simulation of discrete microscopic random dopant distributions effects in sub-0.1 μm MOSFETs, *IEDM Tech. Dig.*, 1993, pp. 705–708.

59. K. Chandrasekaran, Z.M. Zhu, X. Zhou, W. Shangguan, G.H. See, S.B. Chiah, S.C. Rustagi, and N. Singh, Compact modeling of doped symmetric DG MOSFETs with regional approach, *Nanotech Proceedings, WCM*, vol. 3, pp. 792–795, June 1–5, 2006, Boston, MA.

60. Y. Taur and T.H. Ning, *Fundamentals of Modern VLSI Devices*. Cambridge, UK: Cambridge University Press, 1998.

61. A. Ortiz-Conde, F.J. Garcia-Sanchez, and S. Malobabic, Analytic solution of the channel potential in undoped symmetric dual-gate MOSFETs, *IEEE Trans. Electron Devices*, vol. 52, no. 7, July 2005, pp. 1669–1672.

62. M.S. Lundstrom and Z. Ren, Essential physics of carrier transport in nanoscale MOSFETs, *IEEE Trans. Electron Dev.*, vol. 49, January 2002, pp. 133–141.

63. B. Ray and S. Mahapatra, Modeling of channel potential and subthreshold slope of symmetric double-gate transistor, *IEEE Trans. Electron Devices*, vol. 56, no. 2, February 2009, pp. 260–266.

6

Study of Deep Sub-Micron VLSI MOSFETs through TCAD

Srabanti Pandit

CONTENTS

6.1 Introduction

The feature size of metal-oxide-semiconductor (MOS) transistors has been scaled down for higher packing density, reduced cost, and better performance. This reduction in the device dimensions has more or less followed Moore's law, according to which the complexity of device integration is approximately doubled every 18 months. The scaling procedure has pushed the transistor dimensions well below the micrometer scale and into the deep sub-micrometer range [1]. However, in this domain, several fundamental limitations due to the physics of the device lead to the deviation of the scaling process from Moore's prediction. Several physical effects (short channel effects [SCEs], inverse narrow width effects [INWEs], and gate leakage current) critically affect the performances of deep sub-micron MOS transistors [2,3].

The SCEs mainly arise from the increased field at the drain end. The gate starts to lose its control over the channel due to the perturbations caused by the lateral drain field. The INWEs in the narrow devices are primarily caused due to the combined effect of gate fringing field and dopant redistribution phenomena. Further, gate leakage currents arise due to tunneling of carriers through the thin gate oxide layer. To summarize, the channel length and width reduction are associated with physical phenomena that primarily involve the roles of the vertical gate field and/or the lateral drain field. The depletion depths and the electrostatic potentials get altered along the channel length and the channel width. This ultimately leads to an overall degradation of the desired behavior or performance of the devices. The behavioral study of devices includes the study of parameters like threshold voltage roll-off, drain-induced barrier lowering (DIBL) effect, transconductance, subthreshold slope degradation, output resistance, etc.

Synopsys technology computer aided design (TCAD) device simulator is used efficiently in order to study the device behavior. The simulations are based on numerical computations that yield reasonably accurate results.

The rest of the chapter is divided as follows. Section 6.2 presents a brief introduction to the tools of the TCAD simulator used in this chapter. Section 6.3 discusses the device architecture and simulation setup. Section 6.4 deals with the study of SCEs. Section 6.5 covers mobility degradation. In Section 6.6 we study the drain characteristics, and in Section 6.7 we deal with the INWEs. This is followed by a study of an advanced device structure in Section 6.8. Finally, a conclusion is given in Section 6.9.

6.2 Synopsys Technology Computer Aided Design (TCAD) Tool Suite

Synopsys TCAD tool suite is used for the study of the deep sub-micron devices in this chapter. This tool suite includes tools for creating device structures, meshing of the created structure, its simulation, scientific visualization and plotting of simulated data, curve display, and extraction of performance parameters. The tools used here are briefly discussed. The details of the tool suite have been described in Chapter 4, and the respective user guides [4] should be consulted for detailed descriptions.

1. *Sentaurus Structure Editor (SSE)*: This tool is used for device structure creation. The structures are generated or edited interactively using the graphical user interface (GUI). The Synopsys meshing engines

may be configured and called by interfacing through SSE. In addition, SSE generates the necessary input files (the TDR boundary file and mesh command file) for the meshing engines that generate the TDR grid and data file for the device structure. Alternatively, devices can be created in batch mode using scripts. This option is useful for creating parameterized device structures.

2. *Mesh*: This engine helps to mesh the structure created with SSE. The tool produces finite-element meshes for use in semiconductor device simulation. The mesher generates high-quality spatial discretizations for 1D, 2D, and 3D devices using a variety of mesh generation algorithms.

3. *Sentaurus device*: This tool is used to simulate the electrical characteristics of the device. Upon specification of necessary input files, the simulated outputs are generated in separate files. Terminal currents, voltages, and charges are computed based on a set of physical device equations that describe the carrier distribution and the conduction mechanisms.

4. *Tecplot SV*: This tool has extensive 2D and 3D capabilities and is used for scientific visualization and plotting of simulated data.

5. *Inspect*: The electrical characteristics are plotted with the help of this tool. It is basically a curve display and analysis program. The curves are specified at discrete points.

The basic tool flow is illustrated in Figure 6.1. The various files associated with each tool are also shown.

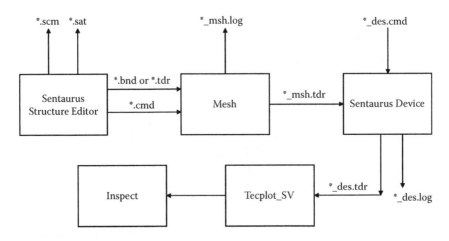

FIGURE 6.1
Basic tool flow using Synopsys TCAD.

6.3 Device Architecture and Simulation Setup

Figure 6.2 shows the schematic cross-section along the length of a typical bulk metal-oxide-semiconductor field-effect transistor (MOSFET) structure that is used for simulation. It consists of an n+ polysilicon gate, a gate oxide, a uniformly doped channel, shallow n+ source-drain extension (SDE) regions, and deep source-drain (DSD) regions.

Here, channel length is along the x-axis and depth of the device is along the y-axis. L_g is the drawn gate length, L_{eff} is the effective channel length, t_{ox} is the gate oxide thickness, and x_j is the depth of the SDE region. For simulation purposes, the real device is created using the Sentaurus Structure Editor (SSE). The default width of the device along the z-axis is 1 μm. The device has $L_{eff} = 65$ *nm*, $t_{ox} = 3$ *nm*, $x_j = 40$ *nm* and power supply $V_{DD} = 1.2V$. The SDE and DSD regions are doped with arsenic with concentrations of 2.5 × 10^{20} cm^{-3} and 3.7 × 10^{20} cm^{-3}, respectively. The channel is doped with boron with a uniform concentration of 1 × 10^{18} cm^{-3}. The constructed structure as obtained from TCAD Structure Editor is shown in Figure 6.3.

The structure is subsequently meshed using the tool 'Mesh'. The meshing strategy ensures that fine elements are generated for the important regions of the device (active region), and coarse elements are generated for the bulk regions. The meshed structure is shown in Figure 6.4 that shows the zoomed-in view of the active region of the transistor. The meshing strategy keeps the problem at a minimum of computer processor time with reasonable accuracy. In one of our devices, in the channel region, the meshing element sizes are 0.00475 μm in x- and y-directions and 1 μm in the z-direction.

The meshed structure is then simulated using the tool 'Sentaurus device'. The hydrodynamic (or the energy-balance) transport model [4] is used as the physical model for simulation purposes. This model serves to describe

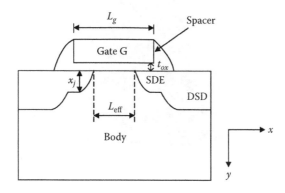

FIGURE 6.2
A typical bulk MOSFET structure.

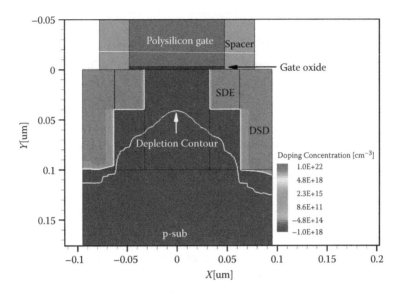

FIGURE 6.3 (See color insert)
Bulk MOSFET structure created using TCAD.

properly the characteristics of the devices in the deep sub-micron regime. The hydrodynamic model consists of a basic set of partial differential equations (Poisson equation and continuity equations) and energy-conservation equations that are solved by considering the carrier temperature to be different

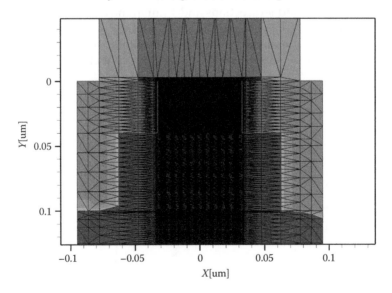

FIGURE 6.4 (See color insert)
A zoomed-in view of the active region of the meshed structure.

from the lattice temperature. The eQCvanDort flag [4] is specified to take into account the quantization effects in the classical device simulation. The OldSlotboom model [4] that takes care of the lattice-temperature dependence of the band gap and band-gap narrowing is used for the determination of the silicon intrinsic carrier concentration. The following mobility models are used: Masetti model [4] in silicon that explains the carrier mobility degradation due to scattering of carriers by the dopant impurity ions; Canali model [4] that explains degradation due to high electric fields, thus taking care of the velocity saturation effect; and Lombardi model that explains the mobility degradation at interfaces due to a transverse electric field [4].

6.4 Short Channel Effects (SCEs)

The short channel effect is the decrease of the threshold voltage of a MOS transistor as the channel length is reduced [5]. The SCE is pronounced under high drain bias.

6.4.1 Threshold Voltage Roll-Off

Figure 6.5 shows the variation of the electrostatic potential along the depth of the device at a particular position along the length of the channel. This figure is obtained using the tool 'Tecplot SV' after the simulated files are

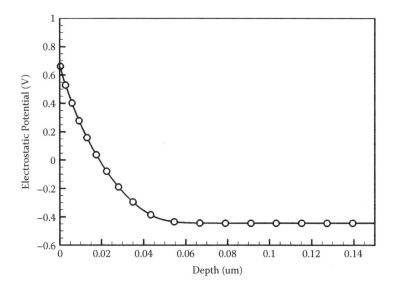

FIGURE 6.5
Electrostatic potential variation along the depth near the middle of the channel.

loaded. The device is sliced using an x-cut tool [4]. It is to be noted that the electrostatic potential values obtained through TCAD are all computed from an arbitrarily defined reference potential. In particular, for silicon, the standard approach is to set the reference potential equal to the Fermi potential of an intrinsic semiconductor. The figure illustrates that with a certain gate bias, the electrostatic potential in the semiconductor is maximum at the surface ($y = 0$) and gradually decreases to zero beyond the depletion depth where the bulk material is neutral.

The field patterns (i.e., the electrostatic potential contours in the depletion region of a long channel and a short channel bulk MOSFET) are shown in Figures 6.6(a) and 6.6(b), respectively. These figures are obtained after the respective structures are simulated using the tool 'Sentaurus device' of TCAD and then visualized using the tool 'Tecplot SV'. The long device (Figure 6.6a) has an effective channel length, L_{eff} of 1 μm, and the short device (Figure 6.6b) has an effective channel length of 65 nm. As seen from Figure 6.6(a), the potential contours are almost parallel to the oxide-silicon interface. The electric field is thus one-dimensional, being along the vertical direction only for almost the entire length of the channel. However, in Figure 6.6(b) the field is two-dimensional (i.e., the components of the electric field along both directions are appreciable). It is also seen that for a given gate bias, the electrostatic potential at a particular depth from the oxide-silicon interface is higher for the shorter device. In other words, the surface potential (electrostatic potential at the surface) of the shorter device is more, with a greater band bending at the oxide-semiconductor interface. This is the key difference between a short channel and a long channel MOS transistor. The depletion width is thus more for the shorter device, as seen from Equation (6.1) [5]:

$$W_d = \sqrt{\frac{2\varepsilon_{Si}\psi_s}{qN_a}} \tag{6.1}$$

where W_d is the depletion width, and ψ_s is the surface potential.

The white colored contour in Figure 6.6(b) is the depletion width contour. The greater depletion depth in the shorter device means that the depletion charge is effectively reduced. This leads to a reduction in the threshold voltage of the shorter device. Equation (6.2) shows the dependence of the threshold voltage on the depletion charge density [5]:

$$V_{th} = V_{FB} + 2\psi_B + \frac{Q_d}{WL_{eff}C_{ox}} \tag{6.2}$$

where V_{th} is the threshold voltage, V_{FB} is the flat-band voltage, ψ_B is the difference between Fermi level and intrinsic level, Q_d is the depletion charge

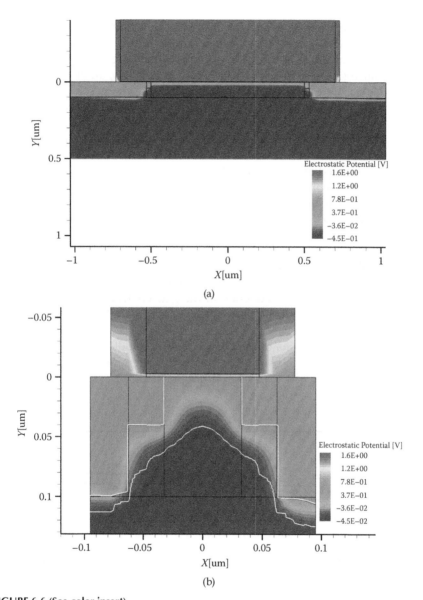

FIGURE 6.6 (See color insert)
Electrostatic potential contours in (a) long channel $L_{eff} = 1\,\mu m$ and (b) short channel $L_{eff} = 65\,nm$ ($V_{gs} = 1.0V, V_{ds} = 0.05V$).

density, and C_{ox} is the gate oxide capacitance per unit area. Thus, as Q_d decreases, V_{th} also decreases.

The two-dimensional field pattern in a short channel device is due to the close proximity of the source and drain regions. In a short channel device, the source-drain distance is comparable to the depletion width in the vertical direction [6]. The source-drain potential exerts a strong influence on the gate potential over a significant portion of the channel length. According to the charge sharing model [5,6], when the drain bias is low, all of the depletion charges beneath the gate are not imaged on the gate charges. Rather, some of them are the terminating centers of the field lines originating near the source and drain junctions. Thus there is an effective reduction in the depletion charge density that ultimately leads to the threshold voltage roll-off. This is in contrast to the long channel devices where almost all of the depletion charges are imaged on all the gate charges.

The threshold voltage roll-off as obtained from the TCAD simulator is shown in Figure 6.7. Different devices of varying effective lengths are simulated. The threshold voltages of the different devices are then determined using the tool 'Inspect' by the constant current (CC) technique [7]. The reference current is taken to be $10^{-6}\,A/\mu m$. It is seen that for long channel lengths the threshold voltage remains almost constant. However, as the channel length is decreased the threshold voltage falls from its long-channel value. This is due to the gradual reduction of the gate control over the channel in the shorter devices. The drain field starts to exert its influence over the gate field in the shorter devices.

The simple assumptions of the charge sharing model, however, render it invalid for high drain and substrate biases. It is therefore unable to explain the drain-induced barrier lowering (DIBL) that is discussed next.

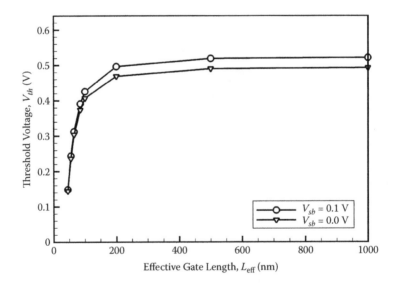

FIGURE 6.7
Threshold voltage roll-off for two different substrate biases $(V_{gs} = 1.0\,V, V_{ds} = 0.05\,V)$.

6.4.2 Drain-Induced Barrier Lowering (DIBL)

For long channels, as seen in Figure 6.6(a), the surface potential is flat over most of the channel region of the device. The surface potential is mainly controlled by the gate voltage and acts as a barrier to the electrons (for n-channel MOSFET). The electrons are not able to surmount this barrier below the threshold condition. However, in case of short devices, the source and the drain fields penetrate deeper into the middle of the channel. This lowers the barrier between the source and the drain. The electrons are then able to overcome the reduced barrier and move toward the drain end. This increases the subthreshold current. The threshold voltage of the short channel device is thus lower than that of a long channel device. As the drain bias is increased the barrier is lowered further, resulting in a further decrease in the threshold voltage. This phenomenon is referred to as the DIBL effect. The surface potential plots as obtained from TCAD simulation for the long and the short channel devices are shown in Figure 6.8 to illustrate the DIBL effect. The shorter device has been studied with two drain biases.

In the figure, the variation of surface potential along the length of the device is obtained by using the y-cut tool of 'Tecplot SV' for each of the devices. The surface potential is then plotted against normalized channel length where 0 is the center of the channel, −1 is the source end, and 1 is the drain end.

It is shown in Figure 6.8 that with a very small drain bias ($V_{ds} = 0.05V$), the surface potential at the source end is nearly equal to the built-in potential

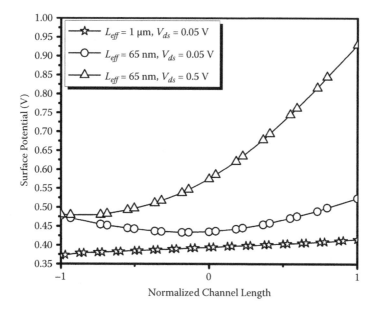

FIGURE 6.8
Surface potential plots as obtained through device simulation.

barrier, V_{bi} of the source-substrate junction. As the drain bias is increased $(V_{ds} = 0.5V)$, the surface potential at the drain end rises to $(V_{bi} + V_{ds})$. Thus the surface potential at the drain end rises by 0.45 V (0.50–0.05 V). Keep in mind that the surface potential values in TCAD are measured relative to a reference. The absolute values are to be calculated accordingly. The DIBL effect for the 65 nm MOSFET as obtained from the simulation results is shown in Figure 6.9. The DIBL coefficient obtained is –0.18.

The threshold voltage roll-off and DIBL can be minimized in three ways: (1) reduction of the oxide thickness to achieve better gate control over the channel, (2) reduction of the depletion width, and (3) reduction of the source-drain junction depth. The depletion width is reduced by increasing the doping concentration in the channel. The source-drain junction depth is effectively reduced by introducing the source-drain extension structure.

Figure 6.10 shows the DIBL effect as obtained from the simulation results of a MOSFET with a lower SDE depth. $L_{eff} = 65\ nm, x_j = 25\ nm, t_{ox} = 2.2\ nm$, and power supply $V_{DD} = 1.2V$. The threshold voltages are calculated using the constant current technique. The DIBL coefficient in this case is calculated to be –0.11.

Figure 6.11 shows the subthreshold characteristics of the 65 nm MOSFET as obtained from TCAD simulation results. It shows plots of the drain current (in log scale) against the gate voltage for two different drain voltages. It is seen that with the increase in the drain bias, the subthreshold current increases. For very short devices, the subthreshold slope degrades because the surface potential is controlled more by the drain than the gate. Eventually, the gate loses all its control, and punch-through takes place when a high

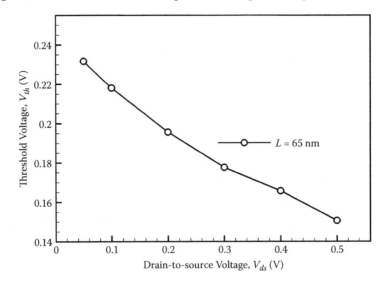

FIGURE 6.9
The DIBL effect, $L_{eff} = 65$nm, $x_j = 40nm$, $t_{ox} = 3\ nm$.

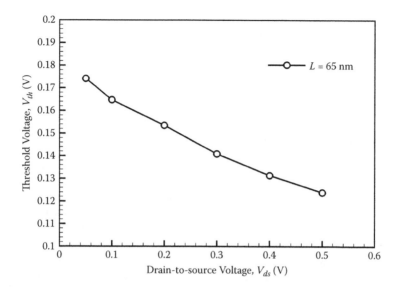

FIGURE 6.10
The DIBL effect, $L_{eff} = 65\ nm$, $x_j = 25\ nm$, $t_{ox} = 2.2\ nm$.

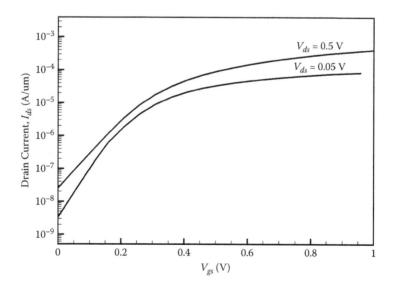

FIGURE 6.11
Subthreshold characteristics for two different drain biases ($L_{eff} = 65\ nm$).

drain current persists irrespective of the gate voltage. The subthreshold swings (SSs) for the low and high drain biases as obtained from Figure 6.11 are 90 mV/decade and 95 mV/decade, respectively.

6.5 Mobility Degradation

The mobility of carriers in the MOSFET channel is significantly lower than that in the bulk silicon [6]. At the surface boundary, lattice (or phonon) scattering is increased due to the presence of crystalline discontinuity. At high vertical fields, surface roughness scattering severely degrades the carrier mobility. Channel mobility is also affected by the oxide and interface traps at the Si-SiO$_2$ interface [6].

Figure 6.12 shows the plot of transconductance, g_m against V_{gs} for the 65-nm device for two different substrate biases under low and high drain biases. Figure 6.12(a) is for a small drain bias, with $V_{ds} = 0.05\,V$, and Figure 6.12(b) is for a comparatively higher drain bias with $V_{ds} = 1.0\,V$. The transconductance plots are obtained by differentiating the curves of the $I_{ds} - V_{gs}$ characteristics. This is done by using the 'diff' command in the 'Inspect' tool. In Figure 6.12(a) it is observed that when the drain bias is low, the transconductance starts to fall off from a high value beyond a certain gate voltage. This is explained as follows. When the vertical electric field is very high (in this case it is greater than 1.6×10^6 V/cm for an oxide thickness of 3 nm and gate bias greater than 0.5 V), the mobility decreases very rapidly due to reasons discussed earlier. Thus, at low drain bias, the drain current and the transconductance are degraded significantly at high gate voltages. Also note that at low V_{ds}, the parasitic source-drain resistance plays a significant role in determining the drain current. The peak drain current and consequently the transconductance decrease in the linear region (low V_{ds}) due to this resistance.

However, if the drain bias is high as in Figure 6.12(b), the transconductance does not fall off. This is due to velocity saturation. The mobility degradation of the carriers is somewhat counterbalanced by the high drift velocity of the carriers under high drain bias. For very short channel transistors, the drain current I_{ds} becomes limited by the velocity saturation effect. This saturated drain current is given as [6]

$$I_{dsat} = C_{ox}Wv_{sat}(V_{gs} - V_{th})$$ (6.3)

Consequently, the transconductance is also fixed at a constant value.

It is interesting to note from Figure 6.12 that the maximum value of the transconductance at low V_{ds} is much lower than that at high V_{ds}. This is explained as follows.

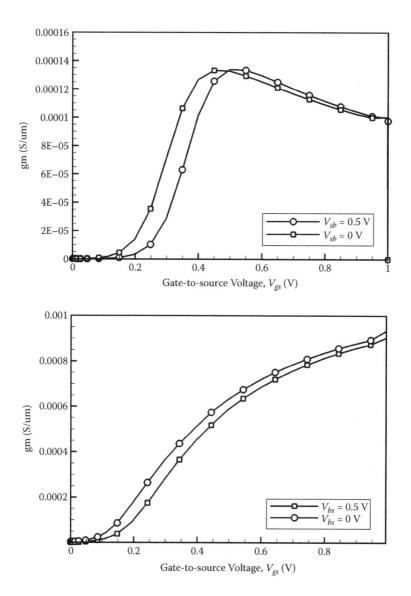

FIGURE 6.12
Transconductance versus Gate-to-source Voltage; (top)Vds = 0.05V and (bottom) Vds = 1.0V.

The drain current, I_{ds} may be written as

$$I_{ds} = \frac{V_{ds}}{R_{ds} + R_{Ch}} \tag{6.4}$$

where R_{Ch} is the channel resistance, and R_{ds} is the source-drain resistance due to the lightly doped SDE regions, or,

$$R_{Ch} = \left. \frac{V_{ds}}{I_{ds}} \right|_{R_{ds}=0}$$

or Equation (6.4) may be written as

$$I_{ds} = \frac{\frac{V_{ds}}{R_{Ch}}}{1 + \frac{R_{ds}}{R_{Ch}}} = \frac{I_{ds}|_{R_{ds}=0}}{1 + \frac{R_{ds} I_{ds}|_{R_{ds}=0}}{V_{ds}}} \tag{6.5}$$

Therefore, when V_{ds} is low, it is due to R_{ds}, that I_{ds} drops from its value, $I_{ds}|_{R_{ds}=0}$. Thus the maximum value of the transconductance is lowered at low drain bias.

6.6 Drain Characteristics

A significant effect that critically affects the I-V characteristics of a deep sub-micron MOS transistor is velocity saturation.

6.6.1 Velocity Saturation

In a short channel device, the saturation of the drain current may take place at a much lower V_{ds} value than the V_{ds} value of the longer device. This limits the saturation current of the device. This is illustrated in Figure 6.13 where two curves are drawn: the upper one is for the long channel, and the lower one corresponds to the shorter device. It is seen that the shorter device experiences an early saturation (i.e., the current saturates at a lower drain voltage). This occurs due to velocity saturation and is explained as follows. In the presence of velocity saturation,

$$\frac{1}{V_{ds_{sat}}} = \frac{m}{V_{gs} - V_{th}} + \frac{1}{\xi_{sat} L_{eff}} \tag{6.6}$$

where ξ_{sat} is the critical electric field beyond which the velocity saturates, and m is the bulk charge factor [8]. Equation (6.6) shows that the short channel $V_{ds_{sat}}$ is an average of $\xi_{sat} L_{eff}$ and long channel $V_{ds_{sat}} (= \frac{V_{GS} - V_{th}}{m})$. Thus short channel $V_{ds_{sat}}$ is smaller than long channel $V_{ds_{sat}}$. Hence the drain current for the shorter MOS transistor saturates earlier compared to the long channel value.

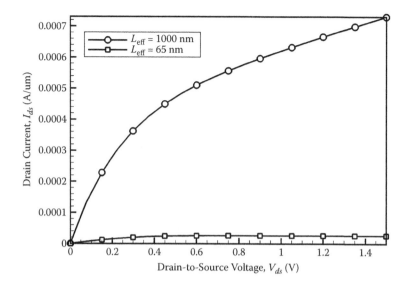

FIGURE 6.13
Drain current versus drain-to-source voltage ($V_{gs} = 1.1$ V).

6.6.2 Output Resistance

Figure 6.14 shows the simulated drain characteristics of a 65 nm MOSFET. As seen in Figure 6.14, for the short channel MOSFET, the drain current increases beyond saturation. There are two reasons for this. First, due to the increase of drain voltage, threshold voltage falls and hence the current increases. This is the DIBL effect. Second, as V_{ds} is increased beyond the saturation voltage, the saturation point (the point along the channel length where carriers attain the saturation velocity) where the surface channel collapses moves slightly toward the source [6]. That is, the conducting channel length deceases. As a result, the current increases beyond the saturation point. This is the channel length modulation effect.

The increase in the current beyond the saturation point implies that the output conductance is finite. A plot of the output resistance against the drain-to-source voltage is shown in Figure 6.15. It is seen from the curves that in the subthreshold region, the drain current is low so that the output resistance is high. In the strong inversion region, the drain current is high so that output resistance is low. In weak inversion, with the increase of drain bias, the drain current increases due to various second-order effects such as channel length modulation, DIBL effect, etc. Thus the output resistance value falls. However, in strong inversion, the increase of drain current with the increase of drain bias due to the above effects is somewhat counterbalanced by the carrier mobility degradation effect due to the applied gate bias. Therefore the overall increase of drain current is small. Hence the output resistance remains nearly constant.

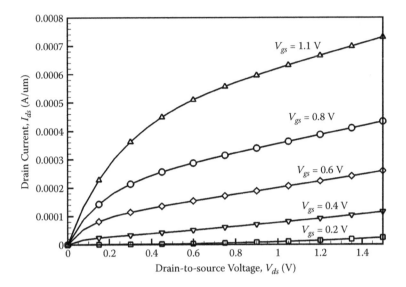

FIGURE 6.14
Drain characteristics of a 65 nm MOSFET.

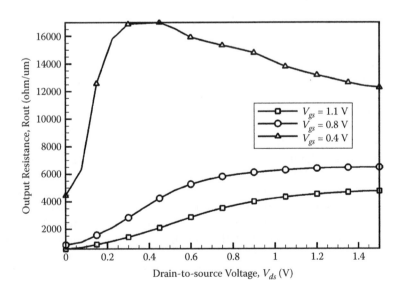

FIGURE 6.15
Output resistance versus drain voltage (L_{eff} = 65 *nm*).

6.7 Inverse Narrow Width Effects (INWEs)

As the device dimension is reduced along its width, we come across effects known as the narrow width effects. In LOCOS isolated devices, threshold voltage increases with a decrease in the width of the device. The deep submicron devices isolated using the shallow trench isolation (STI) process are associated with a decrease in the threshold voltage with a reduction in the channel width. This effect associated with the STI MOSFETs is known as the inverse narrow width effect (INWE) [5]. Narrowing the channel width of a transistor affects its performance to an extent comparable to the effects caused by shortening the channel length [9,10]. Hence the study of MOS transistor performances along the width dimension becomes important for the design of low power complementary metal-oxide-semiconductor (CMOS) circuits and memory cells [11,12].

The INWE is primarily caused due to two reasons. First is the combined effect of gate fringing fields through trench oxide and dopant redistribution phenomenon [13,14]. Second is the effect of STI stress [15]. The effect of the latter is dominant for comparatively larger widths, while the former is dominant for narrower widths [15]. STIMOS transistors with channel widths below 1 μm exhibit threshold voltage roll-off where the INWE effect is the dominant cause.

6.7.1 Gate Fringing Field Effect

Figure 6.16 shows the schematic of the cross-section along the width of the STIMOS device. The field lines originating from the gate terminate on charges under the gate through the thin gate oxide. Some of the field lines also terminate on charges along the trench oxide sidewalls. These are the fringing field lines.

Figure 6.17 shows the cross-section along the width of a typical STIMOS device as obtained by using the appropriate slicing tool of 'Tecplot SV'. The electrostatic potential contours are shown in the figure. It is seen in the figure that the depletion depth (the white contour is the depletion contour) is nonuniform and varies along the width of the device. In wide MOS devices, the enhanced depletion region near the sidewalls caused due to the gate fringing field is a small percentage of the total depletion volume and can be neglected. The depletion depth may then be considered to be uniform throughout the width without resulting in too much of an error in the analysis of the device performances. For devices with small widths, the depletion volume near the sidewalls becomes a large percentage of the total depletion volume. This is referred to as the gate fringing field effect and is illustrated in Figure 6.18.

MOS devices with varying channel widths are created using the Sentaurus Structure Editor and are then simulated using the tool 'Sentaurus device'. The simulated devices are then visualized using the tool 'Tecplot SV'. The depletion depths along the width are noted and then plotted as in Figure 6.18.

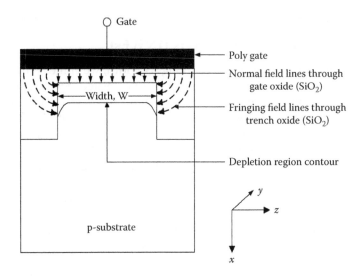

FIGURE 6.16
Cross-section along the width of a trench-isolated MOSFET including gate fringing fields.

It is observed from Figure 6.18 that as the device becomes narrower, the gate fringing increases.

The effect of gate fringing is modeled by a parasitic fringe capacitance. The higher depletion depths at the trench oxide sidewalls are associated with a higher surface potential. Figure 6.19 shows the schematic representation of

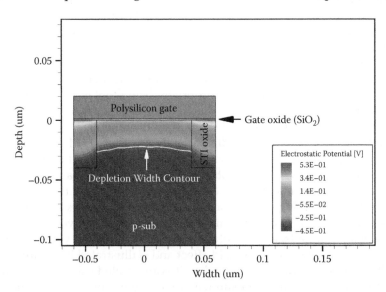

FIGURE 6.17 (See color insert)
Cross-section along the width of a trench-isolated MOSFET.

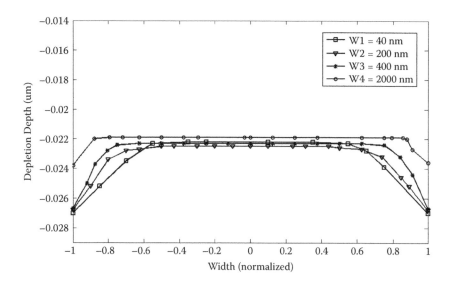

FIGURE 6.18
Gate fringing field effect for STI MOSFETs.

the variation of the depletion depth and the corresponding surface potential along the width of the device.

In Figure 6.19(a) the depletion depth at the trench-oxide sidewall is $d_{w\,max}$, decreases to a value d_w, remains constant at d_w beyond a critical point z_b, and then finally increases again to $d_{w\,max}$ at the other sidewall. Figure 6.19(b) shows the variation of the surface potential ψ_S along the width of the device. ψ_{SM} is the surface potential in the middle of the device; ψ_{ST} is the surface potential at the trench oxide sidewall edges.

Figure 6.20 shows the surface potential profile along the width of a typical simulated device. The gate fringing field through the trench oxide causes the surface potential at the sidewall edges to increase in comparison to that at the middle of the channel width.

The sidewall surface potential ψ_{ST} varies with V_{gs} in a similar fashion as the surface potential at the middle ψ_{SM} does with V_{gs}. A typical device has been simulated for different gate biases, and the results are as shown in Figure 6.21.

The surface potential at the center of the device is related to that at the trench-oxide sidewalls by the relation [13,16]

$$\psi_{ST} = l\psi_{SM} + d \tag{6.7}$$

where l is close to unity, and d is a numerical constant expressed as

$$d = \frac{nkT}{q}$$

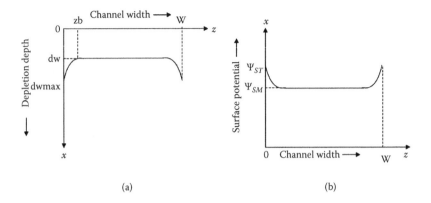

FIGURE 6.19
(a) Variation of depletion depth along the channel width. (b) Variation of surface potential along the channel width.

where $\frac{kT}{q}$ is the volt equivalent of temperature. The value of n lies between 1.2 and 2.3 for devices with largely varying dimensions like 200 nm gate length to 40 nm gate length having substrate doping ranging from 10^{17}/cc to 10^{19}/cc.

6.7.2 Dopant Redistribution

The depletion charge density does not remain constant along the entire width of the channel. This is due to the non-uniformity of the doping concentration

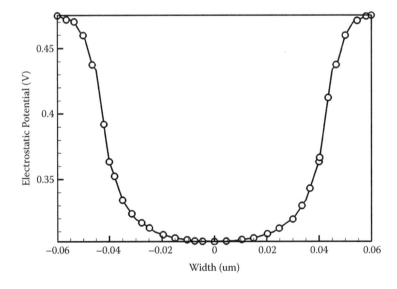

FIGURE 6.20
Plot of surface potential against the width of the device.

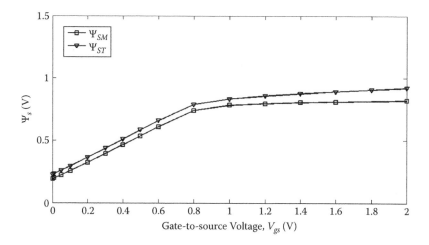

FIGURE 6.21
Simulation results showing variation of ψ_{ST} and ψ_{SM} with V_{gs} ($V_{sb} = 0V$).

along the width of the channel which is caused by dopant redistribution (comprised of dopant segregation [17–19] and transient-enhanced diffusion, TED [20]). After the gate oxidation processing step, dopant atoms in the active channel regions of the transistors start to diffuse out to the adjacent trench oxide/Si interface. This is referred to as the dopant segregation effect [21]. In particular, boron in NMOS transistors gets depleted at the STI edges and segregates to the oxide/Si interface [17]. Boron segregation reduces the B-concentration near the STI edges. If the active area of the device is narrow enough, boron segregation would reduce the boron concentration significantly even at the center of the channel [19]. Further, during the silicon processing step, ion implantation of the dopant impurities dislodges silicon atoms from their lattice sites, which are known as interstitial silicon atoms. A subsequent annealing step repairs some of these lattice damages. The remaining interstitial silicon atoms play an active role in the diffusion of the dopant impurities. High concentrations of these silicon atoms enhance boron diffusion. This phenomenon is known as transient-enhanced diffusion (TED). The TED process leads to a substantial decrease of the boron concentration in proximity to the STI edge relative to the center of the channel [20]. Again if the device is narrow enough, the decrease in the boron concentration may take place significantly even at the center of the channel. The doping profile along the channel width is expressed as [22]

$$N(z) = \left(\frac{N_a - N_t}{2} \right) \left(erf\left(\frac{z + W/2}{2\sqrt{Dt}} \right) - erf\left(\frac{z - W/2}{2\sqrt{Dt}} \right) \right) + N_t \qquad (6.8)$$

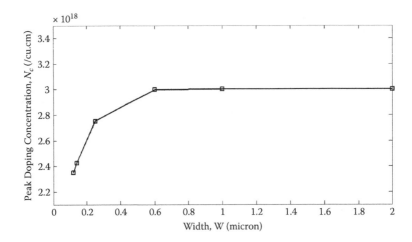

FIGURE 6.22
Variation of the peak channel doping concentration with the channel width.

where N_t is the doping concentration at the trench oxide sidewalls, N_a is the substrate doping concentration, and $2\sqrt{Dt}$ is the diffusion length.

Figure 6.22 shows the plot of the peak concentration against channel width. The figure shows that for large widths, the doping concentration of the channel is equal to N_a. However, as the width decreases, the peak concentration in the channel is reduced due to dopant redistribution.

Figure 6.23 shows the plot of the INWE on the threshold voltage of bulk STI MOSFETs with $L_{eff} = 40\ nm$. Different devices with varying channel widths are created using Sentaurus Structure Editor. The source and drain regions are doped with arsenic. A Gaussian doping profile is considered with a peak concentration of 1×10^{19} cm^{-3}. The peak concentration, peak position, and concentration value at the junction depth are specified to the simulator. The channel is doped with boron with a maximum concentration of 3×10^{18} cm^{-3}. An analytical doping profile, identical to that described by Equation (6.8) has been defined to incorporate the phenomenon of dopant redistribution. The specified parameters are maximum concentration and inflection point. The constructed structure is subsequently meshed and simulated using the respective tools. The meshing strategy ensures that fine elements are generated for the critical parts of the device (active region) and coarse elements for the bulk regions.

6.8 Advanced Device Structures

Scaling of MOSFETs has led to several serious limitations; for example, (1) device performance compromised due to undesirable effects such as threshold voltage roll-off, drain-induced barrier lowering (DIBL), degraded

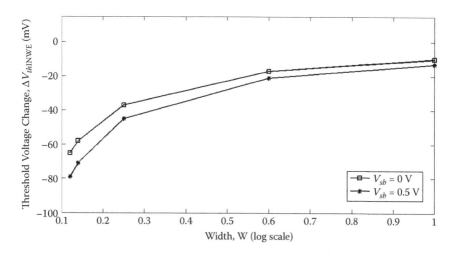

FIGURE 6.23
Inverse narrow width effect (INWE) on the threshold voltage versus channel width ($V_{gs} = 1.0V$, $V_{ds} = 0.05\ V$, $V_{sb} = 0.5\ V$).

subthreshold slope; (2) technological barriers like the manufacturing of optical equipment required for the reduction in the wavelength of light for lithography procedures; (3) gate-oxide thickness reduction leading to quantum mechanical tunneling, which in turn necessitates the introduction of gate dielectric materials with high dielectric constants so that the physical thickness of the dielectric can be increased; and (4) high doping between the source and drain which increases the parasitic capacitance between the source, drain, and the substrate.

All of the above led to the development of the silicon-on-insulator (SOI) technology.

6.8.1 SOI Structures

SOI structures are of two types. These are partially depleted SOI (PDSOI) and fully depleted SOI (FDSOI). In PDSOI, the silicon film is larger than the sum of the gate depletion widths from the front and back ends. These devices exhibit a floating body effect (kink effect). This occurs when carriers of the same type as the body, generated by impact ionization near the drain, are stored in the floating body. This alters the body potential and hence the threshold voltage. A way to minimize this is to use body contacts. However, the body effect advantage is lost in that case. In FDSOI, the silicon film is thin enough that the entire film is depleted before the threshold condition is attained. These devices exhibit a steeper subthreshold slope as compared to the bulk devices. The FDSOI devices were found to perform better than the PDSOI.

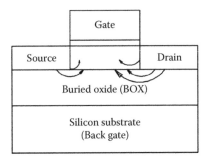

FIGURE 6.24
Conventional thin film SOI MOSFET.

Figure 6.24 shows the schematic of a conventional thin film SOI MOSFET structure. Most of the electric field lines as shown in Figure 6.24 from the source and drain propagate through the buried oxide (BOX) layer before reaching the channel layer. The SOI structures exhibited lower leakage, low junction capacitance, low latch-up, and better subthreshold swing [23–25]. The performance of these devices in regard to SCEs depends on the silicon film thickness, BOX layer thickness, and doping concentrations [26].

In the deep sub-micron regime, the merits offered by the FDSOI devices diminished, and then double gate structures were found to be more promising.

6.8.2 Double Gate (DG) MOSFETs

The DG-MOSFETs are very attractive to improve the performance of CMOS devices and to overcome some of the difficulties faced in the downscaling of MOSFETs into the deep sub-micron regime. Threshold voltage roll-off, DIBL, off-state leakage, etc., are significantly reduced; hence, these devices are preferred in nanoscale circuits [27–29].

Figure 6.25 shows the schematic diagram of a typical DG-MOSFET. It consists of a silicon slab sandwiched between two oxide layers. A metal or a poly-silicon film contacts the two oxide layers [26]. These act as the front and back gate electrodes that create inversion layers near the two Si-SiO$_2$ interfaces

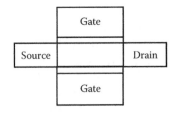

FIGURE 6.25
A typical DG-MOSFET.

upon the application of a suitable bias. Thus there are two MOSFETs that share the same source, drain, and substrate.

The salient features of a DG-MOSFET are control of the SCEs by device geometry and not by doping (channel doping or halo doping) as done in bulk MOSFETs. The two gate electrodes jointly control the carriers, thereby screening the effect of drain field from the channel. Additionally, the thin silicon channel leads to a stronger coupling of the gate potential with the channel potential. The reduced SCEs lead to greater scalability than bulk MOSFETs. The undoped body (intrinsic channel) reduces mobility degradation by eliminating impurity scattering, thereby improving the carrier transport. The random microscopic dopant fluctuations are also avoided [30]. The current drive (or gate capacitance) per unit area is increased [31].

The DG-MOSFETs are again of two types, symmetric DG-MOS and asymmetric DG-MOS [23]. In the former, the two oxide thicknesses are the same, the two gates have the same flat-band voltage, and the gates are connected together. In the latter, the two oxide thicknesses are different.

Figure 6.26 shows the cross-section along the length of a DG-MOSFET structure as seen in the tool 'Tecplot SV' of TCAD.

Figure 6.27 shows the subthreshold characteristics of the simulated DG-MOSFET for two drain biases. The lower curve corresponds to $V_{ds} = 0.05\,V$, and the upper one corresponds to $V_{ds} = 0.5\,V$. The subthreshold swing for $V_{ds} = 0.05\,V$ is calculated to be 75 mV/decade, and that for $V_{ds} = 0.5\,V$ is 80 mV/decade.

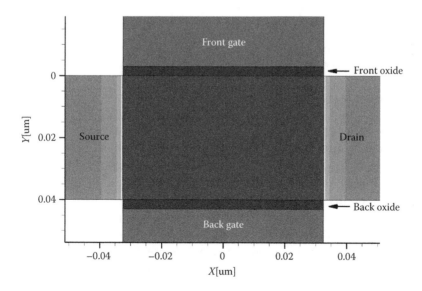

FIGURE 6.26 (See color insert)
A typical DG-MOSFET structure as simulated in TCAD.

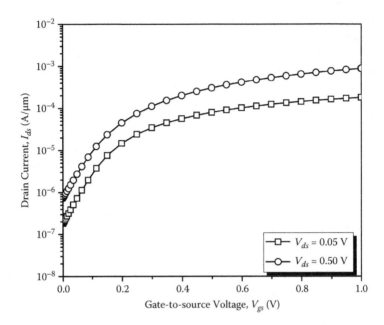

FIGURE 6.27
Subthreshold characteristics of DG-MOSFETs for two different drain biases ($L_{eff} = 65\ nm$).

Figure 6.28 shows the DIBL effect as obtained from TCAD results. The DIBL coefficient is calculated to be −0.10. It is seen that the DG-MOSFETs show a steep subthreshold swing and a high drive current. However, the limitations of DG-MOSFETs in relation to how far it can be scaled come from the SCEs, such as threshold voltage roll-off and DIBL.

The DG-MOSFETS are associated with the following phenomena:

1. The SOI-based DG-FinFETs exhibit *self-heating effects* [8] because the device is thermally insulated from the substrate by the BOX layer. This consequently alters the output characteristics of the device, especially in the sub-micron regime.

2. *Quantum mechanical effects:* Due to the ultra thin nature of the silicon channels, quantization effects are seen. This affects the distribution and mobility of carriers [32].

3. *Volume inversion:* As the silicon film is reduced, the inversion layer is formed not only near the two oxide interfaces, but the entire silicon film gets inverted. The carriers are no longer confined to the interfaces but are distributed throughout the entire silicon volume [26]. This phenomenon is found to increase the mobility of the carriers by reducing their scattering at the oxide and inter-face traps.

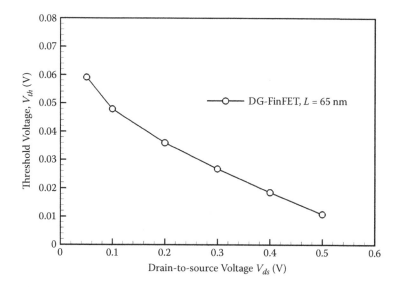

FIGURE 6.28
The DIBL effect ($L = 65$ nm).

4. *Misalignment of top and bottom gates:* The DG-MOSFETs are difficult to fabricate; in particular, the achievement of a perfect vertical alignment of the top and the bottom gates poses a serious problem. Any misalignment between the gates leads to device performance degradation due to overlap capacitance and loss in current drive [3].

6.9 Conclusion

The performances of MOSFETs in the sub-micron regime are limited by several physical phenomena such as the short channel effects along the channel length that mainly consist of the threshold voltage roll-off, drain-induced barrier lowering, and subthreshold slope degradation. Apart from these, mobility degradation, transconductance, and output resistance are also affected. The inverse narrow width effect along the channel width consists of the combined effect of gate fringing and dopant redistribution that leads to a threshold voltage roll-off. All of these effects have been studied for a typical bulk MOSFET with the help of the TCAD device simulator. Additionally, a typical double gate (DG) MOSFET has also been studied with TCAD. The TCAD studies reveal the device behavior accurately.

References

1. International Technology Roadmap for Semiconductors, 2007.
2. Wong, H.S.P., Frank, D.J., Solomon, P.M., Wann, C.H.J., and Welser J.J., Nanoscale CMOS, *Proceedings of the IEEE*, vol. 87, no. 4, pp. 537–570, April 1999.
3. Wong, H.S.P., Beyond the conventional transistor, *IBM J. Res. Dev.*, vol. 46, no. 2/3, pp. 133–168, March/May 2002.
4. TCAD Sentaurus Manuals, Version C-2009.06, Synopsys Inc., Mountain View, CA.
5. Tsividis, Y.P., *Operation and Modeling of the MOS Transistor*, 2nd ed. New York: McGraw-Hill, 1999.
6. Taur, Y., and Ning, T.H., *Fundamentals of Modern VLSI Devices*, Cambridge University Press, United Kingdom, 1998.
7. Conde, A.O., Sanchez, F.J.G., Liou, J.J., Cerdeira, A., Estrada, M., and Yue, Y., A review of recent MOSFET threshold voltage extraction methods, *Microelectronics Reliability*, vol. 42, pp. 583–596, 2002.
8. Cheng, Y., and Hu, C., *MOSFET Modeling and BSIM3 User's Guide*, Norwell, MA: Kluwer Academic, 2002.
9. Agrawal, B., De, V.K., and Meindl, J.D., Three-dimensional analytical subthreshold models for bulk MOSFETs, *IEEE Trans. Electron Devices*, vol. 42, no. 12, pp. 2170–2180, December 1995.
10. Lin, S.C., Kuo, J.B., Huang, K.T., and Sun, S.W., A closed-form back-gate-bias related inverse narrow-channel effect model for deep-submicron VLSI CMOS devices using shallow trench isolation, *IEEE Trans. Electron Devices*, vol. 47, no. 4, pp. 725–733, April 2000.
11. Roy, K., Mukhopadhyay, S., and Mahmoodi-Meimand, H., Leakage current mechanisms and leakage reduction techniques in deep-submicrometer CMOS circuits, *Proceedings of the IEEE*, vol. 91, no. 2, pp. 305–327, February 2003.
12. Lau, W.S. et al., Anomalous narrow width effect in p-channel metal-oxide-semiconductor surface channel transistors using shallow trench isolation technology, *Microelectronics Reliability*, vol. 48, pp. 919–922, June 2008.
13. Pandit, S., and Sarkar, C.K., Modeling the effect of gate fringing and dopant redistribution on the inverse narrow width effect of narrow channel shallow trench isolated MOSFETs, 24th *IEEE Annual Conference on VLSI Design*, pp. 195–200, 2011.
14. Pandit, S., and Sarkar, C.K., Analytical modelling of inverse narrow width effect for narrow channel STI MOSFETs, *International Journal of Electronics*, vol. 99, no. 3, pp. 361–377, 2012.
15. Pacha, C., Martin, B., von Arnim, K., Brederlow, R., Schmitt-Landsiedel, D., Seegebrecht, P., Berthold, J., and Thewes, R., Impact of STI-induced stress, inverse narrow width effect, and statistical VTH variations on leakage currents in 120 nm CMOS, *ESSDERC*, pp. 397–400, 2004.
16. Pandit, S., and Sarkar, C.K., A compact threshold voltage model for narrow channel nano-scale MOSFETs, *IEEE International Conference on Computers and Devices for Communication*, 2009, pp. 1–4.
17. Wang, R.V., Lee, Y.H., Lu, Y.L.R., McMahon, W., Hu, S., and Ghetti, A., Shallow trench isolation edge effect on random telegraph signal noise and implications for flash memory, *IEEE Trans. Electron Devices*, vol. 56, no. 9, pp. 2107–2113, September 2009.

18. Arnaud, F., and Bidaud, M., Gate oxide process impact on RNCE for advanced CMOS transistors, *IEEE Proceedings of ESSDERC* 2002.
19. Nouri, F., Scott, G., Rubin, M., and Manley, M., Narrow device issues in deep-submicron technologies—The influence of stress, TED and segregation on device performance, *IEEE Proceedings of ESSDERC* 2000, pp. 112–115.
20. Ghetti, A., Benvenuti, A., Molteni, G., Albenci, S., Soncini, V., and Pavan, A., Experimental and simulation study of boron segregation and diffusion during gate oxidation and spike annealing,*Technical Digest of IEEE Electron Devices Meeting*, pp. 983–986, 2004.
21. Dabrowski, J., Mussig, H.J., Zavodinsky, V., Baierle, R., and Caldas, M.J., Mechanism of dopant segregation to SiO_2/Si (001) interfaces, *Phys. Rev. B,* vol. 65, pp. 245–305, 2002.
22. Fung, S.K.H., Chan, M., and Ko, P.K., Inverse-narrow-width effect of deep sub-micrometer MOSFETs with Locos isolation, *Solid-State Electronics*, vol. 41, no. 12, pp. 1885–1889, 1997.
23. Lu, H., and Taur, Y., An analytic potential model for symmetric and asymmetric DG MOSFETs, *IEEE Trans. Electron Devices*, vol. 53, no. 5, pp. 1161–1168, May 2006.
24. Sadachika, N., Kitamaru, D., Uetsuji, Y., Navarro, D., Yusoff, M.M., Ezaki, T., Mattausch, H.J., and Mattausch, M.M., Completely surface-potential-based compact model of the fully depleted SOI-MOSFET including short-channel effects, *IEEE Trans. Electron Devices*, vol. 53, no. 9, pp. 2017–2024, September 2006.
25. Trivedi, V.P., and Fossum, J.G., Scaling fully depleted SOI CMOS, *IEEE Trans. Electron Devices*, vol. 50, no. 10, pp. 2095–2103, October 2003.
26. Colinge, J.P., *FinFETs and Other Multi-Gate Transistors*, New York: Springer, 2008.
27. Jung, H.K., and Dimitrijev, S., Analysis of subthreshold carrier transport for ultimate DG MOSFET, *IEEE Trans. Electron Devices*, vol. 53, no. 4, pp. 685–691, April 2006.
28. Masahara, M. et al., Demonstration, analysis, and device design considerations for independent DG MOSFETs, *IEEE Trans. Electron Devices*, vol. 52, no. 9, pp. 2046–2053, September 2005.
29. Iniguez, B., Fjeldly, T.A., Lazaro, A., Danneville, F., and Deen, M.J., Compact-modeling solutions for nanoscale double-gate and gate-all-around MOSFETs, *IEEE Trans. Electron Devices*, vol. 53, no. 9, pp. 2128–2142, September 2006.
30. Wong, M., and Shi, X., Analytical I-V relationship incorporating field-dependent mobility for a symmetrical DG MOSFET with an undoped body, *IEEE Trans. Electron Devices*, vol. 53, no. 6, pp. 1389–1397, June 2006.
31. Colinge, J.P., Multiple-gate SOI MOSFETs, *Solid-State Electronics*, vol. 48, pp. 897–905, 2004.
32. Trivedi, V.P., and Fossum, J.G., Quantum-mechanical effects on the threshold voltage of undoped double-gate MOSFETs, *IEEE Electron Dev. Lett.*, vol. 26, no. 8, pp. 579–582, August 2005.

7

MOSFET Characterization for VLSI Circuit Simulation

Soumya Pandit

CONTENTS

7.1 Introduction

With the continual downscaling of metal-oxide-semiconductor (MOS) transistors to the sub-90 nm regime, several secondary issues related to the transistor device physics, hitherto considered to be insignificant, are found to play significant roles in circuit performances. The circuit designers therefore need proper understanding of the various parameters related to geometry as well as performances of a single MOS transistor and the effect of these on the performances of an overall very large scale integrated (VLSI) circuit. A detailed characterization of the MOS transistor device to be used by them for the design is therefore an essential requirement prior to the design task, especially in sub-90 nm design domain. The objective of this chapter is to present a comprehensive discussion on the characterization of sub-90 nm MOS transistor for VLSI circuit simulation purpose. The discussion is limited to conventional bulk MOS transistor only, because this has been the most widely used device for VLSI circuit simulation, considering the cost and expertise required for fabrication.

The pedagogical approach used in this chapter is that initially the various characteristics are discussed qualitatively. This is followed by introduction of standard mathematical models. The use of these models in a BSIM4 compact model (which is one of the most widely used industry standard compact models) is thereafter discussed briefly. The purpose of this chapter is to present the primary important characteristics of VLSI metal-oxide-semiconductor field-effect transistor (MOSFET) to the integrated circuit (IC) designers in a manner such that they are aware of the internal workings of the SPICE simulation tool. The readers are encouraged to refer to appropriate BSIM manuals and SPICE model user guides for the exact mathematical formulations and the complete modeling works. SPICE simulation results are provided to support the theoretical discussion, as required.

The rest of the chapter is organized as follows. Section 7.2 emphasizes the importance of the use of device models for VLSI circuit simulation. The various categories of device models and the publicly available device models are also discussed. Section 7.3 presents the detailed characterization of the threshold voltage of an n-channel enhancement mode transistor. Section 7.4 discusses the I-V characteristics, supported with results. Sections 7.5 and 7.6 discuss the impact ionization process and the gate dielectric model. Characterizations of MOS capacitances are discussed in Section 7.7. Noise and statistical characterization are discussed in Sections 7.8 and 7.9, respectively. Finally, Section 7.10 presents a summary and conclusion of the chapter.

7.2 Device Models for Circuit Simulation

7.2.1 Necessity of Device Models

An outline of the IC simulation procedure is illustrated in Figure 7.1. The procedure starts with a set of desired specifications for the circuit to be designed. This acts as the input to the procedure. A particular topology of the circuit (transistor-level topology) is then selected by the designer. This choice is primarily based upon the designer's experience. The circuit is described either schematically or through a textual description, referred to as the netlist of the circuit. The circuit is subsequently analyzed through SPICE simulation tool. The SPICE simulation tool internally uses a set of device models for the components of the circuits and solves a set of network theory equations through standard numerical algorithms [1].

The performance parameters of the circuit are extracted and compared with the desired specifications. If the extracted parameters satisfy the desired specifications, the design is completed. Otherwise the simulation procedure is carried out iteratively either by changing the circuit description or circuit parameters and conducting further analyses until the specifications are satisfied.

From 1970 onward, SPICE has been the sole tool used by IC designers for circuit simulation. The development of this tool started at the University

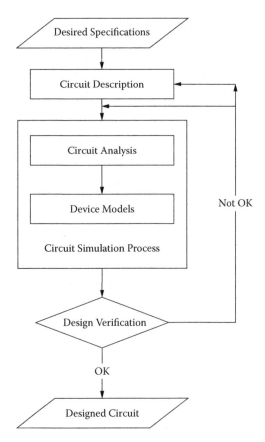

FIGURE 7.1
Outline of the IC simulation flow.

of California, Berkeley, in the late 1960s and continued until the 1990s. The commercial SPICE simulation tools used today by designers (e.g., HSPICE [Synopsys], SPECTRE [Cadence], ELDO [Mentor Graphics]) are all based upon the original SPICE simulation tool developed at the University of California. All of these simulators internally use appropriate device models for faithful description of behaviors of the devices used in the circuit. Thus the use of appropriate device models, not the internal algorithms, is responsible for the success of a circuit simulation program [2]. The accuracy and reliability of a circuit simulation process in predicting the device performances accurately depend upon the accuracy of the internal device models.

7.2.2 Definition and Categories of Device Models

Device models are defined as the link between the physical world (technology, manufacturing, etc.) and the design world (device simulation, timing simulation, etc.) of the semiconductor industry [3]. There are three categories of device models: numerical models, look-up table models, and analytical or compact models. The numerical models are based upon numerical solutions of carrier transport equations, device geometry, and doping profile-related equations. Although these techniques provide accurate results, they are computationally very intensive. Therefore these techniques are not suitable for simulation of large circuits. However, these may be used for exploration of novel device structures and associated performances. The TCAD device simulation tools as discussed in the earlier chapters are based upon this approach. The look-up table approach, on the other hand, uses measured device current and capacitances (and in some cases small signal parameters) as functions of bias voltages and device sizes for characterizing device performances that are subsequently used for circuit simulation purposes. This approach is used when good physical models of any device are not available and is sometimes used in fast circuit simulators. The most popular approach that is used for circuit simulation purpose is the third approach (i.e., the use of analytical or compact models). A compact model is characterized by a set of mathematical equations whose parameters are used as inputs to a SPICE-like circuit simulation program [3]. The physical compact model equations are derived based upon the physics of the device. These equations are expected to reproduce the device characteristics for different device dimensions, range of temperature, process variations, etc. Good physical compact models are usually complex because they consider several physical phenomena. In order to make the equations simple so as to avoid the convergence problem of the circuit simulator, some approximations are sometimes judiciously made keeping the physics intact. Fitting parameters are often introduced to improve the accuracy of the model. Apart from accuracy, a desirable requirement from a compact model is some prediction capability. This helps the designers to predict any statistical behavior of the circuit and to explore circuit performances under migrated technology.

7.2.3 Commercially Used Compact Models

The development of a physical compact model for a MOS transistor began in the 1970s. Since then, more than 100 models, including MOS 1, MOS 2, MOS 3, MOS 9, PCIM, Level 28, ISIM, BSIM1 (Berkeley Short Channel IGFET Model), BSIM2 and BSIM 3, BSIM 4, PSP (Pennsylvania State University's Surface Potential Model), HiSIM (Hiroshima University STARC IGFET Model), and EKV (Enz-Krummenacher-Vittoz) have been reported. Out of them, BSIM3, BSIM 4 PSP, HiSIM, and EKV are mostly used in commercial circuit simulators. The basic idea behind the development of all of these models is to include exact physics-based analytical equations in the model structure and then use fitting parameters to calibrate the equations with measured results. The generations of compact models have evolved to include more and more physical effects of the transistor device and thus to make the models more and more accurate.

In order to ease the task of circuit designers for predicting the circuit performances with technology generation, predictive technology model (PTM) has been developed by Zhao and Cao [4] based on physical models and early stage silicon data. The PTM of bulk CMOS is successfully generated for 130 nm to 32 nm technology nodes, with effective channel length as low as 13 nm. These have been used in the present chapter for device characterization.

7.3 Threshold Voltage Characterization

Accurate characterization of threshold voltage is one of the most important requirements for precise description of transistor behavior and the effect of such behavior on circuit performances. It may be noted that all the commercially available compact models mentioned earlier accurately characterize this important quantity. Consequently, the SPICE simulation tool by using such models properly takes care of the effects of threshold voltages of the individual MOS transistors on the circuit performances. However, it is essential for the designers to properly understand the behavior of this quantity under varying bias conditions and geometry of the device, at least in general, in order to intuitively justify the SPICE simulated performances of the circuits.

7.3.1 Threshold Voltage Characterization for Long Channel MOS Transistor

Throughout this chapter, the discussion is centered on n-channel enhancement mode transistor. In general, the conclusion drawn for an n-channel MOS transistor equally holds true for p-channel MOS transistor, with reversal of polarity of the bias voltages. Historically the source terminal is the conventional voltage reference mostly due to the dominance of digital circuits where source is considered as the reference voltage terminal.

7.3.1.1 Uniform Channel Doping

The schematic diagram of a long channel MOS transistor with uniformly doped substrate is shown in Figure 7.2. The theoretical definition of threshold voltage is based on the *strong inversion* condition. The threshold voltage is defined as the gate voltage when the surface potential or band bending reaches $2\Phi_F$ and the silicon charge (the square root) is equal to the bulk depletion charge for that potential [5]. This definition is valid for long channel MOS transistors. Based on the threshold voltage value, the operating region of a MOS transistor is broadly divided into three parts: First, if the applied gate bias is greater than the threshold voltage ($V_{GS} > V_T$), the inversion charge density is larger than the substrate doping concentration and the

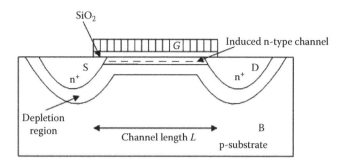

FIGURE 7.2
Cross-sectional view of n-channel MOSFET.

transistor operates in a strong inversion region. In this region, drift current is the dominant carrier transport mechanism. Second, if the applied gate bias is much smaller than the threshold voltage ($V_{GS} > V_T$), the inversion charge density is smaller than the substrate doping concentration and the transistor operates in a weak inversion or sub threshold region. In this region, the diffusion current is the dominant carrier transport mechanism. Last, if the applied gate bias is very close to the threshold voltage, the inversion charge density is close to the substrate doping concentration and the region of operation is called a moderate inversion region. In this region, both diffusion and drift currents are equally important.

The threshold voltage of a long and wide channel MOS transistor with uniform doping is given as [5]

$$V_T = V_{FB} + 2\Phi_F - \frac{Q_B}{C_{ox}} \tag{7.1a}$$

$$V_T = V_{FB} + 2\Phi_F + \frac{\sqrt{4\varepsilon_{Si}qN_A\Phi_F}}{C_{ox}} \tag{7.1b}$$

In (7.1a) and (7.1b), V_{FB} is the flat band voltage, $\Phi_F = \frac{kT}{q}\ln(\frac{N_A}{n_i})$ is the Fermi potential, N_A is the uniform p-type substrate doping concentration, and C_{ox} is the oxide capacitance per unit area. Q_B is the depletion charge per unit area. For NMOS transistor Q_B is negative, and for PMOS transistor Q_B is positive. With the application of substrate bias V_{BS} (<0 for NMOS and >0 for PMOS), the bulk depletion charge region is widened and the threshold voltage is increased as given as [5]:

$$V_T = V_{FB} + 2\Phi_F + \frac{\sqrt{2\varepsilon_{Si}qN_A(2\Phi_F - V_{BS})}}{C_{Ox}} \tag{7.2a}$$

This can be alternatively written as follows:

$$V_T = V_{T0} + \gamma\left(\sqrt{2\Phi_F - V_{BS}} - \sqrt{2\Phi_F}\right) \tag{7.2b}$$

Here the quantity V_{T0} is referred to as the zero substrate bias large geometry threshold voltage. The factor

$$\gamma = \frac{\sqrt{2\varepsilon_{Si}qN_A}}{C_{ox}} \tag{7.3}$$

is referred to as the body-effect parameter. The substrate sensitivity is defined as

$$\frac{dV_T}{dV_{BS}} = -\frac{\gamma}{2\sqrt{2\Phi_F - V_{BS}}} = -\frac{\sqrt{2q\varepsilon_{Si}N_A}}{2C_{ox}\sqrt{2\Phi_F - V_{BS}}} \tag{7.4a}$$

At $V_{BS} = 0$, this is equal to

$$\frac{dV_T}{dV_{BS}} = -\frac{1}{C_{ox}}\sqrt{\frac{q\varepsilon_{Si}N_A}{4\Phi_F}} = -\frac{1}{C_{ox}}\frac{\varepsilon_{Si}}{W_{dm}} = -\frac{C_{dm}}{C_{ox}} = -(m-1) \tag{7.4b}$$

In (7.4b), m is referred to as the body-effect coefficient, W_{dm} is the width of the depletion region, and C_{dm} is the depletion capacitance.

This model is derived based on the assumption that the transistor is of long channel length and width and the substrate doping concentration is uniform. However, in current VLSI MOS transistors, the channel is doped non-uniformly in both vertical and lateral directions. These are collectively referred to as channel engineering. In addition, the channel lengths are very short. Therefore, the basic model needs to be modified. The following sub-sections describe how the vertical and lateral channel non-uniform doping effects are incorporated in industry standard compact models.

7.3.1.2 Vertical Channel Engineering

In the process of making a VLSI MOS transistor, several ion implantation doping process steps are required to adjust the threshold voltage and to suppress the punch-through and hot-carrier effects. The regions underneath the interface are doped with various concentrations along the vertical directions. The schematic diagram of a VLSI MOS transistor using vertical channel engineering is shown in Figure 7.3(a).

A shallow implantation of channel dopants of the same type as substrate is designed to obtain the desired threshold voltage value, and another deep implantation of channel dopants of the same type as substrate is designed to suppress the punch-through and drain-induced barrier lowering (DIBL) effect. A typical high-to-low doping profile in the vertical direction and the corresponding step

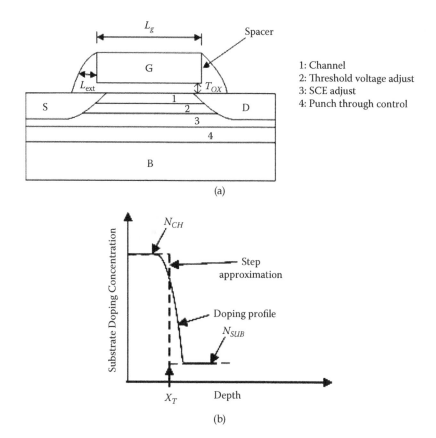

FIGURE 7.3
(a) A MOS transistor illustrating vertical channel engineering. (b) High-to-low channel doping profile.

function approximation are shown in Figure 7.3(b). With non-uniform vertical doping, the body bias coefficient γ becomes dependent on substrate bias.

Contrary to the uniform doping concentration N_A, the doping concentration in the channel region is denoted by N_{CH} and that in the deep substrate region is N_{SUB}. Two body-effect parameters γ_1 and γ_2 are to be defined by substituting appropriate concentrations in (7.3). However, this makes the expression for threshold voltage difficult to compute. Therefore, a compact representation of threshold voltage in the presence of a non-uniform vertical doping profile as used in BSIM4 is given by [6]

$$V_T = V_{T0} + K_1\left(\sqrt{2\Phi_F - V_{BS}} - \sqrt{2\Phi_F}\right) - K_2 V_{BS} \tag{7.5}$$

In (7.5), K_1 and K_2 are the two key parameters responsible for characterizing the vertical non-uniform channel doping effects. The values of these coefficients are determined by fitting (7.5) to measured threshold voltage data.

7.3.1.3 Halo/Pocket Implantation

For suppression of short channel effects, local high doping concentration regions near the source and drain junction edges are generally employed. This is known as lateral channel engineering or halo/pocket implantation. With this type of channel engineering, the doping concentration in the channel along the channel length becomes non-uniform. The schematic diagram of a VLSI MOS transistor using lateral channel engineering is shown in Figure 7.4(a). The lateral non-uniform doping with higher doping

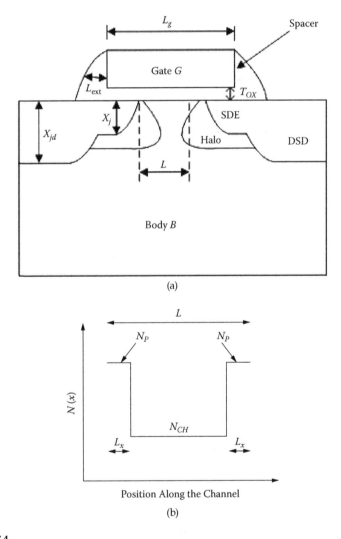

(a)

(b)

FIGURE 7.4
(a) A MOS transistor illustrating lateral channel engineering. (b) Step doping profile approximating the variation of the channel concentration from source to drain side.

concentration near the source/drain regions results in an increase in the average doping concentration in the channel and in turn results in an increase in the threshold voltage. This is sometimes referred to as reverse short channel effect (RSCE), because it helps to compensate charge sharing effects from the source/drain fields [7].

The lateral non-uniform doping concentration is approximated by a step doping profile along the channel, as shown in Figure 7.4(b). The average channel doping is given by [7,8]

$$N_{eff} = \frac{N_{CH}(L - 2L_x) + N_P 2L_x}{L} = N_{CH}\left(1 + 2\frac{L_x}{L}\frac{N_P - N_{CH}}{N_{CH}}\right) \cong N_{CH}\left(1 + \frac{L_{PE0}}{L}\right)$$

(7.6)

In (7.6), N_P is the pocket concentration. L_{PE0} is a fitting parameter extracted, whose value is to be extracted from the measured data. With the introduction of lateral doping, the threshold voltage model in (7.5) is modified as follows [8]:

$$V_T = V_{T0} + K_1\left(\sqrt{2\Phi_F - V_{BS}} - \sqrt{2\Phi_F}\right)\sqrt{1 + \frac{L_{PEB}}{L}} - K_2 V_{BS}$$

$$+ K_1\left(\sqrt{1 + \frac{L_{PE0}}{L}} - 1\right)\sqrt{2\Phi_F}$$

(7.7)

In (7.7) the following BSIM4 model parameters are identified:

- K_1 and K_2 model the effects of a non-uniform vertical channel doping profile on threshold voltage.
- L_{PE0} and L_{PEB} model effects of non-uniform lateral channel doping profile on threshold voltage. L_{PEB} models $V_T(V_{BS})$ dependence.
- At zero substrate bias, only the term L_{PE0} represents lateral non-uniformity, and as the channel length is reduced, due to the fourth term, the threshold voltage increases. This is significant for a short channel MOS transistor.

7.3.2 Threshold Voltage Characterization for Short Channel MOS Transistor

The threshold voltage of a long channel device is found to be independent of the channel length and the applied drain voltage. However, in short channel devices, it has been experimentally found that the threshold voltage of a MOS transistor decreases as the channel length is reduced or the drain bias is increased. This effect is known as the short channel effect [5,7]. In addition, the dependence of the threshold voltage on the body bias becomes weak as the channel length is reduced.

The long channel theory of MOS transistor is based upon the assumption that the depletion charge underneath the gate is controlled by the vertical electric field due to the applied gate bias. However, in a short channel device the channel length is comparable to the MOS depletion width in the vertical direction, and the source-drain potential has a significant effect on the band bending over a major portion of the device. The earlier assumption related to long channel device does not remain valid for short channel devices. The depletion charge under the gate is actually induced by the gate together with the source and the drain. Therefore, the channel charge may be considered to be shared by the gate as well as the source and drain. This is illustrated in Figure 7.5(a). Consequently, smaller gate voltage is required to induce inversion in short channel MOS transistors compared to long channel transistors.

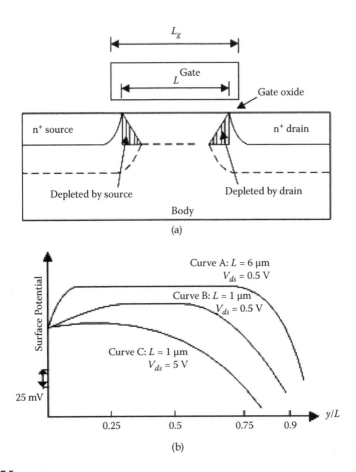

FIGURE 7.5
(a) Sharing of gate depletion charge by source and drain. (b) Variation of surface potential to lateral distance (normalized to the channel length L) for three different cases highlighting the reduction of the source-drain potential barrier for lower channel length and higher drain bias.

This leads to a smaller value of threshold voltage in short channel transistors compared to long channel transistors. This approach of explaining short channel effect is referred to as the charge sharing approach [7].

The decrease in the threshold voltage due to the reduction of channel length and increase in the drain bias can be explained by considering the surface potential. As the channel length decreases and the drain bias increases, the potential barrier (p-type region) between the source and the drain for the carriers (electrons) is lowered. Hence, less gate voltage is required to bring the surface potential to $2\Phi_F$.

In effect the threshold voltage is lowered. This is called DIBL [5,7]. The concept of DIBL is explained in Figure 7.5(b), illustrating the surface potential plots along the channel for three different (long and short channel) devices [5]. It is observed that with small channel length and high drain bias, the potential barrier between the source and the drain is the lowest.

The short channel effect in threshold voltage is incorporated in a BSIM4 compact model as follows [8]:

$$V_T = V_{T0} + K_1\left(\sqrt{2\Phi_F - V_{BS}} - \sqrt{2\Phi_F}\right)\sqrt{1 + \frac{L_{PEB}}{L}} - K_2 V_{BS}$$

$$+ K_1\left(\sqrt{1 + \frac{L_{PE0}}{L}} - 1\right)\sqrt{2\Phi_F} + \Delta V_T \tag{7.8}$$

In (7.8), ΔV_T represents the reduction of threshold voltage due to short channel effect. A simple and accurate model, derived in [9] is

$$\Delta V_T = -\frac{[2(V_{bi} - \psi_s) + V_{DS}]}{2\cosh\left(\frac{L_{eff}}{l_t} - 1\right)} \tag{7.9a}$$

Through several approximations, (7.9a) reduces to [7]

$$\Delta V_T = -[3(V_{bi} - \psi_s) + V_{DS}]e^{-\frac{L}{l_t}} \tag{7.9b}$$

In (7.9a) and (7.9b), V_{bi} is built-in potential of the source/drain (S/D) junction. This is given by

$$V_{bi} = \frac{kT}{q}\ln\left(\frac{N_{DEP}N_{SD}}{n_i^2}\right) \tag{7.10}$$

In (7.10), N_{DEP} is the channel concentration at the edge of the depletion boundary, and N_{SD} is the source-drain doping concentration. In (7.9a), l_t represents the characteristic length given as

$$l_t = \sqrt{\frac{\varepsilon_{Si}t_{ox}W_{dm}}{\varepsilon_{ox}\eta}} \tag{7.11}$$

In (7.11), W_{dm}/η represents the average depletion width in the channel. In order to improve the flexibility of the model equations over different technology generations, several fitting parameters are introduced. The following equations are used in BSIM4 to represent SCE and DIBL, respectively [8]

$$\Delta V_T (SCE) = -\frac{0.5 DVT0}{\cosh(DVT1 \cdot \frac{L_{eff}}{l_t}) - 1} [V_{bi} - \psi_s] \tag{7.12}$$

The characteristic length is given by

$$l_t = \sqrt{\frac{\varepsilon_{Si} t_{ox} W_{dm}}{\varepsilon_{ox}}} (1 + DVT2 \cdot V_{BS}) \tag{7.13}$$

The DIBL effect is modeled in BSIM as follows [8]:

$$\Delta V_T (DIBL) = -\frac{0.5}{\cosh(DSUB \cdot L_{eff}/l_{t0}) - 1} (ETA0 + ETAB \cdot V_{BS}) V_{DS} \tag{7.14}$$

$$l_{t0} = \sqrt{\frac{\varepsilon_{Si} t_{ox} W_{dm}}{\varepsilon_{ox}}} \tag{7.15}$$

$$W_{dm} = \sqrt{\frac{2\varepsilon_{Si} \psi_s}{q N_{DEP}}} \tag{7.16}$$

In these, *DVT0, DVT1, DVT2, DSUB, ETA0,* and *ETAB* are the SPICE BSIM4 model parameters whose values are to be extracted from measured data. *DVT2* and *ETAB* account for the body bias effect on short channel effect and DIBL, respectively.

7.3.2.1 Short Channel Effect Reduction

It is observed that the threshold voltage roll-off is dependent upon the characteristic length l_t as defined in (7.11). It is well known that the roll-off decreases with reduction in S/D junction depth X_j. However, this term has not been included in the model. Following Brews' approach [10], the characteristic length is redefined as

$$l_t \propto \left(X_j t_{ox} W_{dm}^2 \right)^{1/3} \tag{7.17}$$

Therefore, in order to have V_T less sensitive to charge sharing and DIBL effect, X_j, t_{ox}, and W_{dm} should be reduced.

7.3.3 Techniques for Threshold Voltage Extraction

This sub-section outlines three commonly used approaches for extraction of the threshold voltage value from the measured drain current versus gate voltage transfer characteristics. A good review of these techniques including the background theory and others is provided in [11].

7.3.3.1 Constant Current Method

In the constant current method, the threshold voltage is evaluated as the value of the gate voltage, corresponding to a given arbitrary constant drain current I_{DS} and drain voltage $V_{DS} < 100\,mV$. A typical value for this arbitrary constant drain current is $(W_m/L_m) \times 10^{-7}\,A$, where W_m and L_m are the mask channel width and length, respectively. Because of its simplicity, this is the most widely used method in industry. However, this method has a serious drawback of being totally dependent on the arbitrary chosen value of the drain current.

7.3.3.2 Extrapolation in the Linear Region Method

The extrapolation in the linear region method is another very popular method for extracting the threshold voltage from the measured transfer characteristics of a MOS transistor. Both the drain current and the linear transconductance are degraded significantly at high gate voltages because of the decrease of mobility with the increasing normal field. The technique consists of determining the gate voltage intercept (i.e., $I_{DS} = 0$) of the linear extrapolation of the $I_{DS} - V_{GS}$ curve at its maximum first derivative point (i.e., the point of maximum transconductance). This is referred to as the linearly extrapolated threshold voltage V_{ON}. V_{ON} is obtained by subtracting $V_{DS}/2$ from the intercept. The main drawback of this method is that the maximum slope point is often slightly greater ($\sim 2kT/q - 4kT/q$) than the threshold voltage V_T at $\psi_s(\text{inv}) = 2\Phi_F$ due to mobility degradation effects and the presence of significant source and drain series parasitic resistances.

7.3.3.3 Second Derivative Method

In this method, the threshold voltage is determined as the gate voltage at which the derivative of the transconductance $(dg_m/dV_{GS} = d^2 I_D/dV_{GS}^2)$ is maximum.

7.3.4 Simulation Results and Discussion

In this sub-section the discussion on characterization of threshold voltage for VLSI MOS transistors is reviewed through SPICE simulation results. In the subsequent discussion considerably long channel width is considered so as to avoid the narrow channel width effect on the threshold voltage.

7.3.4.1 Simulation Setup

For all the simulation results provided in this chapter, conventional bulk NMOS transistor has been selected. A 45-nm technology node has been selected, and the drawn channel length of the transistor is taken to be 65 nm, if not mentioned otherwise. The supply voltage is taken to be 1 V. The physical oxide thickness is 1.1 nm and the electrical oxide thickness is 1.75 nm, considering poly-depletion effect and inversion layer thickness. The substrate is uniformly doped with concentration equal to $3.24E18/cm^3$. The source/drain concentration is $2E20/cm^3$. The source/drain junction depth is 14 nm. The HSPICE simulation tool has been used to obtain all the simulation results, with BSIM4 as the compact model. The model parameters of the corresponding predictive technology model [4] have been taken for simulation purposes.

7.3.4.2 Threshold Voltage Characterization with Substrate Bias Effect

The variation of threshold voltage of a large geometry MOS transistor ($W = L = 10\,\mu m$) is shown in Figure 7.6(a). It is observed that the threshold voltage increases from its zero substrate bias value as the substrate voltage is increased. The value of zero substrate bias long geometry threshold voltage V_{T0} as observed from Figure 7.6(a) is 0.466 V. The threshold voltage variation with substrate bias follows (7.5). Noting the values of the necessary model parameters from the model file, the theoretical curve is drawn and compared with the simulation results. It is observed that the theoretical curve closely follows the simulation results.

The variation of the substrate sensitivity of threshold voltage with the substrate bias is shown in Figure 7.6(b). The sensitivity obtained from theoretical formulation is also shown in Figure 7.6(b).

7.3.4.3 Threshold Voltage Characterization for Short Channel Transistors

The variation of threshold voltage with channel lengths of a MOS transistor is shown in Figure 7.7(a). It is observed that as the channel length is reduced from 100 nm onwards, the threshold voltage reduces. This is referred to as the threshold voltage roll-off. The effect is more pronounced when the applied drain bias is high. This is demonstrated in Figure 7.7(b). The amount of threshold voltage roll-off as observed from the simulation results for different substrate bias and drain bias are summarized in Table 7.1. It is observed that with the increase in substrate bias, the short channel effect increases. This is easy to understand considering the fact that with the increase in substrate bias, the depletion width increases and consequently the short channel effect increases as suggested in Brew's relation.

The DIBL effect is shown in Figure 7.8. The DIBL coefficient η is defined as the slope of the curve ($\eta = \frac{V_T(V_{DS}=1V)-V_T(V_{DS}=0.1V)}{0.9}$). From simulation results, its

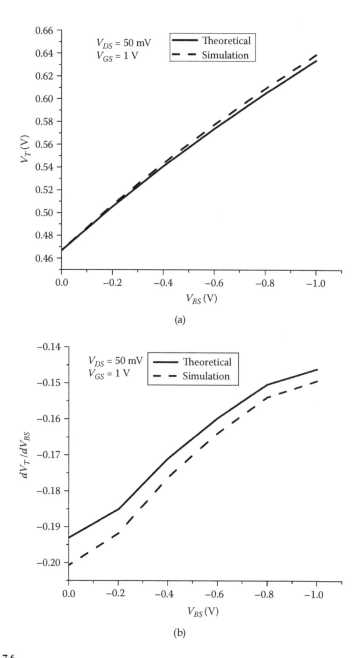

FIGURE 7.6
(a) Variation of threshold voltage with substrate bias for n-channel MOS transistor. (b) Substrate sensitivity of threshold voltage.

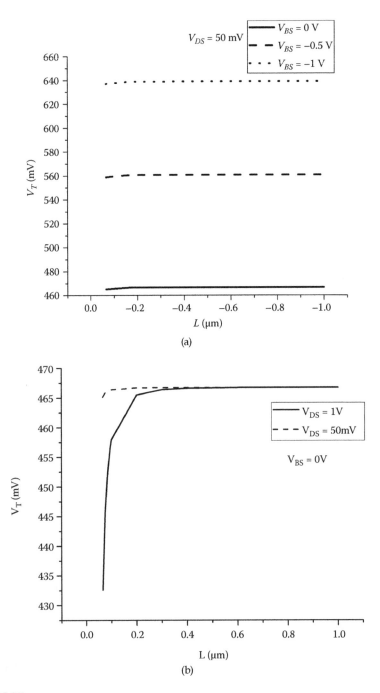

FIGURE 7.7
(a) Threshold voltage roll-off for low drain bias. (b) Simulation results showing that the threshold voltage roll-off increases with increased drain bias.

TABLE 7.1

Amount of Threshold Voltage Roll-Off at Low and High Drain Bias
for Different Substrate Bias

Roll-Off	$V_{DS} = 50$ mV			$V_{DS} = 1$V
	$V_{BS} = 0$ V	$V_{BS} = -0.5$ V	$V_{BS} = -1$ V	$V_{BS} = 0$ V
ΔV_T(mV)	1.55	1.70	1.86	32.16

value is found to be 0.0342, and the theoretical value as calculated from the
model discussed earlier is 0.0346.

7.3.4.4 Threshold Voltage Extraction

The threshold voltage extraction method through extrapolation in the linear
region method is shown in Figure 7.9 and that through the second derivative
method is shown in Figure 7.10. The threshold voltage as extracted from the
constant current method, linear extrapolation method, and second derivative method for low drain bias and different substrate bias are summarized

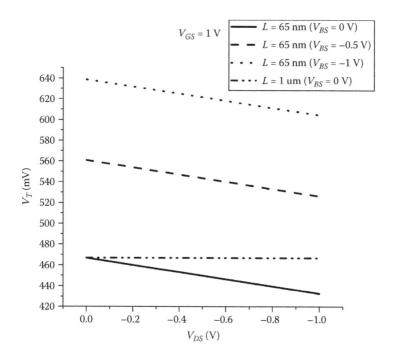

FIGURE 7.8
Simulation results illustrating the DIBL effect.

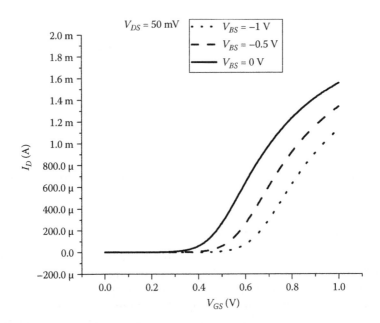

FIGURE 7.9
Threshold voltage extraction: extrapolation in the linear region method.

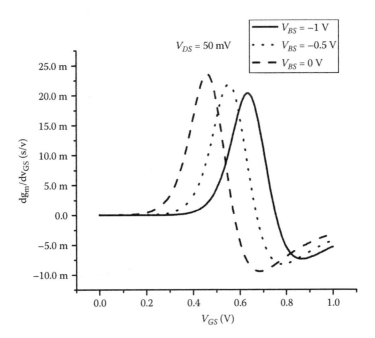

FIGURE 7.10
Threshold voltage extraction: second derivative method, evaluated at $V_{DS} = 50$ mV.

TABLE 7.2

Threshold Voltage Value Extracted through Different Methods

Method	Extracted Value of the Threshold Voltage at V_{DS} = 50 mV. W = 10 μm L = 65 nm		
	V_{BS} = 0 V	V_{BS} = –0.5 V	V_{BS} = –1 V
Constant current	0.457 V	0.559 V	0.645 V
Linear extrapolation	0.409 V	0.502 V	0.583 V
Second derivative	0.462	0.558	0.636

in Table 7.2. For the constant current method, the constant current is taken to 1 μA, the channel width is 10 μm and the channel length is 65 nm.

7.4 I-V Characterization

Precise knowledge of the I-V characteristics of a MOS transistor is a basic requirement for a good VLSI designer. The fundamental current transport equations are introduced, followed by channel charge, mobility, and velocity saturation effects. The I-V models for long and short channel devices are derived, followed by some advanced issues.

7.4.1 Current Density Equations

The total current density is the sum of the drift current density and the diffusion current density, written as

$$J_n = qn\mu_n\xi + qD_n\frac{dn}{dx} \qquad (7.18a)$$

$$J_p = qp\mu_p\xi - qD_p\frac{dp}{dx} \qquad (7.18b)$$

The total conduction current density is thus $J = J_n + J_p$. The diffusion coefficients D_n and D_p for electrons and holes are related to the corresponding mobilities μ_n and μ_p through Einstein's relationship [5]. Thus the current densities are written as follows:

$$J_n = qn\mu_n\xi + kT\mu_n\frac{dn}{dx} \qquad (7.18c)$$

$$J_p = qp\mu_p\xi - kT\mu_p\frac{dp}{dx} \qquad (7.18d)$$

The electric field ξ, which is defined as the electrostatic force per unit charge, is written as $\xi = -d\psi_i/dx$. It may be noted that gradual channel approximation has been assumed, according to which the variation of the electric field in the y-direction (along the channel) is much less than that in the x-direction (perpendicular to the channel). With this the conduction current densities are written as

$$J_n = -qn\mu_n \frac{d\phi_n}{dx} \tag{7.19a}$$

$$J_p = -qp\mu_p \frac{d\phi_p}{dx} \tag{7.19b}$$

The quasi-Fermi potentials ϕ_n and ϕ_p are defined as [5]

$$\phi_n \equiv \psi_i - \frac{kT}{q}\ln\left(\frac{n}{n_i}\right) \tag{7.20a}$$

$$\phi_p \equiv \psi_i + \frac{kT}{q}\ln\left(\frac{p}{n_i}\right) \tag{7.20b}$$

The electron current density at a point (x,y) in the channel is

$$J_n(x,y) = -q\mu_n n(x,y)\frac{dV_{CS}(y)}{dy} \tag{7.21}$$

Here $V_{CS}(y)$ is the quasi-Fermi potential. The total current at any point y along the channel is

$$I_{DS}(y) = qW\int_0^{x_i} \mu_n n(x,y)\frac{dV_{CS}}{dy}dx \tag{7.22}$$

The integration is carried out from $x = 0$ to $x = x_i$, the bottom of the inversion layer where $\psi = \Phi_F$. There is a sign change as the drain current flows in the negative y direction. The inversion charge density is defined as

$$Q_{inv}(y) = -q\int_0^{x_i} n(x,y)dx \tag{7.23}$$

With this the drain to source current is given as

$$I_{DS} = \mu_n \frac{W}{L}\int_0^{V_{DS}} [-Q_{inv}(V)]\cdot dV_{CS} \tag{7.24}$$

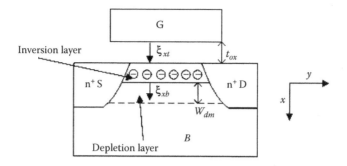

FIGURE 7.11

Inversion layer forms one capacitor with the gate and another capacitor with the body. Surface mobility is a function of the average electric fields at the top and the bottom of the inversion charge layers.

The quasi-Fermi potential $V_{CS}(y) = 0$ at $y = 0$(source) and $V_{CS}(y) = V_{DS}$ at $y = L$(drain).

7.4.2 Channel Inversion Charge Density

With reference to Figure 7.11, the following assumption is made. The inversion layer in the channel of a MOS transistor is a sheet of charge and there is no potential drop or band bending across the inversion layer. This is referred to as the charge sheet approximation [5]. This inversion layer forms a capacitor with the gate, the oxide being the dielectric. Also it forms another capacitor with the body, the depletion layer being the dielectric. Thus the inversion layer is coupled with both gate and substrate of the transistor. The inversion charge density in strong inversion is given by

$$Q_{inv} = -[C_{ox}(V_{GS} - V_{CS}(y) - V_{T0}) + C_{dm}(V_{BS} - V_{CS}(y))] \tag{7.25}$$

Simplification of (7.25) leads to the following expression for inversion charge density:

$$Q_{inv} = -C_{ox}(V_{GS} - mV_{CS}(y) - V_T) \tag{7.26}$$

where

$$m \equiv 1 + \alpha = 1 + \frac{C_{dm}}{C_{ox}} = 1 + \frac{3t_{ox}}{W_{dm}} \tag{7.27}$$

In (7.26), $V_T = V_{T0} - \frac{C_{dm}}{C_{ox}} V_{BS} = V_{T0} - \alpha V_{BS} = V_{T0} - (m-1)V_{BS}$ has been taken using (7.4b).

As defined earlier, m is referred to as the body-effect coefficient.[*] The value of m is typically 1.2, however, it can be taken to be unity for simplified calculations. Clearly the bulk charge factor is closely related to the body-effect parameter, as observed also from (7.4b).

The inversion charge density in the weak inversion region is given by [7]

$$Q_{inv} \simeq -\sqrt{\frac{q\varepsilon_{Si}N_{CH}}{4\Phi_F}}U_T \exp\left(\frac{\psi_s - 2\Phi_F - V_{CS}}{U_T}\right)$$ (7.28)

In (7.28), $U_T = kT/q$ is the thermal voltage, and N_{CH} is the effective channel concentration.

7.4.3 Carrier Mobility Degradation Model

In the inversion layer of a MOS transistor, the current flow is determined by the surface mobility, whose value is much lower than the bulk mobility of the carriers. This is because of several mechanisms of scattering, primarily the phonon or the lattice scattering, the coulombic scattering, and the surface roughness scattering [5]. For good quality interfaces, phonon scattering is the dominant mechanism at room temperature.

The surface mobility is a function of the average of the electric fields at the top and the bottom of the inversion charge layer. These fields are shown in Figure 7.11.

From Gauss's law using the depletion layer as the Gaussian box, it is possible to write

$$\xi_{xb} = \frac{Q_B}{\varepsilon_{Si}} = \frac{C_{ox}(V_T - V_{FB} - 2\Phi_F)}{\varepsilon_{Si}}$$ (7.29a)

Considering the Gaussian box to be a box that encloses both the depletion and the inversion layer, we have

$$\xi_{xt} = \frac{Q_B + Q_{inv}}{\varepsilon_{Si}}$$ (7.29b)

From (7.29a) and (7.26), we have, after substitutions,

$$\xi_{xt} = \frac{C_{ox}}{\varepsilon_{Si}}(V_{GS} - V_{FB} - 2\Phi_F)$$ (7.29c)

It is to be noted that the effect of the lateral field is ignored and $m = 1$ for simplicity.

---

[*] Some authors refer to this as the bulk-charge factor.

The average electric field is defined as

$$\xi_{eff} = \frac{1}{2}(\xi_{xb} + \xi_{xt}) \tag{7.30}$$

Substituting from (7.29a) and (7.29c), and after some simplifications for n⁺ poly-gate n-channel MOS transistor

$$\xi_{eff} = \frac{V_{GS} + V_T + 0.2V}{6t_{ox}} \tag{7.31}$$

Physically, ξ_{eff} means the average electric field experienced by the carriers in the inversion layer. The dependence of the surface mobility of the carriers on this average electric field and hence on the gate bias is given by the following empirical relationship [3]:

$$\mu_s = \frac{\mu_0}{1 + \left(\frac{\xi_{eff}}{\xi_0}\right)^\upsilon} \tag{7.32}$$

In (7.32), μ_0 is the low field surface mobility, υ is a constant whose value is ~1.85 for electrons at the surface, and ~1.0 for holes at the surface. ξ_0 is the critical electric field (~0.9 MV/cm for electrons at surface and ~0.45 MV/cm for holes at the surface). The model proposed in (7.32) fits experimental data well, but because it involves a power function, it is difficult to integrate the model in a circuit simulation program. A Taylor series expansion of (7.32) and retaining only up to three terms, the following expression for the vertical field mobility degradation model is derived:

$$\mu_s = \frac{\mu_0}{1 + U_A\left(\frac{V_{GS} + V_T}{t_{ox}}\right) + U_B\left(\frac{V_{GS} + V_T}{t_{ox}}\right)^2} \tag{7.33}$$

In (7.33), U_A and U_B are two parameters, whose values are to be extracted from the experimental I-V data. The substrate bias dependence of the mobility is incorporated by introducing another parameter U_C in (7.33). With this, the model becomes [6]

$$\mu_s = \frac{\mu_0}{1 + (U_A + U_C.V_{BS})\left(\frac{V_{GS} + V_T}{t_{ox}}\right) + U_B\left(\frac{V_{GS} + V_T}{t_{ox}}\right)^2} \tag{7.34}$$

$$\mu_s = \frac{\mu_0}{1 + \left[U_A\left(\frac{V_{GS} + V_T}{t_{ox}}\right) + U_B\left(\frac{V_{GS} + V_T}{t_{ox}}\right)^2\right](1 + U_C V_{BS})} \tag{7.35}$$

These two different models are incorporated in SPICE simulator by using suitable mobility selector flags. It may be noted that all the mobility degradation models discussed above include the effect of vertical electric field only. It is observed that the mobility in a strong inversion region is a function of the gate bias. In the subthreshold region, the variation of Q_{inv} with V_{GS} cannot be modeled accurately. Therefore, μ_s becomes constant in the subthreshold region.

7.4.4 Carrier Velocity Saturation Model

Carrier velocity is another important physical phenomenon that critically affects the I-V characteristics of a short channel VLSI MOS transistor. If the lateral electric field is small, the drift velocity of the carriers is given by

$$v_d = \mu_s \xi_y \tag{7.36}$$

In (7.36), μ_s is the surface mobility and is independent of the lateral field ξ_y. However, as the lateral field ξ_y becomes high, the carrier velocity no longer follows (7.36). With the increase of the lateral field, the kinetic energy of the carrier increases. When the energy of the carrier exceeds the optical phonon energy, an optical phonon is generated by the carrier, and in this process the carrier loses its velocity significantly. Consequently, the kinetic energy and therefore the drift velocity of the carriers cannot exceed a certain value. This limiting velocity is called the *saturation velocity*. The v_d-ξ_y relationship is illustrated in Figure 7.12. An accurate model for the drift velocity is given by [3]

$$v_d = \frac{\mu_s \xi_y}{\left[1 + \left(\frac{\xi_y}{\xi_{sat}}\right)^\alpha\right]^{1/\alpha}} \tag{7.37}$$

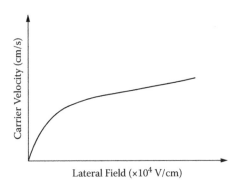

FIGURE 7.12
Carrier velocity saturation.

In (7.37), $\alpha = 2$ for electrons and $\alpha = 1$ for holes. ξ_{sat} is the critical electric field at which the carrier velocity becomes saturated and is linked with the saturation drift velocity of the carrier as $\xi_{sat} = 2v_{dsat}/\mu_s$. For electrons, v_{dsat} varies between ~6–10 × 10^4 m/s and that for holes between ~4–8 × 10^4 m/s. The model presented in (7.37) fits experimental data well but suffers from the drawback that it is computationally difficult to be incorporated in any circuit simulation program. In BSIM, a piece-wise velocity-field relationship is thus suggested. This is as follows [6]:

$$v_d = \frac{\mu_s \xi_y}{1 + \frac{\xi_y}{\xi_{sat}}} \quad \xi_y < \xi_{sat} \quad \xi_{sat} = \frac{2v_{dsat}}{\mu_s}$$

$$v_d = v_{dsat} \quad \xi_y \geq \xi_{sat} \tag{7.38}$$

This model is in accordance with the experimental data.

7.4.5 Basic MOSFET I-V Model

Substituting (7.26) in (7.24) and carrying out the simple integration, we get the following expression as the basic MOSFET I-V model:

$$I_{DS} = \mu_{ns} \frac{W}{L} C_{ox} \left[(V_{GS} - V_T)V_{DS} - \frac{m}{2} V_{DS}^2 \right] \tag{7.39}$$

The following observations are made from (7.39):

- The drain current is proportional to the channel width W, surface electron mobility μ_{ns}, lateral electric field V_{DS}/L, and the average inversion charge density $C_{ox}(V_{GS} - V_T - mV_{DS}/2)$.

- When the applied drain bias is small, the term $V_{DS}^2/2$ can be neglected so that the drain current is linearly proportional to the drain bias V_{DS}, which means that the transistor is acting as a resistor.

- As the drain bias increases, the average inversion charge density decreases and $\frac{dI_{DS}}{dV_{DS}}$ decreases. By differentiating (7.39) w.r.t V_{DS} and putting equal to 0, we get

$$\frac{dI_{DS}}{dV_{DS}} = \mu_{ns} \frac{W}{L} C_{ox} [(V_{GS} - V_T) - mV_{DS}] = 0 \quad \text{at } V_{DS} = V_{DSsat}$$

$$V_{DSsat} = \frac{V_{GS} - V_T}{m} \tag{7.40}$$

- The drain current that flows when the drain bias is saturated is referred to as the drain source saturation current. This is given as

$$I_{DSsat} = \frac{W}{2mL} C_{ox}\mu_{ns} (V_{GS} - V_T)^2 \qquad\qquad V_{DS} > V_{DSsat} \qquad (7.41)$$

- Substituting $V_{DSsat} = \frac{V_{GS}-V_T}{m}$ for $V_{CS}(y)$ in (7.26), it is found that the inversion charge disappears at the drain side. This phenomenon is referred to as pinch-off. After the pinch-off region in the channel, there exists a depletion region. The electrons after reaching the pinch-off region are swept down by the drain bias and thus constant current flows, which is given by (7.41).
- Equations (7.39) and (7.41) form the basic I-V model for a long channel MOS transistor. Because of its simplicity, these equations are widely used by the designers for hand calculations. In addition, these are used in first-generation SPICE models [3].
- The transconductance parameter is defined as

$$g_m = \frac{dI_{DS}}{dV_{GS}}\bigg|_{V_{DS}} = \sqrt{2\mu_{ns}C_{ox}\frac{W}{L}I_{DS}} \qquad (7.42)$$

It is observed that when a MOS transistor operates in the strong inversion region, the transconductance is proportional to the square root of the drain current.

7.4.6 MOSFET I-V Model with Velocity Saturation

Assuming the inversion charge density to be given by (7.26), the drain current from (7.24) is

$$I_{DS} = \mu_{ns}C_{ox}\frac{W}{L} \int_0^{V_{DS}} (V_{GS} - mV_{CS} - V_T)\frac{\frac{dV_{CS}}{dy}}{1 + \frac{1}{\xi_{sat}}\frac{dV_{CS}}{dy}} \qquad (7.43)$$

In (7.43), the velocity saturation effect given by (7.38) has been assumed and $\xi_y = \frac{dV_{CS}}{dy}$. Performing the simple integration, we arrive at the following model for the I-V characteristics:

$$I_{DS} = \mu_{ns}C_{ox}\frac{W}{L}\frac{1}{1+\frac{V_{DS}}{\xi_{sat}L}}\left[(V_{GS} - V_T)V_{DS} - \frac{m}{2}V_{DS}^2\right] \qquad (7.44)$$

Comparing (7.44) and (7.39), we see that

$$I_{DS} = \frac{\text{long channel } I_{DS} \,(7.39)}{1 + \frac{V_{DS}}{\xi_{sat}L}} \tag{7.45}$$

Thus the effect of velocity saturation is to reduce the long channel drain current by the factor $1 + \frac{V_{DS}}{\xi_{sat}L}$. When V_{DS} is small or L is large, this factor reduces to 1 (i.e., velocity saturation becomes negligible). The drain current model in (7.44) is valid before the carrier velocity saturates (i.e., in the linear or triode region).

If the drain voltage (and hence the lateral electric field ξ_y) is sufficiently high, the carrier velocity near the drain saturates. At this stage, the channel may be considered to be split into two portions: one adjacent to the source where the carrier velocity is field dependent, and the other near the drain where the carrier velocity is saturated to v_{dsat}. At the junction between these two portions, the channel voltage is V_{DSsat}, and the lateral electric field is ξ_{sat}. Therefore, the saturation drain current is given as

$$I_{DSsat} = WC_{ox}(V_{GS} - V_T - mV_{DSsat})v_{dsat} \tag{7.46}$$

Comparing (7.46) with (7.44), we arrive at the following expression for saturation drain voltage:

$$\frac{1}{V_{DSsat}} = \frac{m}{V_{GS} - V_T} + \frac{1}{\xi_{sat}L} \tag{7.47}$$

The following observations are made from (7.46) and (7.47):

- When the channel length is large or the gate overdrive voltage $(V_{GS} - V_T)$ is low, $\xi_{sat}L \gg (V_{GS} - V_T)$, $V_{DSsat} \approx (V_{GS} - V_T)/m$, we arrive at (7.40) (i.e., the long channel saturation drain voltage).
- For the very short channel transistor, $\xi_{sat}L \ll (V_{GS} - V_T)$, $V_{DSsat} \approx \xi_{sat}L$ the drain current becomes

$$I_{DSsat} = WC_{ox}v_{dsat}(V_{GS} - V_T - m\xi_{sat}L) \tag{7.48}$$

- Thus, I_{DSsat} is linearly proportional to $(V_{GS} - V_T)$, instead of the long channel quadratic dependence. In addition, I_{DSsat} is proportional to W but less sensitive to L.

- The classical long channel model suggests that drain current saturates when the inversion charge density becomes zero, a phenomenon referred to as pinch-off. However, a more accurate description of the cause of drain current saturation is that the carrier velocity reaches its maximum value v_{dsat} at the drain. Thus instead of the pinch-off region, there is a velocity saturation region next to the drain where the inversion charge density given as $Q_{inv} = C_{ox}(V_{GS} - V_T - mV_{DSsat})$ does not vanish.

- In order to increase I_{DSsat}, there must be an increase in $C_{ox}(V_{GS} - V_T)$. This is achieved by reducing t_{ox}, minimizing V_T, and increasing V_{GS}. The limit of t_{ox} is determined by oxide tunneling leakage and reliability. On the other hand, the lower limit of V_T is determined by the leakage current in the OFF state. The maximum value of V_{GS} is the supply voltage V_{DD}, which is determined by concerns over power consumption and reliability.

- It may be noted that for low power analog circuit operations, gate overdrive voltage $(V_{GS} - V_T)$ is often taken to be nearly 0.1 V and assuming $\xi_{sat} = 6 \times 10^4 V/cm$ and $L = 50\ nm$, $\xi_{sat}L > (V_{GS} - V_T)$, so that the transistor exhibits some long channel characteristics and consequently the long channel model may be used.

7.4.7 Parasitic Source Drain Resistance

It follows from (7.17) that for short channel effect immunity, the source/drain junction depth must be reduced. Therefore, as shown in Figure 7.13, extra processing steps are performed to produce the shallow S/D junction extension between the deep junction and the channel. To further minimize dopant diffusion, the doping concentration in the shallow S/D extension

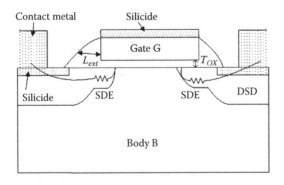
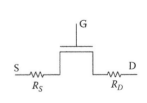

FIGURE 7.13
Source/drain series resistance in the SDE region.

is kept much lower compared to that in the deep S/D region. The shallow S/D extension along with light doping leads to parasitic S/D resistance. This is shown in Figure 7.13. The parasitic source and drain resistances are important device parameters that critically affect the MOS transistor performances. The drain current in the linear region with high gate bias is severely degraded due to this resistance, because channel resistance is lowest under such a bias condition. This is modeled in the presence of this resistance as follows:

$$I_{DS} = \frac{I_{DS0}}{1 + \frac{R_S I_{DS0}}{V_{DS}}} \tag{7.49}$$

In (7.49), I_{DS0} is the intrinsic current expression given by (7.44). R_S is the parasitic source resistance. It appears from (7.49) that the effect of the series resistance is lowest in the saturation region, where V_{DS} is high. A second effect of this resistance is the increase of V_{DSsat}, as follows:

$$V_{DSsat} = V_{DSsat0} + I_{DSsat}(R_S + R_D) \tag{7.50}$$

In (7.50), R_S and R_D are the parasitic source and drain resistances, respectively.

Reducing the value of these resistances is thus an important task for both circuit and device designers. A popular option is to cover the drain and source regions with a low resistivity material such as titanium or tungsten. This process is called *silicidation* and it effectively reduces the sheet resistance of the S/D regions by a factor of 10. Another option to be used by the circuit designer is to make the transistor wide. With a process that includes silicidation and proper attention to layout, parasitic resistance may be reduced.

7.4.8 Output Resistance

A typical I-V curve and its output resistance are shown in Figure 7.14. The drain current in the output I-V curve is divided into two parts: (1) the linear region, in which the drain current increases with the drain voltage, and (2) the saturation region, in which the drain current weakly depends upon the drain voltage. However, the output resistance curve reveals more detailed information about the various physical mechanisms involved in the saturation region. The output resistance is the reciprocal of the derivative of the I-V curve, and it is shown in Figure 7.14. The physical causes of such variation of the output resistance are the influences of drain voltage on the threshold voltage and a phenomenon called channel length modulation. The output resistance is divided into four regions.

The first region is the linear or triode region. In this region, the output resistance is very small because of the strong dependence of drain current on drain voltage. The other three regions belong to the saturation region,

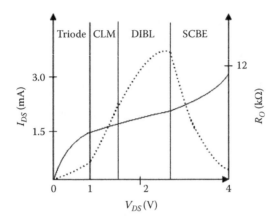

FIGURE 7.14
Typical behavior of MOSFET output resistance.

namely channel length modulation (CLM) region, DIBL region, and sub-strate current induced body effect (SCBE) region. The SCBE results in a dra-matic decrease in output resistance in the high drain bias region.

The drain current has a weak dependence on the drain voltage in the satura-tion region. Therefore, by Taylor series expansion of I_{DS} at $V_{DS} = V_{DSsat}$, we have

$$I_{DS}\left(V_{GS}, V_{DS}\right) = I_{DS}\left(V_{GS}, V_{DSsat}\right) + \frac{\partial I_{DS}\left(V_{GS}, V_{DS}\right)}{\partial V_{DS}}\left(V_{DS} - V_{DSsat}\right)$$

$$\equiv I_{DSsat}\left(1 + \frac{V_{DS} - V_{DSsat}}{V_A}\right) \tag{7.51}$$

In (7.51) we have

$$I_{DSsat} = I_{DS}(V_{GS}, V_{DSsat}) = W\upsilon_{dsat}C_{ox}(V_{GS} - V_T - mV_{DSsat}) \tag{7.52}$$

$$V_A = I_{DSsat}\left(\frac{\partial I_{DS}}{\partial V_{DS}}\right)^{-1} \tag{7.53}$$

Here V_A is called the Early voltage and is introduced for the analysis of out-put resistance in the saturation region only. It is assumed that a specific Early voltage parameter can be computed independently for each of the different regions of the output characteristics, namely the CLM region, DIBL region, and SCBE region. These can be calculated analytically; however for accuracy it is better to determine them from measurement results.

It is instructive for IC designers to use the following set of equations for all sorts of hand analysis works.

$$I_{DS} = \mu_{ns}C_{ox}\frac{W}{L}\frac{1}{1+\frac{V_{DS}}{\xi_{sat}L}}\left[(V_{GS}-V_T)V_{DS}-\frac{m}{2}V_{DS}^2\right] \qquad V_{DS}<V_{DSsat}$$

$$I_{DS} = Wv_{dsat}C_{ox}(V_{GS}-V_T-mV_{DSsat})\left(1+\frac{V_{DS}-V_{DSsat}}{V_A}\right) \qquad V_{DS}>V_{DSsat}$$

$$V_{DSsat} = \frac{\xi_{sat}L(V_{GS}-V_T)}{m\xi_{sat}L+(V_{GS}-V_T)}$$

$$(7.54)$$

The parameters V_T and V_A are to be extracted from measurements. For simplicity, the value of the body-effect coefficient m may be considered to be unity.

7.4.9 Subthreshold I-V Model

When the applied gate voltage is smaller than the threshold voltage V_T, ideally the drain current is zero. However, this does not happen in real devices and the drain current remains at a non-negligible level for several tenths of a volt below V_T. This occurs because the inversion charge density does not drop to zero abruptly. Subthreshold characteristics are significant in scaled CMOS digital applications, because they describe how a transistor switches OFF. Thus, the subthreshold region, immediately below V_T, in which $\Phi_F < \psi_s < 2\Phi_F$ is referred to as the weak inversion region. Contrary to the strong inversion region, in which the drift current dominates, subthreshold conduction is dominated by the diffusion current, arising due to a gradient in minority carrier concentration. In weak inversion, an n-channel MOS transistor operates as an n-p-n bipolar transistor, where the source acts as the emitter, the substrate as the base, and the drain as the collector.

The inversion charge density is repeated here from (7.28)

$$Q_{inv} \simeq -\sqrt{\frac{q\varepsilon_{si}N_{CH}}{4\Phi_F}}U_T\exp\left(\frac{\psi_s-2\Phi_F-V_{CS}}{U_T}\right) \qquad (7.55)$$

In order to derive a relationship between ψ_s and V_{GS}, the following expansion is made [12]:

$$V_{GS} \approx V_{GS}\big|_{\psi_s=1.5\Phi_F} + \frac{\partial V_{GS}}{\partial \psi_s}\bigg|_{\psi_s=1.5\Phi_F}(\psi_s-1.5\Phi_F) \qquad (7.56a)$$

It is known that $V_{GS} = V_T$ when $\psi_s = 2\Phi_F$. Therefore, from (7.56a), we have

$$V_T = V_{GS}\big|_{\psi_s=1.5\Phi_F} + \frac{\partial V_{GS}}{\partial \psi_s}\bigg|_{\psi_s=1.5\Phi_F} 0.5\Phi_F \qquad (7.56b)$$

Therefore, from (7.56a) and (7.56b), we have

$$V_{GS} = V_T + \frac{\partial V_{GS}}{\partial \psi_s}\bigg|_{\psi_s=1.5\Phi_F} (\psi_s - 2\Phi_F) \qquad (7.56c)$$

In addition, we have the following relationship:

$$V_{GS} = V_{FB} + \psi_s + \gamma\sqrt{\psi_s} \qquad (7.57)$$

From (7.57),

$$\frac{\partial V_{GS}}{\partial \psi_s} = 1 + \frac{\gamma}{2\sqrt{\psi_s}}\bigg|_{\psi_s=1.5\Phi_F} \qquad (7.58)$$

Substituting the value of γ from (7.3) and considering,

$$C_{dm} = \frac{\varepsilon_{Si}}{W_{dm}} \qquad (7.59)$$

with W_{dm} as given in (7.16), the following relationship is achieved:

$$\frac{\partial V_{GS}}{\partial \psi_s} \approx 1 + \frac{C_{dm}}{C_{ox}} \equiv n \qquad (7.60)$$

From (7.56c), (7.58), and (7.60), we get

$$V_{GS} = V_T + n(\psi_s - 2\Phi_F) \qquad (7.61)$$

(7.61) is widely used for developing compact models in the subthreshold region. Here n is referred to as subthreshold swing factor. It may be noted that the subthreshold swing factor n is not equal to the body-effect coeffient m, although very closely related. Substituting (7.61) in (7.55), we get

$$Q_{inv} \simeq -\sqrt{\frac{q\varepsilon_{Si}N_{CH}}{4\Phi_F}} U_T \exp\left(\frac{V_{GS} - V_T - nV_{CS}}{nU_T}\right) \qquad (7.62)$$

Substituting this in (7.24) and performing the simple integration, the following expression for the subthreshold drain current is obtained:

$$I_{DS} = \mu_n \frac{W}{L} \sqrt{\frac{q\varepsilon_{Si}N_{CH}}{4\Phi_F}} U_T^2 \exp\left(\frac{V_{GS}-V_T}{nU_T}\right)\left[1-\exp\left(\frac{-V_{DS}}{U_T}\right)\right] \qquad (7.63)$$

This can be alternatively written as

$$I_{DS} = I_0 \exp\left(\frac{V_{GS}-V_T}{nU_T}\right)\left[1-\exp\left(\frac{-V_{DS}}{U_T}\right)\right] \qquad (7.64)$$

In (7.64), I_0 is defined as

$$I_0 = \mu_0 \frac{W}{L}\sqrt{\frac{q\varepsilon_{Si}N_{CH}}{4\Phi_F}} U_T^2 \qquad (7.65)$$

In (7.64), n is referred to as the subthreshold swing factor. From experimental data it has been found that the subthreshold swing factor is a function of channel length and the interface state density. The interface traps located at the oxide-silicon interface exchange carriers with the silicon. The charge trapped in them depends on the value of the surface potential ψ_s. This is modeled by an incremental capacitance in parallel with the depletion capacitor. In addition, there are coupling capacitors between the drain/source and channel. With these considerations, the subthreshold swing factor as defined in (7.60) is modified in the BSIM compact model as follows [5,8]:

$$m = 1 + \text{NFACTOR}\frac{C_{dm}}{C_{ox}} + \frac{C_{it}+C_{DSC}}{C_{ox}} \qquad (7.66)$$

The parameter NFACTOR as introduced in (7.66) is for compensating any error while calculating the depletion width capacitance. The value of this parameter is close to unity. C_{it} is the interface trap capacitance per unit area. The capacitance C_{DSC} is sensitive to the body bias as well as to the drain bias that is incorporated in the model as follows:

$$C_{DSC} = (C_{DSC} + C_{DSCD}V_{DS} + C_{DSCB}V_{BS})\frac{0.5}{\cosh(DVT1\cdot\frac{L}{l_t})-1} \qquad (7.67)$$

This should be included in (7.66) for computing the exact value of the subthreshold swing factor. The various capacitance parameters in (7.67) are to be extracted.

7.4.10 Characterization of Poly-Silicon Gate Depletion Effect

The use of poly-silicon gates is considered as an advantage in modern CMOS technology. This is because the source and drain regions can be self-aligned to the gate, thus eliminating parasitic from overlay errors. The poly-silicon is usually heavily doped to behave almost similar to that of metal. However, in many process technologies, it is not possible to dope the poly-silicon gate to arbitrarily high concentrations. Therefore, a thin depletion layer is formed at the interface between the poly-silicon and the gate oxide, with the application of gate voltage.

7.4.10.1 Reduction of Gate-Source Voltage

Although the depletion region is very thin, its effect cannot be ignored in the deca-nanometer MOSFETs, because the gate oxide thickness is also very small. This is especially critical with the dual n^+-p^+ poly-silicon gate process in which the gates are doped by ion implantation. The effect of the presence of such a depletion region is that the voltage drop across the gate oxide and the substrate is reduced, because part of the gate voltage will be dropped across the depletion region in the gate. Consequently, the effective gate voltage is reduced.

Figure 7.15 shows an n-channel MOS transistor with a depletion region in the n^+ poly-silicon gate. Let us assume that the doping concentration in the poly gate near the interface is N_p, and the potential drop across the depletion region in the poly-silicon gate is ψ_p. From the depletion approximation, the depletion charge density in poly-silicon is

$$Q_p = \sqrt{2q\varepsilon_{Si}N_p} \cdot \sqrt{\psi_p} = C_{ox}\gamma_p\sqrt{\psi_p} \qquad (7.68)$$

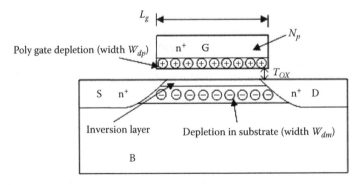

FIGURE 7.15
Poly gate depletion phenomenon.

In (7.68), ψ_p is the potential drop across the poly-silicon gate, and γ_p is given by

$$\gamma_p = \frac{\sqrt{2q\varepsilon_{Si}N_p}}{C_{ox}} \tag{7.69}$$

The potential balance equation is given by

$$V_{GS} = V_{FB} + \psi_p + \psi_{ox} + \psi_s \tag{7.70}$$

The potential drop across the oxide ψ_{ox} in (7.70) is given by

$$\psi_{ox} = \frac{Q_p}{C_{ox}} = \gamma_p\sqrt{\psi_p} \tag{7.71}$$

While writing (7.71), it has been considered that the normal component of electrical displacement is continuous across the interface. From (7.70) and (7.71), the following quadratic equation can be derived:

$$(V_{GS} - V_{FB} - \psi_s - \psi_p)^2 \frac{1}{\gamma_p^2} - \psi_p = 0 \tag{7.72}$$

Solving (7.72), and taking the positive root, the effective gate voltage is found to be

$$V_{GS\_eff} = V_{GS} - \psi_p = V_{FB} + \psi_s + \frac{\gamma_p^2}{2}\left(\sqrt{1 + \frac{4(V_{GS} - V_{FB} - \psi_s)}{\gamma_p^2}} - 1 \right) \tag{7.73}$$

Substituting the value of γ_p from (7.69) in (7.73),

$$V_{GS\_eff} = V_{FB} + \psi_s + \frac{q\varepsilon_{Si}N_p t_{ox}^2}{\varepsilon_{ox}^2}\left(\sqrt{1 + \frac{2\varepsilon_{ox}^2 (V_{GS} - V_{FB} - \psi_s)}{q\varepsilon_{Si}N_p t_{ox}^2}} - 1 \right) \tag{7.74}$$

It is observed that if $t_{ox} = 30A^0$, the effective gate voltage can be reduced by up to 10% due to the poly-silicon gate depletion effect. In a BSIM compact model, N_p is denoted by N_{GATE} and is considered as a model parameter.

7.4.10.2 Effect on Threshold Voltage

Let us now investigate the effect of poly-silicon gate depletion on the threshold voltage of a MOS transistor. The condition for charge balance is

$$Q_G = -(Q_B + Q_{inv}) \tag{7.75}$$

When the gate is positively biased, the positive charge on the gate is supported by the depletion charge due to the donor ions at the poly-Si/SiO$_2$ interface of an n^+ poly-silicon gate. From (7.68),

$$\psi_p = \frac{Q_p^2}{\gamma_p^2 C_{ox}^2} = \frac{Q_G^2}{\gamma_p^2 C_{ox}^2} = \frac{(Q_B + Q_{inv})^2}{\gamma_p^2 C_{ox}^2} \tag{7.76a}$$

The potential drop across the oxide is

$$\psi_{ox} = \frac{Q_G}{C_{ox}} \tag{7.76b}$$

Substituting the values of ψ_p and ψ_{ox} from (7.76a) and (7.76b) in (7.70),

$$V_{GS} = V_{FB} + \psi_s + \frac{(Q_B + Q_{inv})^2}{\gamma_p^2 C_{ox}^2} - \frac{(Q_B + Q_{inv})}{C_{ox}} \tag{7.77}$$

Considering the fact that at threshold, $\psi_s = 2\Phi_F$ and $Q_{inv} \rightarrow 0$, the threshold voltage in the presence of the poly depletion effect is given by

$$V_{T0p} = V_{T0} + \frac{Q_B^2}{\gamma_p^2 C_{ox}^2} \tag{7.78}$$

Thus it is observed that due to the poly-silicon gate depletion effect, the threshold voltage is increased by an amount

$$\frac{Q_B^2}{\gamma_p^2 C_{ox}^2} \sim \frac{\gamma^2}{\gamma_p^2} 2\Phi_F$$

7.4.10.3 Effect on Oxide Thickness

Because a depletion layer is present in the gate, it may be thought that a poly-silicon gate capacitor is added in series with the oxide capacitor. The effect of this is that the oxide dielectric thickness is increased, such that the effective oxide thickness is

$$\hat{t}_{ox} = t_{ox} + \frac{\varepsilon_{ox}}{\varepsilon_{Si}} W_{dp} = t_{ox} + \frac{W_{dp}}{3} \tag{7.79}$$

In (7.79), W_{dp} is the poly-silicon gate depletion width that is related to the potential drop in the depletion region through

$$W_{dp} = \sqrt{\frac{2\varepsilon_{Si}\psi_p}{qN_p}} \tag{7.80}$$

7.4.10.4 Electrical Oxide Thickness

It may be noted in this connection that the charge sheet approximation considered in all calculations presented so far is not true in a fine sense. The assumption that the inversion layer is infinitely thin therefore needs rigorous consideration. To properly calculate the shape of the inversion region, Poisson's equation has to be solved simultaneously with Schrödinger's equation, which governs the behavior of tightly confined particles. The average location of the inversion charge below the Si-SiO$_2$ interface is called the *inversion layer thickness* t_{inv}. Then the effective (often called the electrical oxide thickness, in compact model terminology) that determines the capacitive coupling between the gate and the channel charge becomes

$$\hat{t}_{ox} = t_{ox} + \frac{\varepsilon_{ox}}{\varepsilon_{Si}}(W_{dp} + t_{inv}) = t_{ox} + \frac{W_{dp}}{3} + \frac{t_{inv}}{3} \tag{7.81}$$

The solution to the poly-silicon depletion effect is to dope the poly-silicon heavily. However, very heavy doping may cause dopant penetration from the gate through the oxide into the substrate. Poly-silicon gate depletion effect is eliminated in advanced MOS technology by replacing the gate material with a pure metal. NMOS and PMOS transistors may require two different metals (with metal work functions close to those of n^+ and p^+ poly-silicon) in order to achieve the optimal threshold voltage [13].

7.4.11 Simulation Results and Discussion

7.4.11.1 Variation of Drain Current, Transconductance, and Output Resistance with Gate Voltage

The variations of drain current with applied gate bias and low drain bias for three different substrate biases are shown in Figure 7.16. It is observed from the figure when the gate bias is just above threshold voltage, the current increases at a faster rate, compared to high gate bias. This is because of the carrier mobility degradation phenomenon occurring at higher gate bias. This is also demonstrated through Figure 7.17, which shows the variation of the transconductance with applied gate bias for low drain bias. The same graphs plotted for high drain bias are shown in Figures 7.18 and 7.19, respectively. It is observed that with high drain bias, the mobility degradation phenomenon affecting the drain current is somewhat counterbalanced due to the high drift velocity of the carriers. The variation of output resistance with applied gate bias is shown in Figure 7.20. It is observed that when the transistor operates in the subthreshold region, the output resistance is very high. This is simple to explain by considering the fact that in the subthreshold region, a very small amount of drain current flows. On the other hand, in a strong inversion region, the output resistance drops and the value remains fairly constant with increase of gate bias.

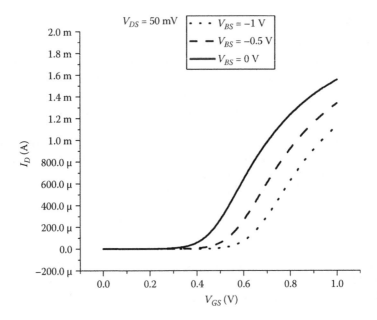

FIGURE 7.16
Variation of drain current with applied gate bias for three different substrate biases and low drain bias.

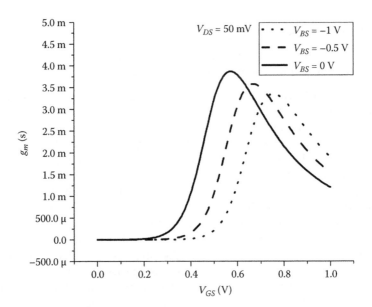

FIGURE 7.17
Variation of transconductance with applied gate bias for three different substrate biases and low drain bias.

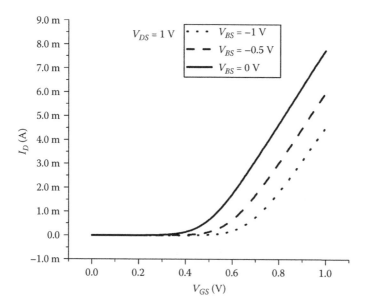

FIGURE 7.18
Variation of drain current with applied gate bias for three different substrate biases and high drain bias.

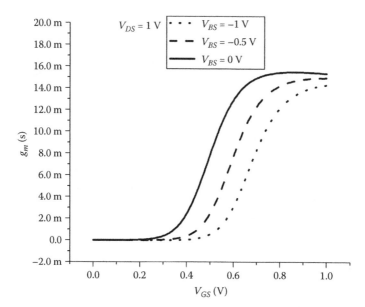

FIGURE 7.19
Variation of transconductance with applied gate bias for three different substrate biases and high drain bias.

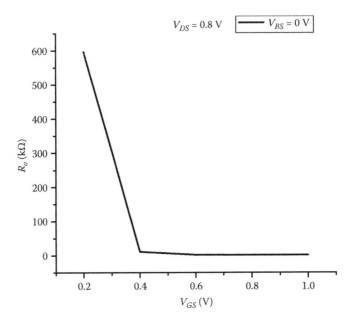

FIGURE 7.20
Variation of output resistance with gate bias.

7.4.11.2 Subthreshold Characteristics

The subthreshold characteristics for low drain bias and high drain bias are shown in Figures 7.21 and 7.22, respectively. The three important performance parameters, related to switching behavior of a MOS transistor, extracted from the subthreshold characteristics are I_{ON}, I_{OFF}, and subthreshold slope S, respectively. The ON and OFF currents are defined as the drain-source current flowing through the transistor when the applied gate bias is either high or zero, respectively. The subthreshold slope is determined as $S = (d(\log_{10} I_{DS})/dV_{GS})^{-1} = 2.3nkT/q$ (i.e., the amount of gate voltage required to change the drain current by an order of magnitude). The values of these three parameters are summarized in Table 7.3. From the results, the value of the subthreshold swing factor n for the two drain biases are calculated and shown in Table 7.3. Thus, the simulation results clearly demonstrate that high drain bias (i.e., DIBL effect) deteriorates the subthreshold characteristics of a MOS transistor.

7.4.11.3 Variation of Transconductance Generation Efficiency, Intrinsic Gain, Linearity, and Cutoff Frequency

An important analog performance parameter of MOS transistor is the transconductance generation efficiency that is measured as g_m/I_{DS}, which

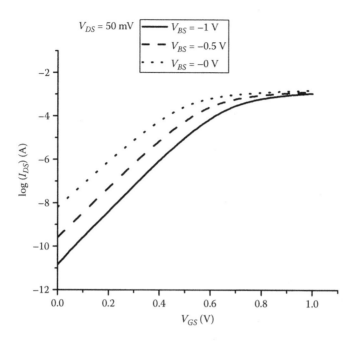

FIGURE 7.21
Subthreshold characteristics for low drain bias.

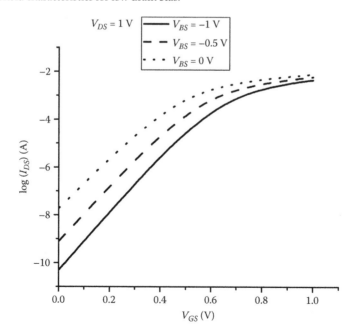

FIGURE 7.22
Subthreshold characteristics for high drain bias.

TABLE 7.3

I_{ON}, I_{OFF}, and Subthreshold Slope S for an NMOS Transistor of $L = 65$ nm and $W = 10$ μm

Drain Bias	I_{ON} @$V_{GS} = 1V$	I_{OFF}@$V_{GS} = 0V$	S	m
$V_{DS} = 50$ mV	1.56 mA	6.52 nA	94 mV/decade	1.596
$V_{DS} = 1$ V	7.74 mA	18.4 nA	95.3 mV/decade	1.618

measures the amount of transconductance generated per unit drain current. The variation of this parameter with applied gate bias is shown in Figure 7.23. It is observed that the transconductance generation efficiency is highest when the transistor works in a weak inversion region. As the gate bias increases, such that the transistor moves on to a strong inversion region, the value of this parameter reduces. This is because in the weak inversion region, a very small amount of current flows through the transistor so that the ratio of transconductance to drain current is high. In a strong inversion region, the drain current increases so that the ratio falls. Theoretically, the maximum value of this factor is found to be $1/nU_T$,

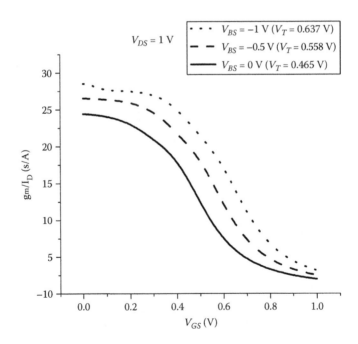

FIGURE 7.23
Variation of transconductance generation efficiency with gate bias for high drain bias and different substrate biases.

and from simulation results, this is 24.41 V^{-1} at $V_{BS} = 0$ V. The effect of substrate bias on the transconductance generation efficiency is also observed. As the substrate bias increases (i.e., as the substrate becomes more reverse biased), the depletion depth increases. Therefore, the depletion capacitance reduces and hence the subthreshold swing factor also reduces. Thus the ratio (g_m/I_{DS}) increases.

The intrinsic voltage gain of a MOS transistor is defined as $g_m R_0$, where R_0 is the output resistance. The variation of output resistance and transconductance with applied gate bias is recalled in Figure 7.24(a). It is observed that at low V_{GS}, when the transistor operates in weak inversion, the transconductance is low, but the output resistance is very high. The variation of the intrinsic gain with the applied gate bias is shown in Figure 7.24(b). Consequently, the intrinsic gain is high at the weak inversion region. As the gate bias is increased, so that the transistor starts to operate in the strong inversion region, the output resistance falls. Thus although the transconductance increases, the intrinsic gain falls. The effect of output resistance plays a significant role in determining the intrinsic gain of a MOS transistor.

Non-linearity of a device is manifested by the presence of higher-order harmonics at the output signal. The linearity of a MOS transistor is quantified in this work through the parameter VIP3. This is the extrapolated gate voltage amplitude, at which the third harmonics of the drain current become equal to the fundamental tone of the drain current [14]. This is mathematically defined as

$$VIP3 = \sqrt{24 \frac{g_m}{|g_{m3}|}}$$

Here

$$g_{m3} = \frac{\partial^3 I_{DS}}{\partial V_{GS}^3}$$

The variation of g_{m3} with applied gate bias is shown in Figure 7.25(a). The VIP3 peak shown in Figure 7.25(b) is due to the second-order interaction effect and can be explained as a cancellation of the third non-linearity coefficient (i.e., g_{m3}) (see Figure 7.25a) by device internal feedback around a second-order non-linearity [14]. It is observed that the linearity of the device is poor when it operates in a weak inversion region. Therefore, for better linearity performance, the device must work in a strong inversion region.

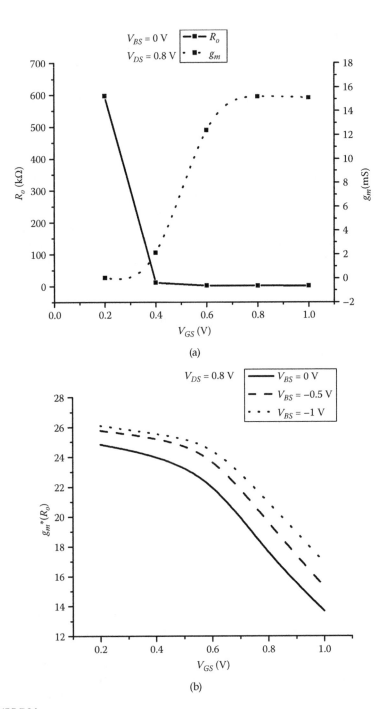

FIGURE 7.24
(a) Variation of transconductance and output resistance with gate bias. (b) Variation of intrinsic gain with applied gate bias for different substrate biases.

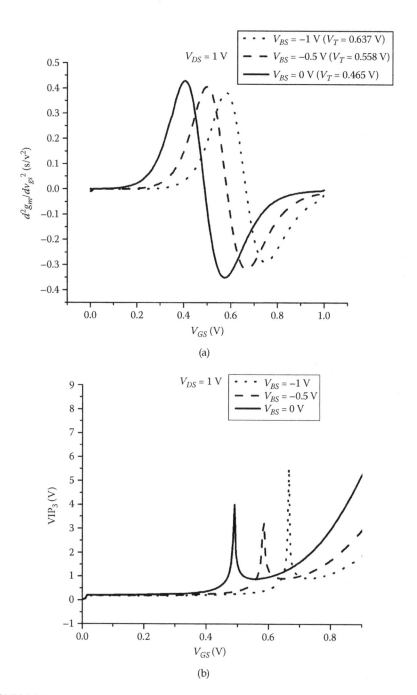

(a)

(b)

FIGURE 7.25
(a) Variation of second derivative of transconductance with applied gate bias. (b) Variation of linearity parameter with applied gate bias.

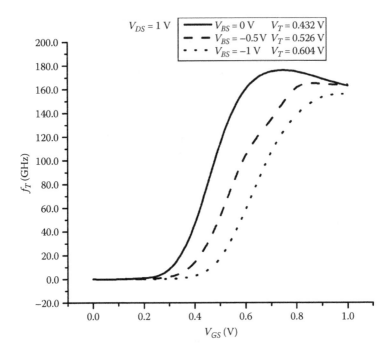

FIGURE 7.26
Variation of cutoff frequency parameter with applied gate bias.

The variation of the cutoff frequency with the applied gate bias is shown in Figure 7.26. The cutoff frequency is also referred to as the unity current gain frequency. This frequency determines the bandwidth of the circuit. It is observed that the value of the cutoff frequency is very small when the device operates in a weak inversion region, and it increases as the gate bias increases so that it operates in a strong inversion region. It may be noted that in a weak inversion region, the intrinsic gain of a MOS transistor is high but the bandwidth is low. Therefore, operation in a moderate inversion region is often preferred for low-power, high-performance analog applications.

7.4.11.4 Variation of Drain Current and Output Resistance with Drain Bias

Variations of the drain current and output resistance with applied drain bias for three different gate biases are shown in Figure 7.27 and Figure 7.28, respectively. It is observed from the graphs that in the subthreshold region, the drain current is small so that the output resistance is high. On the other hand, in the strong inversion region, the amount of drain current is high so that the output resistance is low. For the weak inversion case, with the increase of drain bias, the drain current increases due to various second-order effects such as channel length modulation, DIBL effect, etc. Therefore, the output

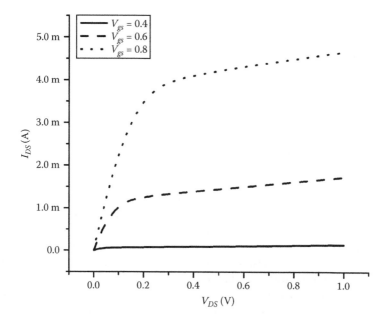

FIGURE 7.27
Variation of drain current with applied drain bias for three different gate biases.

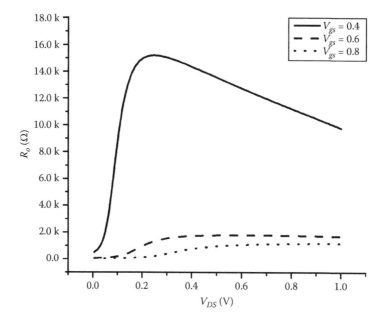

FIGURE 7.28
Variation of output resistance with applied drain bias.

resistance value falls. However, for strong inversion, the increase of drain current with the increase of drain bias due to the above effects is counterbalanced somewhat due to better gate control at high gate bias. Therefore, the resultant rate of increase of drain current is small. Consequently, the output resistance remains nearly constant.

7.5 Hot Carrier Effects Due to Impact Ionization

If the drain voltage (and hence the lateral electric field ξ_y) is sufficiently high, the carrier velocity near the drain saturates. The length of the high field region is a function of channel length, oxide thickness, and gate and drain bias. Even by considering the scaling of supply voltage, the electric field in the high field region is strong enough ($10^4 V/cm$). Consequently, the electrons gain enough energy and collide with the bound electrons in the valence band to create impact ionization of silicon lattice atoms in scaled MOS transistors. These highly energetic electrons are referred to as *hot electrons* because if their kinetic energy is expressed as kT_e, then T_e becomes as high as 1000 K, which is much higher than the lattice temperature. As a result of the impact ionization process, electron-hole pairs are generated. Among these pairs, the electrons are collected by the drain which increases the drain current. On the other hand, the holes are pushed toward the source, which in turn are directed toward the substrate due to the action of the vertical electric field. This results in an impact ionization induced substrate current as illustrated schematically in Figure 7.29.

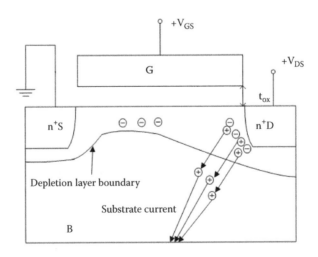

FIGURE 7.29
Substrate current due to impact ionization.

The empirical relationship describing the impact ionization rate is given as [5]

$$\alpha_i(y) = A_i \exp\left(\frac{-B_i}{\xi_y}\right) \tag{7.82}$$

In (7.82), $\alpha_i(y)$ is the number of ionization events per unit length, and A_i and B_i are ionization constants. Thus the substrate current is given by

$$I_{sub} = I_{DS} \int_0^{\Delta L} \alpha_i(y) dy = A_i I_{DS} \int_0^{\Delta L} e^{\frac{-B_i}{\xi_y}} dy \tag{7.83}$$

The velocity saturation region is bounded by $y = 0$(saturation point) to $y = \Delta L$(drain). Also along the surface, the quasi-Fermi level $V(y)$ increases from V_{DSat} at $y = 0$ to V_{DS} at $y = \Delta L$. From pseudo-two-dimensional analysis, an exponential relationship between the lateral field ξ_y and the lateral channel distance can be derived [5], which is given below

$$\xi_y(y) = \xi_{sat} \cosh\left(\frac{y}{l_t}\right) \tag{7.84}$$

In (7.84), ξ_{sat} is the critical field for velocity saturation, and l_t is the characteristic length of the exponentially rising electric field and is given as [5]

$$l_t = \sqrt{\frac{\varepsilon_{Si}}{\varepsilon_{ox}} t_{ox} X_j} \approx \sqrt{3 t_{ox} X_j} \tag{7.85}$$

The peak electric field is reached at the drain [5], where

$$\xi_{max} = \xi_y(y = \Delta L) = \sqrt{\left(\frac{V_{DS} - V_{DSsat}}{l_t}\right)^2 + \xi_{sat}^2} \tag{7.86}$$

In the saturation region, generally $\xi_{max} \gg \xi_{sat}$, so that ξ_{max} can be approximated as

$$\xi_{max} = \frac{V_{DS} - V_{DSsat}}{l_t} \tag{7.87}$$

This field can be as high as mid-10^5 to 10^6 V/cm and leads to impact ionization and other hot carrier effects. From (7.84), we find after using necessary trigonometric identity,

$$\frac{d\xi_y(y)}{dy} = \frac{1}{l_t} \sqrt{\xi_y^2(y) - \xi_{sat}^2} \tag{7.88}$$

Substituting (7.88) in (7.83), with appropriate change of limits, we get

$$I_{sub} = A_i l_t I_{DS} \int_{\xi_{sat}}^{\xi_{max}} \frac{e^{-\frac{B_i}{\xi_y}}}{\sqrt{\xi_y^2 - \xi_{sat}^2}} d\xi_y = \frac{A_i l_t I_{DS} \xi_{max}}{B_i} \exp\left(-\frac{B_i}{\xi_{max}}\right) \qquad (7.89)$$

From (7.87) and (7.89), we get

$$I_{sub} = \frac{A_i}{B_i}(V_{DS} - V_{DSsat}) \exp\left(-\frac{B_i l_t}{V_{DS} - V_{DSsat}}\right) I_{DS} \qquad (7.90)$$

This is used to calculate the substrate current in MOS transistors. It may be noted that I_{sub} strongly depends on the effective channel length because the drain saturation current strongly depends on the effective channel length. In addition, the drain current I_{DS} depends on the source-drain series resistance.

The substrate current causes an ohmic potential drop in the substrate. This leads to substrate bias that causes the threshold voltage to drop. This triggers a positive feedback effect that further enhances the drain current. The substrate current induced body bias effect (SCBE) results in a current increase that is much larger than I_{sub}.

7.5.1 Hot Carrier Injection (HCI)

For high enough electric field, some of the electrons or holes gain sufficient energy from the electric field to cross the interface barrier and enter the SiO_2 layer. The electrons thus trapped in the oxide change the threshold voltage, typically the threshold voltage for NMOS transistors increases and that for the PMOS transistor decreases. The probability of carrier injection is more for hot electrons compared to hot holes. This is because of the smaller effective mass of electrons and because the Si-SiO$_2$ interface energy barrier is larger for holes (~4–6 eV) than for electrons (~3.1 eV). This hot carrier injection phenomenon leads to a long-term reliability problem, or *aging problem*, where a circuit might degrade or fail after being in use for some time.

Present-day CMOS technologies use specially engineered lightly doped drain and source regions which introduces additional series resistance and reduces the peak electric field in the transistor. This prevents carriers from reaching the critical values necessary to become hot. However, drain current and thus device performances are traded off as a result. Therefore, an important design consideration is to operate the circuit at a voltage far enough below the breakdown condition.

7.6 Characterization of Gate Dielectric

The thickness of the gate dielectric (SiO_2 is the preferred gate insulator in the semiconductor industry because of the excellent compatibility of SiO_2 with silicon and established performance record) is reduced from 300 nm for the 10 µm technology to 1.2 nm for the 65 nm technology. Scaling down of SiO_2 thickness is essential for two reasons. First, with the scaling down of the oxide thickness, the gate capacitance C_{ox} increases. This increases the transistor ON-current which leads to an increase of the circuit speed. The second reason is short channel effect immunity, as discussed earlier.

However, if the oxide becomes too thin, the electric field in the oxide becomes so high that it may cause dielectric breakdown. For oxide thickness less than 1.5 nm, tunneling leakage current becomes the most serious limiting factor that prevents the use of such thin film. The reduction of gate oxide thickness (has reached only a few atomic layers) results in an increase in field across the oxide. The high electric field leads to tunneling of electrons from the strongly inverted surface to gate and also from gate to the inverted surface through the gate oxide, resulting in gate oxide tunneling. When the electrons tunnel into the conduction band of the oxide layer, the resulting tunneling is referred to as Fowler-Nordheim tunneling [5]. On the other hand, if the electrons tunnel directly through the forbidden energy gap of the SiO_2 layer, the resulting tunneling is referred to as direct tunneling.

In order to avoid the tunneling problem, there is an intense search of alternative dielectrics with high-κ (permittivity) which has properties very close to SiO_2 but offers the opportunity to use a higher thickness for the gate insulator. The gate capacitance of a MOS transistor using an arbitrary dielectric material with thickness T_d is given by

$$C_G = \frac{\varepsilon_0 \kappa_d A}{T_d} \tag{7.91}$$

In (7.92), ε_0 is the permittivity of free space, κ_d is the relative permittivity of dielectric material, A is the area of the conducting plates, and T_d is the gate dielectric thickness. The thickness of the high-κ dielectric insulator is derived from the relation

$$T_{ox} = \text{Effective oxide thickness (EOT)} = \frac{\kappa_{SiO_2}}{\kappa_d} T_d \tag{7.92}$$

HfO_2 has a relative permittivity (κ) of ~24, six times larger than that of SiO_2(κ_{SiO_2} ~ 3.9). Therefore, a 6-nm HfO_2 film has effective oxide thickness (EOT) of 1 nm, in the sense both films produce the same oxide capacitance. However, the HfO_2 film is physically much thicker compared to SiO_2 film. Therefore the leakage current in the HfO_2 film is several orders of magnitude smaller than that through SiO_2. Some other popular high-κ dielectric insulator materials are ZrO_2 and Al_2O_3. However, the uses of high-κ dielectric insulator materials pose several problems in IC manufacturing. These include chemical reactions between these materials and the silicon substrate, lower surface mobility, and more oxide trapped charges. In order to reduce these problems to some extent, a thin SiO_2 interfacial layer is inserted between the silicon substrate and the high-κ dielectric insulator.

In SPICE simulation, high-κ gate dielectric can be modeled as SiO_2 with an equivalent oxide thickness. Alternatively, the value of the gate dielectric constant parameter (*EPSROX*) can be specified.

7.7 Capacitance Characterization

A VLSI circuit operates both under DC conditions (when the terminal voltages do not change with time) and time-varying conditions. The time-varying operation of the circuit is largely influenced by the various capacitors present in a MOS transistor. Therefore, proper characterization of the various capacitances of a MOS transistor is an essential task for IC designers. The capacitance model is based on the quasi-static approximation, which implies that the potential and charge density at any given point in the channel of the transistor follow the time-varying terminal voltages immediately without any delay. In other words, under quasi-static approximation, it is assumed that the time-varying terminal voltages do not change appreciably within the "transit time" duration of the device. The various intrinsic and extrinsic capacitors present within a MOS transistor are identified in the following sub-section.

7.7.1 Capacitance Components in a MOS Transistor

The various capacitors present within an n-channel MOS transistor are identified in Figure 7.30. For characterizing the various capacitances, the MOS transistor capacitors are divided into two types: intrinsic and extrinsic. The intrinsic region is identified as the region between the metallurgical source and the drain junction where the gate to S/D region is at flat band voltage. The capacitances involved within the intrinsic region are referred to as the intrinsic capacitances. The extrinsic capacitances that are basically the parasitic capacitances are further divided into five components: (1) the outer fringing capacitances between the poly-silicon gate and the S/D region: C_{FO}; (2) the inner fringing capacitances between the poly-silicon gate and the S/D region: C_{FI}; (3) the overlap capacitances between the gate and the heavily doped S/D regions (as well as the bulk region), C_{GSO}, C_{GDO} (C_{GBO}); (4) the overlap capacitances between the gate and the lightly doped S/D regions C_{GSOL}, C_{GDOL}; and (5) the source/drain junction capacitances C_{JS} and C_{JD}. The intrinsic capacitances

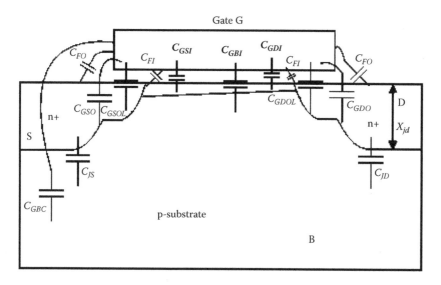

FIGURE 7.30
Identification of intrinsic and extrinsic capacitors present in an n-channel MOS transistor.

are shown bold in Figure 7.29. These are gate-to-source capacitance C_{GSI}, gate-to-bulk capacitance C_{GBI}, and gate-to-drain capacitance C_{GDI}.

7.7.2 Characterization of Intrinsic Capacitances (Meyer's Approach)

The simplest approach for characterizing the gate capacitances was developed by Meyer [15,16]. The simplified intrinsic capacitance model treats the intrinsic MOS capacitances as three lumped capacitances, gate-to-source capacitance C_{GS}, gate-to-drain capacitance C_{GD}, and gate-to-bulk capacitance C_{GB}. The gate capacitances are attributed entirely to the changes in the gate charge, which is written in the following compact formulation:

$$C_{GZ} = \frac{\partial Q_G}{\partial V_{GZ}}$$ (7.93)

In (7.93), C_{GZ} represents the capacitance between the gate and the terminal Z (S/D), while V_{GZ} represents the corresponding voltage difference. It is also assumed that all capacitances are reciprocal (e.g., $C_{GD} = C_{DG}$). From charge neutrality condition, we have

$$Q_G = -(Q_{inv} + Q_B)$$ (7.94)

Here Q_{inv} is the inversion charge density, and Q_B is the bulk charge density. In strong inversion, the channel charge density along the channel is given by (7.26) which is repeated here for convenience:

$$Q_{inv} = -C_{ox}(V_{GS} - V_T - V_{CS}(y))$$ (7.95)

Here m is taken to be unity for simplicity. The drain-to-source current is given by (7.24) which is repeated here for convenience:

$$I_{DS} = \mu_s \frac{W}{L} \int_0^{V_{DS}} [-Q_{inv}(V_{CS})] dV_{CS} \tag{7.96}$$

The drain-to-source current I_{DS} is obtained by integration (7.96) and is written as

$$I_{DS} = \frac{\mu_s W C_{ox}}{L} \left(V_{GS} - V_T - \frac{1}{2} V_{DS} \right) V_{DS} \tag{7.97}$$

Using $V_{GD} = V_{GS} - V_{DS}$, (7.97) is transformed to

$$I_{DS} = \frac{\mu_s W C_{ox}}{2L} [(V_{GS} - V_T)^2 - (V_{GD} - V_T)^2] \tag{7.98}$$

Considering the variation of the charges along the channel length, (7.94) is transformed to

$$Q_G = -W \int_0^L Q_{inv}(y) dy - W \int_0^L Q_b(y) dy = -W \int_0^L Q_{inv}(y) dy - Q_B \tag{7.99}$$

Using (7.96), (7.97), and (7.99) and performing the integration, we get

$$Q_G = \frac{2}{3} W L C_{ox} \left[\frac{(V_{GD} - V_T)^3 - (V_{GS} - V_T)^3}{(V_{GD} - V_T)^2 - (V_{GS} - V_T)^2} \right] - Q_B \tag{7.100}$$

7.7.2.1 Intrinsic Capacitances in the Linear Region

The intrinsic capacitances C_{GS}, C_{GD}, and C_{GB} in the linear region are determined by using the following relationships:

$$C_{GS} = \frac{\partial Q_G}{\partial V_{GS}} \bigg|_{V_{GD}, V_{GB}} \tag{7.101}$$

$$C_{GD} = \frac{\partial Q_G}{\partial V_{GD}} \bigg|_{V_{GS}, V_{GB}} \tag{7.102}$$

$$C_{GB} = \frac{\partial Q_G}{\partial V_{GB}} \bigg|_{V_{GS}, V_{GD}} \tag{7.103}$$

Therefore, by differentiating (7.100) as per the relationships (7.101) through (7.103), the various gate capacitances are determined as follows:

$$C_{GS} = \frac{2}{3} WLC_{ox} \left[1 - \frac{(V_{GD} - V_T)^2}{(V_{GS} - 2V_T + V_{GD})^2} \right] \tag{7.104}$$

$$C_{GD} = \frac{2}{3} WLC_{ox} \left[1 - \frac{(V_{GS} - V_T)^2}{(V_{GS} - 2V_T + V_{GD})^2} \right] \tag{7.105}$$

$$C_{GB} = 0 \tag{7.106}$$

The fact that the capacitance C_{GB} is zero at the strong inversion region may be explained by the fact that the inversion layer in the channel from the source to the drain screens the silicon bulk from the gate charge.

7.7.2.2 Intrinsic Capacitances in the Saturation Region

In the saturation region, the drain voltage is V_{DSsat}, which is given by $V_{DSsat} = V_{GS} - V_T$, assuming long channel MOS transistor. Thus the gate-to-drain voltage becomes

$$V_{GD} = V_{GS} - V_{DSsat} = V_T \tag{7.107}$$

Substituting (7.108) in (7.101), we get

$$Q_G = \frac{2}{3} WLC_{ox}(V_{GS} - V_T) - Q_B \tag{7.108}$$

Therefore, the various intrinsic gate capacitances in the saturation region are obtained as follows:

$$C_{GS} = \frac{2}{3} WLC_{ox} \tag{7.109}$$

$$C_{GD} = 0 \tag{7.110}$$

$$C_{GB} = 0 \tag{7.111}$$

The physical explanation for (7.111) is the same as that provided for (7.106). The physical explanation for (7.110) is that in the saturation region the channel is pinched off, thereby the channel is electrically isolated from the drain. The gate charge is not influenced by the change in drain voltage, and thus the capacitance C_{GD} vanishes.

7.7.2.3 Intrinsic Capacitances in the Subthreshold Region

In the subthreshold region, the inversion charge is negligible compared to the bulk depletion charge, so that the charge neutrality condition is given by

$$Q_G = -Q_B = C_{ox}\gamma\sqrt{\psi_{sa}} \tag{7.112}$$

In (7.112), ψ_{sa} is the surface potential in the subthreshold region, which is given as [7]

$$\psi_{sa} = \left(-\frac{\gamma}{2} + \sqrt{\frac{\gamma^2}{4} + V_{GB} - V_{FB}} \right)^2 \qquad (7.113)$$

Substituting (7.113) in (7.112) and performing an integration as done in (7.100), the total gate charge in the subthreshold region is given by

$$Q_G = -\frac{1}{2} WLC_{ox}\gamma^2 \left[1 - \sqrt{1 + \frac{4}{\gamma^2}(V_{GB} - V_{FB})} \right] \qquad (7.114)$$

Therefore, by differentiating (7.114) as per the relationships (7.101) through (7.103), the various gate capacitances are determined as follows:

$$C_{GS} = 0 \qquad (7.115)$$

$$C_{GD} = 0 \qquad (7.116)$$

$$C_{GB} = \frac{WLC_{ox}}{\sqrt{1 + \frac{4}{\gamma^2}(V_{GB} - V_{FB})}} \qquad (7.117)$$

7.7.2.4 Intrinsic Capacitances in the Accumulation Region

In the accumulation region, $V_{GS} < V_{FB}$, the MOS structure behaves like a simple parallel plate capacitor and the capacitances are as follows:

$$C_{GS} = 0 \qquad (7.118)$$

$$C_{GD} = 0 \qquad (7.119)$$

$$C_{GB} = C_{ox} \qquad (7.120)$$

7.7.2.5 Charge-Based Approach

It may be noted that Meyer's approach for characterizing the intrinsic capacitances of a MOS transistor is simple and is widely used by the IC designers for first-hand estimation of the various MOS capacitances. However, this approach for characterization does not provide good results for some circuits such as MOS charge pump, static RAM, and switched capacitor circuits. Therefore, an alternative approach is used for characterizing the MOS capacitances in today's compact models. This is the charge-based approach for capacitance characterization. In this approach, the

emphasis is put on the accurate characterization of charges of each terminals (Q_D, Q_S, Q_G, Q_B) of the MOS transistor. The calculation of total inversion charge in the channel is fairly easy. However, it is difficult to precisely
characterize the charges on the source and the drain terminals. The inversion charge must be partitioned to the source and drain in a suitable manner. Several charge partitioning approaches have been suggested for the
saturation region. They are 50/50, 40/60, and 0/100 and are distinguished
in the compact models through a model parameter $X_{PART} = Q_D/Q_S$ as the
charge partitioning ratio. The simplest way is to partition the channel
charge and assign 50% of the inversion charge to the source and the rest
to the drain, which corresponds to $(X_{PART} = 0.5)$, which can be written as
$Q_S = Q_D = 0.5Q_{inv}$. When $X_{PART} > 0.5$, the 0/100 charge partitioning scheme
is chosen which implies that $Q_S = Q_{inv}, Q_D = 0$. When $X_{PART} < 0.5$, the 40/60
charge partitioning scheme is chosen. The 40/60 partition scheme, also
known as the Ward Dutton partitioning scheme [17], is physically correct
as demonstrated through 2D device simulation results and experiments.

7.7.2.6 Effect of Poly-Silicon Gate Depletion Effect and Finite Inversion Charge Layer Thickness

The poly-silicon gate depletion effect as discussed earlier needs to be considered while characterizing the intrinsic capacitances. This is implemented by
replacing V_{GS} in all model equations by V_{GS_eff} as defined in (7.74). The effect
of finite inversion charge thickness can be characterized by a capacitance in
series with the gate oxide capacitance C_{ox}. This results in reduced effective
gate oxide capacitance:

$$C_{ox\_eff} = \frac{C_{ox}C_C}{C_{ox} + C_C} \qquad (7.121)$$

In (7.121), C_C is the correction term added due to the inversion layer of thickness t_{inv}.

7.7.3 Characterization of Extrinsic Capacitances

The extrinsic components of MOS transistor capacitances are categorized
broadly into three types: (1) gate overlap capacitances in source/drain and
bulk region $(C_{GSO}/C_{GDO}, C_{GSOL}/C_{GDOL}, C_{GBO})$; (2) inner and outer fringing
capacitances $(C_{FI}$ and $C_{FO})$; and (3) source/drain junction capacitances $(C_{JS}$
and $C_{JD})$. These capacitances at a given operating bias condition are required
to be characterized.

7.7.3.1 Characterization of Fringing and Overlap Capacitances

Characterization of overlap capacitance in a MOS transistor device is especially important when the amount of overlap becomes significant compared

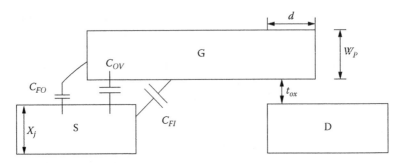

FIGURE 7.31
Overlap and fringing capacitances.

to the electrical channel length. As a crude estimate, the overlap capacitance is determined as follows [3]:

$$C_{OV} = \frac{\varepsilon_{ox} d}{t_{ox}} \tag{7.122}$$

Here d is the amount of gate-to-drain/source overlap. However, when d is small, the fringing effect is significant. Let us consider the approximate structure, shown in Figure 7.31 for precise characterization of the overlap and fringing capacitances [18].

The overlap capacitance consists of the following three components: (1) outer fringing capacitance C_{FO} between the gate and the source/drain, (2) direct overlap capacitance C_{OV} between the gate and the source/drain, and (3) inner fringing capacitance C_{FI} on the channel side between the gate and the side wall of the source/drain junction. These capacitances are calculated using conformal technique with appropriate boundary conditions [18]. These capacitances for unit width of the device are given as follows:

$$C_{FO} = \frac{\varepsilon_{ox}}{\theta} \ln\left(1 + \frac{W_p}{t_{ox}}\right) \tag{7.123}$$

$$C_{FI} = \frac{\varepsilon_{ox}}{\beta} \ln\left(1 + \frac{X_j \sin\beta}{t_{ox}}\right) \tag{7.124}$$

$$C_{OV} = \frac{\varepsilon_{ox}(d + \Delta)}{t_{ox}} \tag{7.125}$$

In (7.123), θ is the slope angle for the poly-silicon gate. For the vertical edge of the poly-silicon gate, $\theta = \pi/2$. In (7.124), β is given by

$$\beta = \frac{\pi\varepsilon_{ox}}{2\varepsilon_{Si}} \tag{7.126}$$

In (7.125), Δ is a correction factor to account for some higher-order effects and is given as follows:

$$\Delta = \frac{t_{ox}}{2}\left(\frac{1-\cos\theta}{\sin\theta} + \frac{1-\cos\beta}{\sin\beta}\right) \tag{7.127}$$

The total overlap capacitance per unit width of the device is thus given by the sum of (7.123), (7.124), and (7.125).

In addition to the above overlap capacitances, there is another overlap capacitance in the channel width direction, which results in an overlap capacitance between the gate and the substrate. This is given as

$$C_{CGBO} = C'_{CBO} \cdot L \tag{7.128}$$

Here in (7.128), C'_{GBO} is the gate-to-bulk overlap capacitance per unit length.

7.7.3.2 Characterization of Junction Capacitances

The junction capacitances arise from the depletion charge between the source or drain and the substrate. These are usually reverse-biased. Therefore, with the variation of source or drain voltages, the depletion charge increases or decreases accordingly. The depletion capacitance per unit area of an abrupt p-n junction is [5]

$$C_j = \frac{\varepsilon_{Si}}{W_{dj}} = \sqrt{\frac{\varepsilon_{Si}qN_A}{2(V_{bi}+V_R)}} = \left[\frac{\varepsilon_{Si}qN_A}{2(V_{bi}+V_R)}\right]^m \tag{7.129}$$

In (7.129), W_{dj} is the depletion layer width, N_A is the impurity concentration of the lightly doped side, ψ_{bi} is the built-in potential, and V_R is the reverse bias voltage across the junction. In (7.129), m is the grading coefficient and its value is ½ for abrupt p-n junction. For zero bias, C_{j0} is defined as

$$C_{j0} = \left[\frac{q\varepsilon_{Si}N_A}{2V_{bi}}\right]^m \tag{7.130}$$

With this, (7.129) can be algebraically manipulated as [3]

$$C_j = C_{j0}\left(1+\frac{V_R}{V_{bi}}\right)^{-m} \tag{7.131}$$

The junction capacitance has two components: bottom component and sidewall/perimeter component. The total junction capacitance is thus written as [3]

$$C_j = C_{jb}A + C_{jsw}P \qquad (7.132)$$

In (7.132), C_{jb} is the bottom component of the junction capacitance per unit area, A is the total junction area, C_{jsw} is the sidewall component of the junction capacitance per unit length, and P is the total junction perimeter. Using (7.131), the bottom component and perimeter component are defined as follows:

$$C_{jb} = C_{j0b}\left(1 + \frac{V_R}{V_{bi}}\right)^{-m_b} \qquad (7.133)$$

$$C_{jsw} = C_{j0sw}\left(1 + \frac{V_R}{V_{bisw}}\right)^{-m_{sw}} \qquad (7.134)$$

In (7.133), C_{j0b} is the zero-bias sidewall capacitance per unit area, and m_b is the grading coefficient for the bottom component. In (7.134), C_{j0sw} is the zero-bias sidewall capacitance per unit length, and m_{sw} is the grading coefficient for the sidewall.

7.7.4 Simulation Results and Discussion

The channel length is taken to be 65 nm and channel width is 10 μm. The oxide capacitance per unit area is 0.0197 F/m². The simulation results include all sorts of extrinsic capacitances and intrinsic capacitances, which are calculated as per the charge-based approach. The charge partitioning ratio is taken to be 0, which means that 40/60 charge partitioning scheme has been considered. Here, we present an intuitive understanding of the graphs. This is important for VLSI designers.

The variations of gate-to-source capacitor C_{GS} with V_{DS} for three different values of V_{GS} are shown in Figure 7.32. It is observed that in the subthreshold region, the gate-to-source capacitance value is very low. This can be explained by the fact that in the subthreshold region, the inversion charge is negligibly small. According to Meyer's approach, $C_{GS} = 0$ in the subthreshold region. However, there will be extrinsic components that contribute to this capacitance. In the linear region, with small V_{DS}, as V_{GS} increases, the inversion charge increases. Therefore, the capacitance value increases. In the saturation region, because of the pinch-off phenomenon, the inversion charge is solely due to the gate-source voltage and the capacitance value is maximum. This behavior is followed by the simulation results.

The variations of gate-to-drain capacitor C_{GD} with V_{DS} for three different values of V_{GS} are shown in Figure 7.33. In the subthreshold region, the inversion charge is negligibly small so that C_{GD} is ideally zero. In the saturation region, due to the pinch-off phenomenon, there is no capacitive coupling

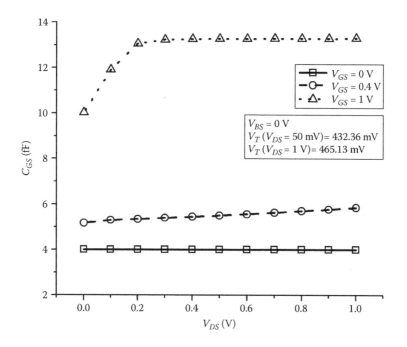

FIGURE 7.32
Variation of gate-to-source capacitance with applied drain bias for three different gate biases.

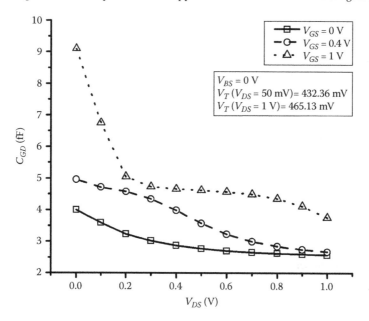

FIGURE 7.33
Variation of gate-to-drain capacitance with applied drain bias for three different gate biases.

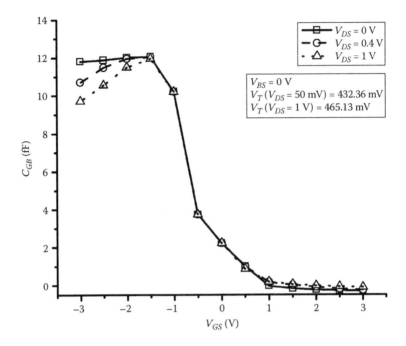

FIGURE 7.34
Variation of gate-to-body capacitance with applied gate bias for three different drain biases.

between the drain and the gate so that C_{GD} is very small. In the linear region, as V_{DS} reduces, the inversion charge increases so that the C_{GD} increases. This behavior is followed by the simulation results.

The variations of the gate-to-bulk capacitance C_{GB} with gate bias for three different drain biases are shown in Figure 7.34. This capacitance has a non-zero value only in the subthreshold region. This is because in this region, the inversion charge is very small. The capacitance is determined by the series combination of oxide capacitance and the depletion capacitance. In the accumulation region, the intrinsic gate-to-bulk capacitance is determined solely by oxide capacitance.

Therefore, Meyer's approach [15] of characterizing the intrinsic capacitance qualitatively explains the variations of the capacitances with bias conditions.

7.8 Noise Characterization

Noise in a MOS transistor is caused by small random fluctuations in signals (currents and voltages), caused due to phenomena generated within the device. Proper characterization of noise in a MOS transistor is essential for analog and RF IC design. The two most important noise components of a MOS transistor are thermal noise and flicker noise. The following subsections deal with each of them individually.

7.8.1 Characterization of Thermal Noise in MOS Transistor

In conventional resistors, the thermal noise is generated due to the random thermal motion of the electrons. This motion does not depend upon the presence or absence of direct current, because the drift velocities of electrons in a conductor are much less than the thermal velocities of the electrons. In a resistor R, the thermal noise is represented by a series voltage generator or a shunt current generator to a noiseless ideal resistor. The noise spectral density is given by [3,5]

$$\overline{v_n^2} = 4kTR\Delta f \tag{7.135}$$

$$\overline{i_n^2} = 4kT\frac{1}{R}\Delta f \tag{7.136}$$

In (7.135) and (7.136), k represents Boltzmann's constant. From (7.135) and (7.136), it is observed that the noise spectral density is independent of frequency f. This characteristic is called *white noise*.

The intrinsic thermal noise of a MOS transistor originates from the channel resistance due to random thermal motion of the carriers. The channel of a MOS transistor may be considered to be divided into several resistive segments, and each of these segments contributes to thermal noise. The corresponding noise spectral density thus follows (7.135). The resistor R is replaced by $(2/3)g_m$, in a saturation region where g_m is the gate transconductance of the device. It follows, therefore, from (7.135) that

$$\overline{v_n^2} = \frac{8kT}{3}g_m\Delta f \tag{7.137}$$

However, (7.137) is inadequate, especially in the linear region $V_{DS} \approx 0$ where the transconductance is zero, so that the calculated noise spectral density becomes zero, which however is not true in practice. Thus (7.137) is modified as follows:

$$\overline{v_n^2} = \frac{8kT}{3}(g_m + g_{ds} + g_{mb})\Delta f \tag{7.138}$$

In (7.138), g_{ds} and g_{mb} are output conductance and body transconductance, respectively. However, a more rigorous approach for characterizing thermal noise is given below which is widely used in SPICE compact models [3,6,7].

Consider an infinitesimally small section of the noiseless channel of a MOS transistor of length dy. Let the resistance of this small section be dR and the channel voltage produced by this resistance is dV_{CS}. For a channel current I_{DS}, these are related as

$$dV_{CS} = I_{DS}dR = -W\mu_s Q_{inv}\frac{dV_{CS}}{dy}dR \tag{7.139}$$

From (7.139) it follows that

$$dR = -\frac{dy}{W\mu_s Q_{inv}} \tag{7.140}$$

The noise spectral density due to the thermal noise generated by this small resistance dR is given by

$$\overline{v_n^2} = 4kTdR\Delta f = -4kT\frac{dy}{W\mu_s Q_{inv}}\Delta f \tag{7.141}$$

The power spectral density for the elemental noise voltage is from (7.141)

$$dS_{V_C} = 4kTdR = -4kT\frac{dy}{W\mu_s Q_{inv}} \tag{7.142}$$

From this elemental noise voltage, the elemental noise current power spectral density is

$$dS_{I_D} = g_C^2 dS_{V_C} \tag{7.143}$$

The conductance for the elemental channel segment is determined as follows:

$$g_C = \frac{dI_{DS}}{dV_{CS}} = -\frac{d}{dV_{CS}}\left[\frac{W}{L}\mu_s\int_0^{V_{DS}}Q_{inv}(V_{CS})dV_{CS}\right] = -\mu_s\frac{W}{L}Q_{inv} \tag{7.144}$$

Substituting g_c from (7.144) and dS_{V_C} from (7.142) into (7.143), we get

$$dS_{I_D} = -\left(-\mu_s\frac{W}{L}Q_{inv}\right)^2 4kT\frac{dy}{\mu_s W Q_{inv}} = -4kT\frac{\mu_s}{L^2}WQ_{inv}dy \tag{7.145}$$

Integrating over the entire channel length, the total noise current power spectral density is given by

$$S_{I_D} = -4kT \frac{\mu_s}{L^2} \int_0^L Q_{inv} W \, dy = -4kT \frac{\mu_s}{L^2} Q_{INV} \tag{7.146}$$

In (7.146), $Q_{INV} = Q_{inv} WL$ represents the total inversion charge under the gate. The thermal noise power spectral density is often expressed in the following manner, referred to as the Klaassen-Prins equation for thermal noise [19]:

$$S_{I_D} = \frac{4kT}{L^2 I_{DS}} \int g^2(V_{CS}) dV_{CS} \tag{7.147}$$

It is to be noted that (7.146) is used in a BSIM compact model with appropriate substitution of Q_{INV}.

7.8.2 Characterization of Flicker Noise in MOS Transistor

The flicker noise in the drain current or gate voltage of a MOS transistor is important to characterize precisely because it deteriorates the signal-to-noise ratio of several analog circuits. It also increases the phase noise of oscillators in RF applications. For proper characterization of flicker noise, the underlying physical mechanism of the flicker noise must be understood. This is briefly discussed below.

7.8.2.1 Physical Mechanisms of Flicker Noise

The conductivity of a conductor due to drift motion of the carriers is given by

$$\sigma = qn\mu \tag{7.148}$$

In (7.148), n represents the carrier concentration, and μ represents the carrier mobility. It appears from (7.148) that any fluctuation in the carrier density or mobility leads to fluctuation of the current flowing through the conductor. There are several different theories for explaining the physical cause of flicker noise. These are broadly classified into three different categories [20]: (1) carrier density fluctuation model, (2) mobility fluctuation model, and (3) correlated carrier and mobility fluctuation model.

According to the carrier density fluctuation model [20], the flicker noise is caused by random trapping and de-trapping of mobile carriers by the interface traps at the Si-SiO$_2$ interface. The interface traps dynamically exchange carriers with the channel causing fluctuation in the surface potentials, giving rise to fluctuation in the inversion charge density. The carrier density fluctuation model is observed to successfully explain the flicker noise spectrum in n-channel MOS transistors. According to the mobility fluctuation model [20], on the other hand, the flicker noise is caused due to fluctuation in the carrier mobility, caused due to phonon scattering. The mobility fluctuation model successfully explains the flicker noise spectrum in p-channel MOS transistor. According to the correlated carrier and mobility fluctuation model [21,22], also referred to as the unified flicker noise model, when an interface trap captures an electron from the inversion layer, it becomes charged and reduces the carrier mobility due to Coulombic scattering. Thus according to this model, both the carrier number and the carrier mobility fluctuate due to trapping and de-trapping of the carriers by the interface traps. The unified model shows good matching with experimental results.

7.8.2.2 Empirical Approach for Characterization of Flicker Noise

The power spectral density of the flicker noise spectrum is given by [21]

$$S_{I_D} = \frac{KF \cdot I_{DS}^{AF}}{f \cdot C_{ox} \cdot WL} \tag{7.149}$$

In (7.149), KF is the flicker noise coefficient, and AF is the flicker noise exponent. The value of the parameter AF lies in the range of 0.5 to 2. The constant KF is proportional to the interface trap density, which is technology-specific. The lack of systematic approach in determining the empirical parameters limits the use of this model. However, two significant observations are made. First, the flicker noise is dominant at low frequency. Because of its dependence on frequency as $(1/f)$, flicker noise is sometimes referred to as the $(1/f)$ noise. At frequencies above 100 MHz, the flicker noise spectrum becomes negligible compared to that of the thermal noise. Second, the flicker noise spectrum reduces as the gate area is increased. Third, for PMOS transistors, it has been found that the value of the flicker noise coefficient is smaller compared to NMOS transistors; therefore, PMOS transistors are used in designing low noise circuits, at least at the first stage.

7.8.2.3 Characterization of Flicker Noise through Physics-Based Model

Consider a section of the channel with width W and length Δy. The drain current is given by

$$I_{DS} = W\mu_s qN\xi_y \tag{7.150}$$

In (7.150), μ_s is the carrier mobility, q is the electron charge, N is the number of channel carriers per unit area, and ξ_y is the lateral channel field. Fluctuation of local drain current is given by [21,22]

$$\frac{\delta I_{DS}}{I_{DS}} = -\left(\frac{1}{\Delta N}\frac{\delta \Delta N}{\delta \Delta N_t} \pm \frac{1}{\mu_s}\frac{\delta \mu_s}{\delta \Delta N_t}\right)\delta \Delta N_t \tag{7.151}$$

In (7.151), $\Delta N = NW\Delta y$, $\Delta N_t = N_t W\Delta y$, where N_t is the number of occupied traps per unit area, and N is the inversion carrier density. The $\pm$ sign in the mobility term of (7.151) denotes whether the trap is neutral or charged when filled.

Let us first evaluate the first term on the right-hand side of Equation (7.151). The ratio of fluctuations in carrier number to fluctuations in occupied trap number $R = \delta \Delta N/\delta \Delta N_t$ is close to unity in strong inversion but assumes a smaller value in other bias conditions. A general expression of R is therefore written as follows:

$$R = \frac{\delta \Delta N}{\delta \Delta N_t} = -\frac{C_{inv}}{C_{ox} + C_{inv} + C_{dm} + C_{it}} \tag{7.152a}$$

In (7.152a), C_{inv}, C_{dm}, and C_{it} are inversion layer, depletion layer, and interface trap capacitances, respectively. A more concise form of R is as follows:

$$R = -\frac{N}{N + N*} \tag{7.152b}$$

In (7.152b), $N* = (kT/q^2)(C_{ox} + C_{dm} + C_{it})$ and the typical value of this quantity is 1–5E10/cm^{-2}.

Let us now evaluate the first term on the right-hand side of Equation (7.151). The carrier mobility is related to the oxide trap density as follows:

$$\frac{1}{\mu_s} = \frac{1}{\mu_B} + \frac{1}{\mu_{SR}} + \frac{1}{\mu_{Ph}} + \frac{1}{\mu_{Cit}} = \frac{1}{\mu_n} + \alpha_{sc}N_t \tag{7.153}$$

In (7.153), $\mu_{Cit} = 1/\alpha_{sc}N_t$ is the mobility limited by Coulombic scattering of the mobile carriers at trapped charges near the Si-SiO$_2$ interface, and $\mu_B, \mu_{SR}, \mu_{Ph}$ represents the mobility limited by ionized impurity scattering, surface roughness scattering, and phonon scattering, respectively. The scattering coefficient α_{sc} is a function of the local carrier density due to the screening effect as well as the distance of the trap from the interface. From experimental results, it has been found that μ_{Cit} increases with the inversion carrier density due to the screening effect. The relationship is given as follows [23]:

$$\mu_{Cit} = \mu_{CO} \frac{\sqrt{N}}{N_t} \tag{7.154a}$$

$$\alpha_{sc} = \frac{1}{\mu_{CO}\sqrt{N}} \tag{7.154b}$$

However, in the original unified mobility model [21,22], the scattering parameter is considered to be independent of the inversion carrier density. The reduction of α_{sc} with an increase of N is understood as follows. As the inversion carrier density increases, the screening length and the scattering cross section due to the screening by minority carriers reduce and hence the scattering parameter increases. In a weak inversion region, screening due to minority carriers becomes less significant compared to that by majority carriers. Because the majority carrier concentration does not change much in the weak inversion region, the scattering cross section remains almost constant with inversion carrier density. Consequently, in the weak inversion region, α_{sc} saturates to a particular value and (7.153b) is no longer valid [23]. By differentiating (7.153) and substituting in (7.151), we arrive at

$$\frac{\delta I_{DS}}{I_{DS}} = -\left[\frac{R}{N} \pm \alpha_{sc}\mu_s\right] \frac{\delta \Delta N_t}{W \Delta y} \tag{7.155a}$$

This can be written as

$$\delta I_{DS} = -\left[\frac{R}{N} \pm \alpha_{sc}\mu_s\right] \frac{I_{DS}}{W \Delta y} \delta \Delta N_t \tag{7.155b}$$

The power spectrum density of the local current fluctuation is obtained from (7.155b) as follows:

$$S_{\Delta I_{DS}}(y, f) = \left(\frac{I_{DS}}{W \Delta y}\right)^2 \left(\frac{R}{N} \pm \alpha_{sc}\mu_s\right)^2 S_{\Delta N_t}(y, f) \tag{7.156}$$

In (7.156), $S_{\Delta N_t}(y, f)$ is the power spectrum density of the fluctuations in the number of occupied traps over the area $W \Delta y$ and is given by

$$S_{\Delta N_t}(y, f) = N_t(E_{fn}) \frac{kTW \Delta y}{\gamma f} \tag{7.157}$$

In (7.157), E_{fn} is the electron quasi-Fermi level, and γ is the attenuation coefficient of the electron wave function in the oxide. Substituting (7.157) in (7.156), we get

$$S_{\Delta I_{DS}}(y,f) = \left(\frac{I_{DS}}{W\Delta y}\right)^2 \left(\frac{R}{N} \pm \alpha_{sc}\mu_s\right)^2 N_t(E_{fn})\frac{kTW\Delta y}{\gamma f} \qquad (7.158)$$

The total drain current noise power spectral density is given as

$$S_{I_{DS}}(f) = \frac{1}{L^2}\int_0^L S_{\Delta I_{DS}}(y,f)\Delta y \, dy \qquad (7.159)$$

Substituting (7.158) in (7.159) and changing the variables of integration by using (7.150), we write

$$S_{I_{DS}}(f) = \frac{qkTI_{DS}\mu_s}{\gamma f L^2}\int_0^{V_{DS}} N_t\left(E_{fn}\right)\left(1 \pm \alpha_{sc}\mu_s \frac{N}{R}\right)^2 \frac{R^2}{N} dV \qquad (7.160a)$$

This can be written in a compact way as follows:

$$S_{I_{DS}}(f) = \frac{qkTI_{DS}\mu_s}{\gamma f L^2}\int_0^{V_{DS}} N_t^*\left(E_{fn}\right)\frac{R^2}{N} dV \qquad (7.160b)$$

In (7.160b), $N_t^*(E_{fn})$ is the equivalent oxide trap density that produces the same noise power in absence of mobility fluctuations and is given as

$$N_t^*(E_{fn}) = N_t(E_{fn})\left(1 \pm \alpha_{sc}\mu_s \frac{N}{R}\right)^2 \qquad (7.160c)$$

In BSIM implementation of the unified noise model, three additional parameters are introduced to fit the noise measurement results:

$$N_t^*(E_{fn}) = A + BN + CN^2 \qquad (7.161)$$

In (7.161), A, B, and C are technology-dependent model parameters. In (7.160b), the integration variable is changed as follows:

$$S_{I_{DS}}(f) = \frac{q^2 kTI_{DS}\mu_s}{\gamma f L^2 C_{ox}}\int_{N_S}^{N_D} N_t^*(E_{fn})\frac{R^2}{N} dN \qquad (7.162)$$

In (7.162), N_S and N_D represent the inversion charge density at the source end and drain end, respectively. These can easily be computed from the inversion charge densities formulae discussed earlier for linear, saturation, and sub-threshold regions. Without going into the detailed mathematical derivations (lengthy but elementary), the drain current noise power at the three regions of operations are written as follows [6]:

Linear Region

$$S_{I_{DS}}(f) = \frac{q^2 kT I_{DS}\mu_s}{m\gamma f L^2 C_{ox}}\left[A\ln\left(\frac{N_S + N^*}{N_D + N^*}\right) + B(N_S - N_D) + \frac{1}{2}C\left(N_S^2 - N_D^2\right)\right] \quad (7.163a)$$

Saturation Region

$$S_{I_{DS}}(f) = \frac{q^2 kT I_{DS}\mu_s}{m\gamma f L^2 C_{ox}}\left[A\ln\left(\frac{N_S + N^*}{N_D + N^*}\right) + B(N_S - N_D) + \frac{1}{2}C(N_S^2 - N_D^2)\right]$$

$$+ \Delta L\frac{kT I_{DS}^2}{\gamma f WL^2}\frac{A + BN_D + CN_D^2}{(N_D + N^*)^2} \quad (7.163b)$$

In (7.163a) and (7.163b), N_S and N_D are evaluated as follows:

$$qN_S = C_{ox}(V_{GS} - V_T) \quad (7.163c)$$

$$qN_D = C_{ox}(V_{GS} - V_T - mV_{DSsat}) \quad (7.163d)$$

In (7.163b), the second term in the flicker noise power spectrum density estimates the noise arising in the velocity saturation region. In the subthreshold region, it is reasonable to assume that $N \ll N^*$ and $N_t^*(E_{fn}) = A + BN + CN^2 \approx A$. Thus the flicker noise power in the subthreshold region is simplified to [6]

$$S_{I_{DS}}(f) = \frac{AkT I_{DS}^2}{WL\gamma f N^{*2}} \quad (7.163e)$$

7.8.3 Simulation Results and Discussion

This section presents simulation results of drain current noise spectra and input referred noise voltage of n-channel MOS transistors and p-channel MOS transistors using HSPICE, utilizing 65-nm PTM technology. The

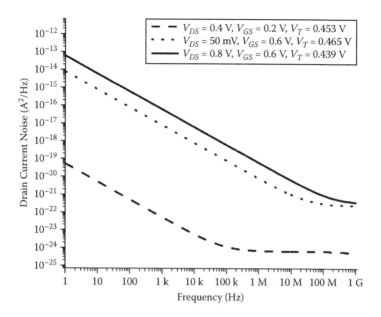

FIGURE 7.35
Drain current noise power spectrum of an n-channel MOS transistor, operating at three different regions of operations.

channel length and width of the transistor in all cases are taken to be 65 nm and 10 μm, respectively. The model selector flags are *fnoimod* = 1 and *tnoimod* = 1. Figure 7.35 shows a typical drain current noise spectrum measured in three different regions of operations for an n-channel MOS transistor. It is observed that noise spectrum shows $1/f^k$ dependency with the exponential factor k close to unity. This is consistent with the assumption regarding the uniform spatial distribution of the oxide traps near the interface. It is observed that in the weak inversion region, the drain current noise of the transistor is lower compared to that in the strong inversion region. This is explained by the fact that noise power spectrum is directly proportional to the drain current, and in the weak inversion region the drain current is very small. The measured drain current noise power at 100 Hz is plotted as a function of gate bias for three different drain biases in Figure 7.36(a). The bias dependence of the input referred noise power is plotted in Figure 7.36(b). At the measured frequency, the thermal noise is negligible compared with the flicker noise. It is observed that the dependence of input referred noise power on the bias point is not significant in both linear and saturation regions. The short channel behavior and DIBL effects are also reflected in the noise power spectrum. The corresponding simulation results for a p-channel MOS transistor are shown in Figures 7.37 and 7.38(a),(b). It is observed that the p-channel transistor has a noise level lower than the n-channel MOS

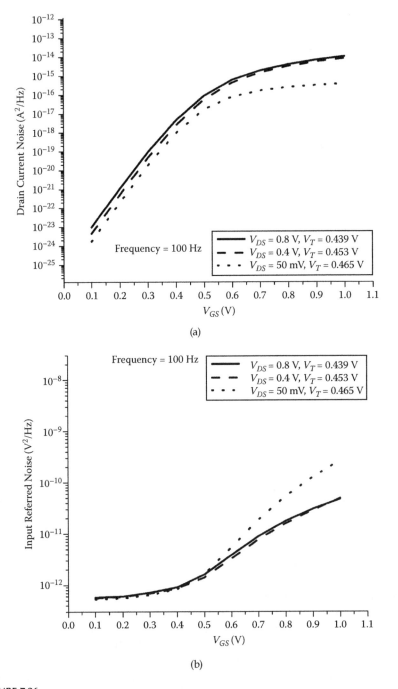

FIGURE 7.36

(a) Bias dependence of drain current noise power of an n-channel MOS transistor. (b) Bias dependence of input referred noise power of an n-channel MOS transistor.

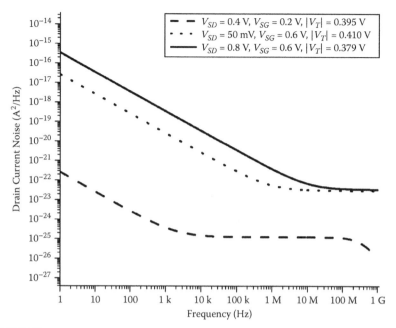

FIGURE 7.37
Drain current noise power spectrum of a p-channel MOS transistor, operating at three differ-ent regions of operations.

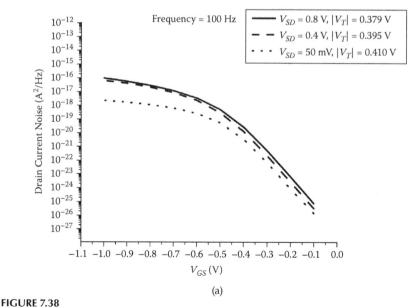

(a)

FIGURE 7.38
(a) Bias dependence of drain current noise power of a p-channel MOS transistor. (b) Variation of input referred noise spectrum for PMOS transistor operating in the subthreshold region. (*continued*)

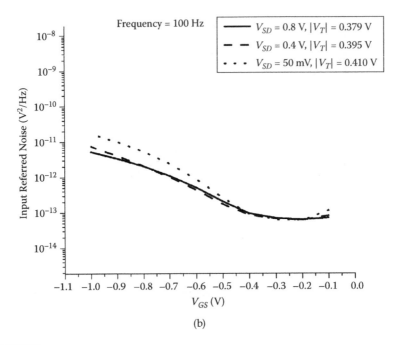

FIGURE 7.38

(*continued*) (a) Bias dependence of drain current noise power of a p-channel MOS transistor. (b) Variation of input referred noise spectrum for PMOS transistor operating in the subthreshold region.

transistor by one or two orders of magnitude. This is because of different oxide trap density near the conduction and valence band edges, different tunneling barriers for the electrons and holes, and different electron and hole mobilities resulting in different degrees of mobility fluctuations.

7.9 Statistical Characterization

With the scaling of MOS transistors to sub-90 nm regime, the effects of statistical variations of process parameters on the performances of VLSI circuits have become critical. The increasing impacts of the within-die variability on the performances of VLSI circuits have posed significant challenges to the conventional VLSI design methodologies. The commercially available computer aided design tools are used to determine the nominal design parameters of a circuit, such that the nominal response of the circuit meets the desired performance specifications. However, after fabrication, the actual circuit response always shows deviations from the nominal value due to process variations. Therefore, a paradigm shift to the conventional deterministic design methodology is required. Statistical design methodology has become indispensable for the present VLSI circuits. Precise characterization of process variability is essential for variability-aware statistical circuit design.

7.9.1 Classification of Process Variability

From a circuit design perspective, the process variations are classified into two types: intra-die and inter-die variations. The intra-die variations are defined as the parametric changes of identical MOS transistors across a short distance, while the inter-die variations refer to such changes for identical MOS transistors separated by longer distance or fabricated at different times [24]. Thus intra-die variations are deviations occurring within a die. On the other hand, the inter-die variations are deviations occurring from die-to-die, wafer-to-wafer, and lot-to-lot. The wafer-to-wafer variations are caused usually by some change in machine conditions along time of manufacturing apparatus. The die-level variations typically originate from lithography steps, because pattern exposure is performed die-to-die. Imperfections in reticles or non-uniformity in the lens system sometimes causes die-level variations. The intra-die variations are significant in sub-90 nm technology and are of critical concern to the IC designers. The intra-die variations have two components: systematic and random. The systematic variations include the variations caused due to optical proximity corrections, phase-sifting mask and layout-induced strain, and well proximity effect [24]. On the other hand, the random variations include variations due to random discrete dopant (RDD), line edge roughness (LER), line width roughness (LWR), oxide thickness variation (OTV), poly-silicon gate/metal gate granularity, and interface roughness. In this chapter, we restricted ourselves to the local component of intra-die process variations, which is of serious concern in the nanoscale regime.

7.9.2 Sources of Random Intra-Die Process Variations and Their Effects

In this subsection, the major sources of process variations are identified and their effects on the device performances are discussed.

7.9.2.1 Random Discrete Dopant (RDD)

With the scaling of the transistor in sub-90 nm technology, it has been found that the number of dopant atoms within the channel of a transistor becomes discrete and is a statistical quantity. In 1 μm technology, the number of dopant atoms in the channel is near about 5000, whereas that in a 32-nm technology node is less than 100. The random fluctuations of the number and position of the dopant atoms in the channel of a MOS transistor cause device-to-device variations of electrical performances which is referred to as the random discrete dopant effect [25].

RDD is considered to be the major contributor to performance mismatch of identical MOS transistors placed very close to each other. Considering uniform channel doping, the effect of RDD on threshold voltage fluctuation for a large geometry MOS transistor is given by [25]

$$\sigma_{V_T,RDD} = \frac{q}{C_{ox}}\sqrt{\frac{N_A W_{dm}}{3LW}}$$ (7.164)

In (7.164), W_{dm} is the depletion depth. For non-uniform doping, N_A in (7.164) is to be replaced by N_{EFF}, which is calculated as follows [26]:

$$N_{EFF} = 3\int_0^{W_{dm}} N(x)\left(1-\frac{x}{W_{dm}}\right)^2 \frac{dx}{W_{dm}}$$ (7.165)

In (7.165), $N(x)$ is the charge density along depth. It is observed from (7.164) that with the scaling of CMOS technology, the device area LW decreases, so that the threshold voltage variability caused due to RDD increases. However, RDD decreases with scaling of oxide thickness. It is found that RDD is a major contributor, over 60%, to the threshold voltage mismatch.

7.9.2.2 Line Edge Roughness (LER)

Line edge roughness is the second most important source of process variability. The cause of this phenomenon is statistical variation in the incident photon count during lithographic exposure and the absorption rate and molecular composition of the photo resist [27]. As a result, roughness occurs along the gate of the MOS transistor, causing variations in gate length along the width of the transistor. This is schematically shown in Figure 7.39. Experimentally it has been demonstrated that the LER is of the order of 4 nm and does not scale down with device scaling. This is because of the fact that the lithographic gap, defined as the difference between the wavelength of the light used for patterning and the minimum feature size is increasing until extreme ultraviolet technology is available [28]. LER is considered to be a dominant source of process variability for short channel length devices beyond 45 nm technology node, where the amount of LER becomes a significant fraction compared to the channel length. Because the LER affects the channel length of the transistor, it leads to threshold voltage variations and variations of all other performances that depend on the channel length. Moreover, LER and RDD are statistically independent. The threshold voltage mismatch due to LER depends on the variations of the width of MOS transistors and is given by [28,29]

$$\sigma_{V_T,LER} \propto \frac{1}{W} < \sigma V_{T,RDD}$$ (7.166)

$$\sigma_{V_T,total} = \sqrt{(\sigma_{V_T,RDD})^2 + (\sigma_{V_T,LER})^2}$$ (7.167)

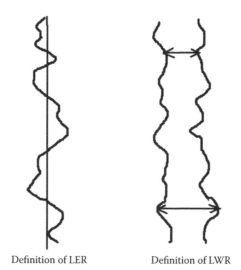

Definition of LER   Definition of LWR

FIGURE 7.39
Microscopic view of channel length along width illustrating LER and LWR.

7.9.2.3 Oxide Thickness Variation (OTV)

Another important source of random intra-die process variability is atomic-scale oxide thickness variation. With the downscaling of the physical gate oxide thickness up to 1 nm, it becomes equivalent to approximately five inter-atom spacing. Through experimentation, it has been observed that the oxide thickness roughly varies by one or two atomic spacing [30]. The oxide thickness variation occurs primarily due to interface roughness. The oxide thickness variations also lead to variations of all performances related to oxide thickness either implicitly or explicitly. The threshold voltage variations due to oxide thickness are also statistically independent of RDD and LER, so that (7.167) is transformed to

$$\sigma_{V_{T,total}} = \sqrt{\left(\sigma_{V_{T,RDD}}\right)^2 + \left(\sigma_{V_{T,LER}}\right)^2 + \left(\sigma_{V_{T,OTV}}\right)^2} \qquad (7.168)$$

7.9.3 Characterization of Process Variability

Accurate characterization of process variability is a challenging task. The present subsection highlights the conventional approach and briefly discusses the new approaches. The development of an accurate characterization procedure is an important research topic.

7.9.3.1 Design Corner Approach

From an IC designer's point of view, the collective effects of process variations are lumped into their effects on the performances of a circuit. These define the

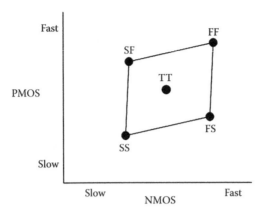

FIGURE 7.40
Design corners.

design or process corners. The term *corner* refers to an imaginary box that surrounds the guaranteed performance of the circuits, as shown in Figure 7.40. The corners for analog applications are slow NMOS and slow PMOS (SS) to characterize the worst-case speed and fast NMOS and fast PMOS (FF) to characterize the worst-case power. The corners for digital applications are fast NMOS and slow PMOS (FS) to characterize the worst-case logic 1 and slow NMOS and fast PMOS (SF) to characterize the worst-case logic 0. The typical (TT) case characterizes the nominal design of the transistors. The corner parameters are generated by deviating the selected process-sensitive SPICE model parameters by a fixed number n of standard deviation σ. For example, an arbitrary SPICE parameter s_i of the typical model file is represented through

$$s_i = s_i^0 \pm n\sigma \tag{7.169}$$

In (7.169), n is selected to set the fixed lower and upper limits of the worst-case models. The direction of the deviation from the mean/typical value s_i^0 depends on whether increasing or decreasing the parameter makes the performance worse. This is usually determined by the sensitivity analysis, by computing the derivative of the performance with respect to the chosen SPICE parameter, and by considering the sign of the derivative.

The advantage of this design corner approach is that the corner models are supplied to the designers so that the circuit can be simulated at each of the process corners for statistical characterization of the effects of process variabilities on circuit performances. However, this approach has two serious limitations. First, it has the significant risk of over- or underestimation of the process variations and their impact on the design. Overestimation makes the task of designing the circuits difficult such that the performances meet the specifications at all the corners. On the other hand, underestimation may lead to manufacturability problems and eventual loss in yield. The second problem is that while generating the corner parameters, the correlations between the device parameters are ignored. This approach therefore does not provide adequate information about the robustness of the design.

7.9.3.2 Monte Carlo Simulation Approach

The Monte Carlo simulation technique is a stochastic technique widely used for statistical characterization of the performance parameter variations due to process variability. The Monte Carlo approach allows direct estimation of the yield of a VLSI circuit. In this technique, the crucial SPICE parameters are sampled from a pre-defined statistical distribution conforming to the process specification. This forms a large database of process-related SPICE parameters, and for each sample of the database, the performance parameters of the circuit are simulated through SPICE simulation. The statistical distributions of the performance parameters corresponding to each sample of the process database are estimated by determining the mean and the standard deviation. The method is very general and accurate for statistical characterization. The problem with the Monte Carlo-based approach is that hundreds of simulation runs have to be performed and depending upon the complexity of the circuit, the entire procedure may take several hours. However, some strategies are available to reduce the sample size, such as variance reduction techniques, stratified sampling, etc.

7.9.3.3 Statistical Corner Approach

The basic idea of the statistical corner model approach is to make the design corner approach more realistic by adding a realistic value of the standard deviation of the corresponding model parameter to its nominal value following (7.158). The value of each σ is obtained from the distribution of a large set of production data. The production data are actual measurement data collected over multiple dies, wafers, and lots. In the absence of actual measurement data, which are fairly common for new process technology, the production data may consist of calibrated TCAD simulation results. The electrical test data, whatever the collection procedure, are mapped to the appropriate SPICE parameters either directly or through some extraction procedure [31]. A statistical corner-based approach is thus more realistic compared to the conventional design corner approach discussed earlier and is faster compared to the Monte Carlo approach.

7.9.4 Simulation Results and Discussion

This section presents SPICE simulation results for statistical characterization of three important performance parameters of an n-channel MOS transistor.

These are (1) threshold voltage V_T, (2) OFF current I_{OFF}, and (3) subthreshold slope S. The intra-die variations studied are random discrete dopants, line edge roughness, and oxide thickness variations. The Monte Carlo simulation technique has been utilized using 45-nm PTM model file.

7.9.4.1 Statistical Characterization of RDD

The effects of RDD on device performances are studied through SPICE simulation by varying the threshold voltage SPICE parameter *VTHO*. The reason behind such a selection is that with the variation of dopant number and hence the doping concentration within the channel, the long channel threshold voltage is primarily affected. In order to characterize the amount of variation of this parameter, there are two approaches. The first is to extract from TCAD simulation results, and the second is to calculate it from (7.164). Because the present work attempts to provide only the philosophy of the characterization procedure, the latter approach is performed, although the first one is preferred for accurate characterization.

The effective channel length and width of the chosen MOS transistor are 37.5 nm and 120 nm, respectively. The substrate doping concentration is 3.24E10/cm³, and the electrical oxide thickness is 1.75 nm. Substituting the necessary values, $\sigma_{V_{T,RDD}}$ is calculated from (7.158) and is found to be 17.71 mV. The same amount of variation is taken for the SPICE parameter *VTHO*. For Monte Carlo simulation, a set of 1000 samples has been chosen. The distribution of the SPICE parameter *VTHO* is considered to be Gaussian. The simulation is performed both at low drain bias and high drain bias—that is, $V_{DS} = 50mV$ and $V_{DS} = 1V$.

The variations of the gate characteristics of the transistor due to RDD at low and high drain bias are shown in Figures 7.41(a) and 7.41(b), respectively. The performance samples are characterized by four measures: mean, standard deviation, skew factor, and kurtosis. These are summarized in Tables 7.4 and 7.5 for low drain bias and high drain bias, respectively. The distributions of the samples for the high drain bias case are shown in Figures 7.42(a) through 7.42(c). The distribution of threshold voltage is Gaussian, whereas that for I_{OFF} and S are log-normal.

7.9.4.2 Statistical Characterization of Line Edge Roughness (LER)

Line edge roughness is the distortion of the gate edge. In order to characterize the distortion of the gate edge, a simplified model of a rough line as shown in Figure 7.43 is considered. The roughness in the gate edge is characterized by high-frequency roughness and low-frequency roughness. The gate is divided into segments with characteristic width W_C, which characterizes the change at which the low frequency part changes the gate length. Within this portion, only high-frequency roughness is present. Assuming

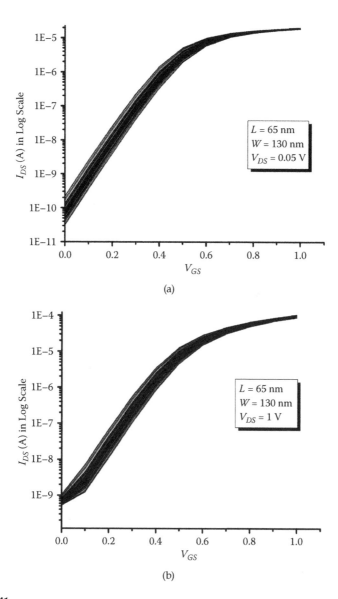

FIGURE 7.41
(a) Effect of RDD process variations on the gate characteristics of the MOS transistor at $V_{DS} = 50$ mV. (b) Effect of RDD process variations on the gate characteristics of the MOS transistor at $V_{DS} = 1$ V.

TABLE 7.4

Summary of RDD, LER, and OTV on Subthreshold Slope, OFF Current, and Threshold Voltage at V_{DS} = 50 mV

Statistical Parameters	RDD			LER			OTV			All		
	S	I_{OFF}	V_T	S	I_{OFF}	V_T	S	I_{OFF}	V_T	S	I_{OFF}	V_T
STDEV (σ)	0.568 mV/ decade	38.8 pA	18.45 mV	4.992e-3 mV/ decade	1.89 pA	1.42 mV	0.749 mV/ decade	4.888 pA	0.743 mV	0.932 mV/ decade	39.11 pA	18.52 mV
MEAN (m)	96.45 mV/ decade	84.7 pA	0.464 V	96.424 mV/ decade	78.40 pA	0.463 V	96.39 mV/ decade	78.35 pA	0.463 V	96.45 mV/ decade	84.97 pA	0.462 V
SKEW	0.546578	1.255	-0.009	0.246	0.135	-0.026	-0.029	0.107	-0.045	0.918	1.060	-0.012
KURT	0.225871	2.004	-0.177	-1.879	-0.296	-0.302	-0.166	-0.135	-0.182	0.802	1.464	-0.142

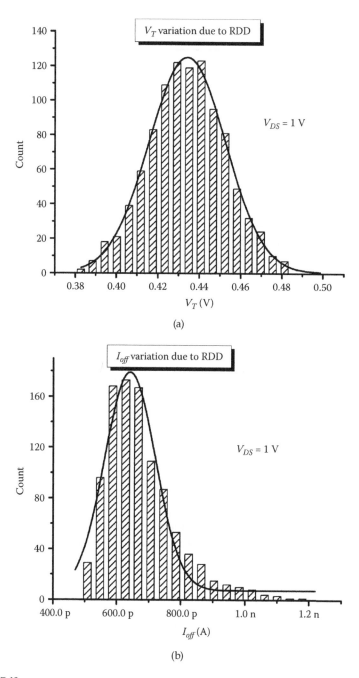

(a)

(b)

FIGURE 7.42
(a) Statistical distributions of threshold voltage variations occurring due to RDD at high drain bias. (b) Statistical distributions of I_{OFF} variations occurring due to RDD at high drain bias. (c) Statistical distributions of S variations occurring due to RDD at high drain bias. (*continued*)

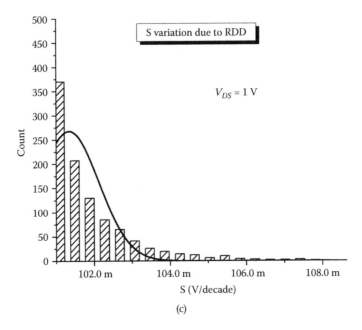

(c)

FIGURE 7.42
(*continued*) (a) Statistical distributions of threshold voltage variations occurring due to RDD at high drain bias. (b) Statistical distributions of I_{OFF} variations occurring due to RDD at high drain bias. (c) Statistical distributions of S variations occurring due to RDD at high drain bias.

fluctuations of two gate edges are uncorrelated, random variation of channel length due to LER is calculated as follows [29]:

$$\sigma L = \sqrt{\frac{2}{1 + W/W_C}} \sigma_{LER} \qquad (7.170)$$

The edge locations of two different segments are uncorrelated and have a standard deviation σ_{LER}. In the present work, effective channel width $W = 120\ nm$,

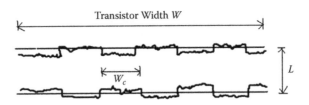

FIGURE 7.43
Simplified model for estimating LER of the gate.

$W_C = 30\,nm$, and $3\sigma_{LER} = 4\,nm$. Substituting these, in (7.170), $\sigma_L = 0.843\,nm$. In BSIM, the effective channel length is defined in simple form as

$$L_{eff} = L_{drawn} + XL - 2LINT \qquad (7.171)$$

In (7.171), XL is the channel length offset due to mask/etch effect, and $LINT$ is the channel length offset parameter. In the present work, $XL = -20\,nm$ and $LINT = 3.75\,nm$. The effects of LER on device performance are simulated in HSPICE Monte Carlo analysis by varying the value of the parameter XL. For Monte Carlo simulation, a set of 1000 samples has been chosen. The distribution of the SPICE parameter XL is considered to be Gaussian. The simulation is performed both at low drain bias and high drain bias (i.e., $V_{DS} = 50\,mV$ and $V_{DS} = 1V$). The effects of LER on the chosen device performances are summarized in Tables 7.4 and 7.5, respectively. The effect of LER on sub-threshold slope is not significant due to lack of any direct functional relationship between the two. However, at high drain bias, the depletion width changes due to DIBL effect so that fluctuations in subthreshold slope are observed. The distributions of the samples for the high drain bias case are shown in Figures 7.44(a) through 7.44(c).

7.9.4.3 Statistical Characterization of OTV

The oxide thickness variation is induced by atom-level interface roughness between silicon and gate dielectric. The minimum magnitude of oxide thickness variation is the height of one silicon atom layer, which is 2.71 A^0. The effects of OTV on the chosen device performances are summarized in Tables 7.4 and 7.5, respectively. The distributions of the samples for the high drain bias case are shown in Figures 7.45(a) through 7.45(c).

7.9.4.4 Statistical Characterization of Simultaneous Variations

In real devices, the various sources of process variations simultaneously affect the device and circuit performances. This can also be simulated in HSPICE. The cumulative effects of RDD, LER, and OTV on the chosen device performances are summarized in Tables 7.4 and 7.5. It is observed from the simulation results that (7.168) is valid. The relative contributions of the different process variations on the device performances are shown in Figures 7.46(a) and 7.46(b). It is observed that in all cases, RDD is a dominant source of process variations. Therefore, mitigation of RDD is an important challenge for the device designers for advancement of nanoscale VLSI circuits. In addition, proper characterization of the amount of process variability (RDD, LER, OTV) is also extremely important, which is not an easy task. An elegant approach for this is to use backward propagation of the variance method through which these are

TABLE 7.5

Summary of RDD, LER, and OTV on Subthreshold Slope, OFF Current, and Threshold Voltage at $V_{DS} = 1$ V

Statistical Parameters	RDD			LER			OTV			All		
	S	I_{OFF}	V_T	S	I_{OFF}	V_T	S	I_{OFF}	V_T	S	I_{OFF}	V_T
STDEV σ	1.122 mV/decade	109.2 pA	18.27 mV	0.165 mV/decade	14.1 pA	2.904 mV	0.68 mV/decade	125.2 pA	0.061 mV	1.45 mV/decade	167.58 pA	18.48 mV
MEAN m	101.9 mV/decade	676.1 pA	0.434 V	101.434 mV/decade	659 pA	0.433 V	101.7 mV/decade	684.6 pA	0.433 V	101.48 mV/decade	702.30 pA	0.432 V
SKEW	2.123	1.249	-0.010	0.116	0.298	-0.111	1.998	0.992	-0.197	0.888	0.932	-0.004
KURT	5.339	1.979	-0.176	-0.341	-0.198	-0.215	4.846	1.152	0.243	0.978	1.098	-0.117

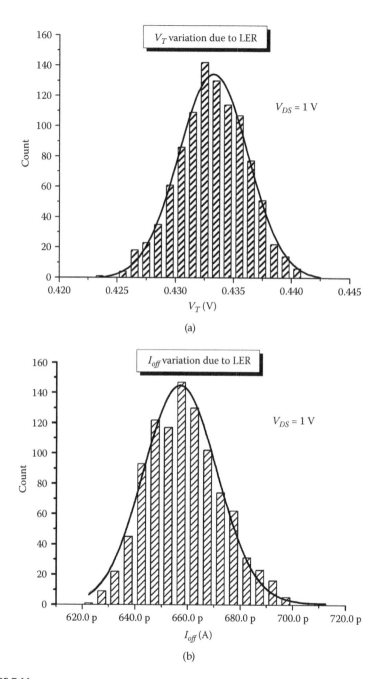

FIGURE 7.44

(a) Statistical distributions of threshold voltage variations occurring due to LER at high drain bias. (b) Statistical distributions of I_{OFF} variations occurring due to LER at high drain bias. (c) Statistical distributions of S variations occurring due to LER at high drain bias. (*continued*)

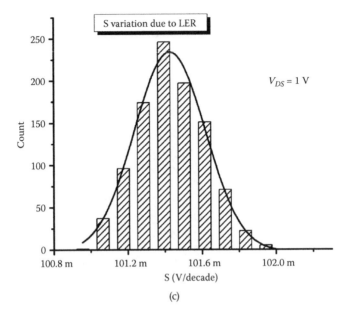

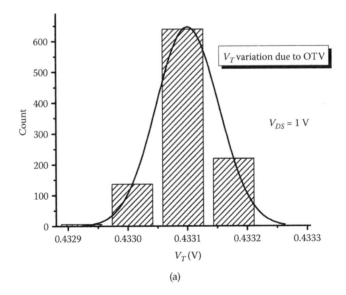

FIGURE 7.44

(*continued*) (a) Statistical distributions of threshold voltage variations occurring due to LER at high drain bias. (b) Statistical distributions of I_{OFF} variations occurring due to LER at high drain bias. (c) Statistical distributions of S variations occurring due to LER at high drain bias.

FIGURE 7.45

(a) Statistical distributions of threshold voltage variations occurring due to OTV at high drain bias. (b) Statistical distributions of I_{OFF} variations occurring due to OTV at high drain bias. (c) Statistical distributions of S variations occurring due to OTV at high drain bias. (*continued*)

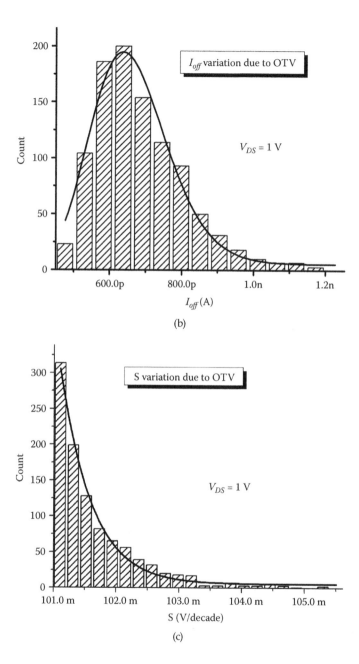

FIGURE 7.45
(*continued*) (a) Statistical distributions of threshold voltage variations occurring due to OTV at high drain bias. (b) Statistical distributions of I_{OFF} variations occurring due to OTV at high drain bias. (c) Statistical distributions of S variations occurring due to OTV at high drain bias.

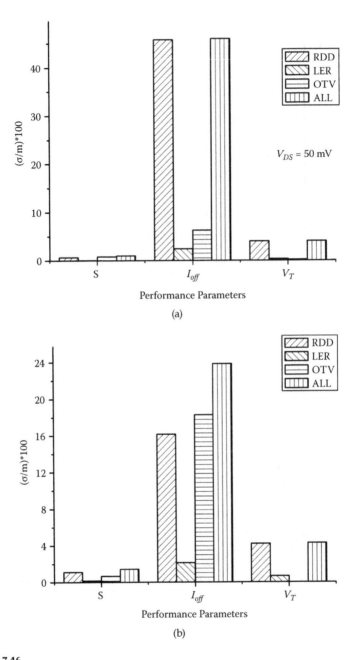

FIGURE 7.46
(a) Contributions of different sources of process variabilities on the performance parameters at $V_{DS} = 50$ mV. (b) Contributions of different sources of process variabilities on the performance parameters at $V_{DS} = 1$ V.

estimated from the effects of these process variabilities on circuit performances [32].

7.10 Summary and Conclusion

This chapter presents a comprehensive overview about characterizing a MOS transistor to be used in VLSI circuit simulation. The issues discussed along with the approaches mentioned are considered by the compact models used in commercial circuit simulation packages. However, the objective is to make the designers aware of the various issues related to the present-day VLSI MOS transistors, such that these are taken care of by the IC designers while designing and optimizing any VLSI circuit. Moreover, with better control over the physics of the circuit operations, the design procedure becomes more perfect and the design effort and time reduce drastically. In the sub-90 nm design domain, several challenges related to circuit performances can be solved at the device design level. This offers an additional flexibility to the designers for designing an optimal circuit without adding any extra circuit components, thus making the circuit area and power efficient. Technology-aware circuit design and device-circuit co-design are important areas of research in nanoscale VLSI circuit design, as presented here in a comprehensive manner.

References

1. K.S. Kundert, *The Designer's Guide to SPICE and SPECTRE®*, Kluwer Academic, Dordrecht, 2003.
2. R. Rohrer, Growing SPICE, *IEEE Solid State Circuit Mag.*, vol. 3, no. 2, pp. 30–35, June 2011.
3. A.B. Bhattacharya, *Compact MOSFET Models for VLSI Design*, IEEE Press, Wiley, Singapore, 2009.
4. W. Zhao and Y. Cao, New generation of predictive technology model for sub-45 nm early design exploration, *IEEE Trans. Electron Devices*, vol. 53, no. 11, pp. 2816–2823, November 2006.
5. Y. Taur and T.H. Ning, *Fundamentals of Modern VLSI Devices*, Cambridge University Press, United Kingdom, 2008.
6. Y. Cheng and C. Hu, *MOSFET Modeling and BSIM3 User's Guide*, Kluwer Academic, Dordrecht, 2002.
7. Y. Tsividis, *Operation and Modeling of the MOS Transistor*, 2nd ed., Oxford University Press, Oxford, 1999.

8. M.V. Dunga, X. Xi, J. He, W. Liu, K.M. Cao, X. Jin, J.J. Ou, M. Chan, A.M. Niknejad, and C. Hu, BSIM4.6.0 MOSFET model, 2006. Available at http://www-device. eecs.berkeley.edu/bsim/Files/BSIM4/BSIM460/doc/BSIM460_Manual.pdf.

9. Z.H. Liu, C. Hu, J.H. Huang, T.Y. Chan, M.C. Jeng, P.K. Ko, and Y.C. Cheng, Threshold voltage model for deep-submicrometer MOSFETs, *IEEE Trans. Electron Devices*, vol. 40, pp. 86–95, January 1993.

10. J.R. Brews, W. Fichtner, E.H. Nicollian, and S.M. Sze, Generalized guide for MOSFET miniaturization, *IEEE Electron Device Lett.*, vol. EDL-1, no. 1, pp. 2–4, January 1980.

11. A. Ortiz-Conde, F.J.G. Sanchez, J.J. Liou, A. Cerdeira, M. Estrada, and Y. Yue, A review of recent MOSFET threshold voltage extraction methods, *Microelectron. Reliability*, vol. 42, pp. 583–596, 2002.

12. R.M. Swanson and J.D. Meindl, Ion implanted complementary MOS transistors in low voltage circuits, *IEEE J. Solid State Circuits*, vol. SC-7, no. 2, pp. 146–193, April 1972.

13. Y.C. Yeo, Q. Lu, P. Ranade, H. Takeuchi, K.J. Yang, I. Polishchuk, T.-J. King, C. Hu, S.C. Song, H.F. Luan, and D.-L. Kwong, Dual-metal gate CMOS technology with ultrathin silicon nitride gate dielectric, *IEEE Electron Device Lett.*, vol. 22, no. 4, pp. 227–229, May 2001.

14. P.H. Woerlee, M.J. Knitel, R. Langevelde, D.B.M. Klaassen, L.F. Tiemeijer, A.J. Scholten, and A.T.A. Zegers-van Duijnhoven, RF-CMOS performance trends, *IEEE Trans. Electron Devices*, vol. 48, no. 8, pp. 1776–1782, August 2001.

15. J. E. Meyer, MOS models and circuit simulation, *RCA Rev.*, vol. 32, pp. 42–63, 1971.

16. M.A. Cirit, The Meyer model revisited: Why is charge not conserved?, *IEEE Tran. Computer Aided Design*, vol. 8, no. 10, pp. 1033–1037, October 1989.

17. D.E. Ward and R.W. Dutton, A charge-oriented model for MOS transistor capacitances, *IEEE J. Solid State Circuits*, vol. 13, no. 5, pp. 703–708, October 1978.

18. R. Shrivastava and K. Fitzpatrick, A simple model for the overlap capacitance of a VLSI MOS device, *IEEE Trans. Electron Devices*, vol. ED-29, no. 12, pp. 1870–1875, December 1982.

19. F.M. Klassen and J. Prins, Thermal noise of MOS transistors, *Phillips Research Report*, vol. 22, pp. 505–514, 1967.

20. M. Haartman and M. Ostling, *Low Frequency Noise in Advanced MOS Devices*, Springer, New York, 2007.

21. K.K. Hung, P.K. Ko, C. Hu, and Y.C. Cheng, A unified model for the flicker noise in metal-oxide-semiconductor field-effect transistors, *IEEE Trans. Electron Devices*, vol. 37, no. 3, pp. 654–665, March 1990.

22. K.K. Hung, P.K. Ko, C. Hu, and Y.C. Cheng, A physics based MOSFET noise model for circuit simulators, *IEEE Trans. Electron Devices*, vol. 37, no. 5, pp. 1323–1333, May 1990.

23. E.P. Vandamme and L.K.J. Vandamme, Critical discussion on unified 1/f noise models for MOSFETs, *IEEE Trans. Electron Devices*, vol. 47, no. 11, pp. 2146–2152, November 2000.

24. S.K. Saha, Modeling process variability in scaled CMOS technology, *IEEE Design and Test of Computers*, vol. 27, no. 2, pp. 8–16, March–April 2010.

25. P.A. Stolk, F.P. Widdershoven, and D.B.M. Klassen, Modeling statistical dopant fluctuations in MOS transistors, *IEEE Trans. Electron Devices*, vol. 45, no. 9, pp. 1960–1971, September 1998.

26. K. Takeuchi, T. Fukai, T. Tsunomura, A.T. Putra, A. Nishida, S. Kamohara, and T. Hiramoto, Understanding random threshold voltage fluctuation by comparing multiple fabs and technologies, *Proc. IEDM 2007*, Technical Digest IEEE Electron Device Meeting, pp. 467–470.

27. K. Bernstein, D.J. Frank, A.E. Gattiker, W. Haensch, B.L. Ji, S.R. Nassif, E.J. Nowak, D.J. Pearson, and N.J. Rohrer, High performance CMOS variability in the 65-nm regime and beyond, *IBM J. Res. Dev.*, vol. 50, no. 4/5, July/September 2006.

28. C. Millar, S. Roy, and A. Asenov, Understanding LER-induced MOSFET V_T variability—Part I: Three-dimensional simulation of large statistical samples, *IEEE Trans. Electron Devices*, vol. 57, no. 11, pp. 2801–2807, November 2010.

29. J.A. Croon, G. Storms, S. Winkelmeier, et al., Line edge roughness: Characterization, modeling and impact on device behavior, *Proceedings of IEDM 2002*, Technical Digest IEEE Electron Device Meeting, pp. 307–310.

30. A. Asenov, S. Kaya, and J.H. Davies, Intrinsic threshold voltage fluctuations in decanano MOSFETs due to local oxide thickness variations, *IEEE Transactions on Electron Devices*, vol. 49, no. 1, pp. 112–119, January 2002.

31. J.C. Chen, C. Hu, C.-P. Wan, P. Bendix, and A. Kapoor, E-T based statistical modeling and compact statistical circuit simulation methodologies, *Proceedings of IEDM*, Technical Digest IEEE Electron Device Meeting, 1996.

32. C.C. McAndrew, I. Stevanovic, X. Li, and G. Gildenblat, Extensions to backward propagation of variance for statistical modeling, *IEEE Design and Test of Computers*, vol. 27, no. 2, pp. 36–43, March–April 2010.

8

Process Simulation of a MOSFET Using TSUPREM-4 and Medici

Atanu Kundu

CONTENTS

8.1 Introduction

The objective of this chapter is to fabricate a 5 μm 2D n-MOSFET (n-type metal-oxide-semiconductor field-effect transistor) using process simulator TSUPREM-4 [1] and device simulator Medici [2]. SUPREM is the acronym of Stanford University Process Engineering Modeling. Taurus TSUPREM-4 is for the version IV, which is a 2D simulation program. TSUPREM-4 is a computer program for the simulation of the fabrication steps required for the manufacture of silicon integrated circuits and for other integrated circuits (ICs). TSUPREM-4 simulates the changes in semiconductor structure which take place after various processing steps used during the actual fabrication procedure.

As the device dimensions have been reduced to micro or nano level, the specialization and application of technology computer aided design (TCAD) tools in new device creation for future technology generations are indispensable to harness the ever-increasing complexity and challenges of the "ever-shrinking transistors." One of the main advantages of TCAD tools is visualization. For deep sub-micron devices, it is possible to visualize the evolution of the actual cross-sections of the structure during various process simulation steps in order to obtain better insight into the IC processing steps. TSUPREM-4, a popular commercial TCAD process simulator tool, allows verifying the entire structure after every realistic silicon wafer processing step via hands-on simulation, without the need for high-cost IC processing facilities. Moreover, these TCAD tools after calibration exhibit impressive predictive power with required accuracy, which can be utilized to speed up the technology integration and transfer to volume manufacturing. Therefore it is possible to experiment and explore the impact of process flow modifications at virtually no cost. This results in the possibility of manufacturing high-yield profitable product with short product development life cycles, which is absolutely necessary given the huge costs of nanoscale integrated circuit fabrication lines.

8.2 Why Silicon?

- Silicon can be easily oxidized to form high-quality silicon-dioxide (SiO_2) insulator, which is used as a masking or barrier material for selective doping steps required for IC fabrication.

- The SiO_2 layer is essential in metal-oxide-semiconductor (MOS) device structure, and high-quality Si-SiO_2 interface formation is possible to form the gate of MOSFET.

- Silicon also has wider band-gap than germanium, which means that silicon devices can operate at higher temperatures than their germanium counterparts.

- Silicon is available abundantly in nature as its primary constituent is ordinary sand. So, silicon provides a very low cost source of semiconductor device or IC fabrication material.

Actual industrial device fabrication consists of several steps that have been followed in fabrication procedure. The process starts with initializing a <100> silicon wafer of 5 μm, and it requires proper meshing of the device.

During the selective doping procedure, several portions of the wafer need to be masked or covered during various steps of the fabrication. By convention, the extension .tl1 is used for the mask layout files used by TSUPREM-4. This mask file named '*t.tl1*' has been used here as an input file for the entire device fabrication where nine different mask names have been assigned for different fabrication steps.

To run any program in TSUPREM-4, the linux environment is required and one has to type '*TSUPREM4*' followed by the filename having *.inp* extension in the terminal. The file in which the script is written is named MOSFET. inp. To run this script file, the command would be '*TSUPREM4 MOSFET.inp*'. With this script file another file is essential to execute the program: the mask file. The mask file is of extension .tl1. Here the mask file name is t.tl1, which contains the name of the masks with their length. Mask file has to be linked with the MOSFET.inp file as an input file in the beginning of its script file by the command *MASK IN.FILE = t.tl1*. Now it is possible to call any of its mask names when required. Here masks used in the t.tl1 file are of names *gateoxet, nbl, Nplus, contact, metal1, metal2, metal3*, and *gateunderdoping*. All lengths are by default in micrometers. The first statement of the mask file 1e3 or 1000 signifies the length mentioned here divided by 1e3 to convert it into micrometers. For example, in the mask named *gateoxet*, only one length is mentioned here from (1600–4100); that is why '1' is mentioned after the mask name, like *gateoxet 1*. (1600–4100) μm signifies that it is the length 1.6 to 4.1 μm of the 0 to 5 μm device as it is divided by 1e3. Similarly for mask name *contact*, there are three lengths, so 3 is mentioned after the mask name *contact*, like *contact 3*, and different lengths are (300–1100), (2550–2850), and (4400–4850).

8.3 Initial Meshing of the Wafer

TSUPREM-4 is used to simulate 2D structures [2]. In the TSUPREM-4 coordinate system, the distance from the surface of the wafer into the silicon is positive (y-axis). The x-axis values are numbered from left to right. The device performance is mainly dependent on *vertical* grid-spacing, and grid spacing is crucial for predictive technology simulation [3,4]. The *'mesh'* statement generates and controls the automatic simulation grids for TSUPREM-4. Meshing refinement is chosen in such a way that meshing density is very high on the top side of the wafer as the device structure will be grown there, so that carrier flow and any other changes occurring due to the terminal voltages will be found out accurately. The plot shown in Figure 8.1 shows the mesh generation of the 5μm wafer, where the *MASK IN.FILE = t.tl*1 statement signifies the input file name is *t* and extension is *tl*1. The output figure will be plotted with name *'Field,Poly,Contact'*. From Figure 8.1, it can be seen that a denser grid is chosen in the areas where a lot of activity and precision of information are important.

The *'grid.fac'* parameter multiplies all grid spacing specifications in the horizontal and vertical directions. By default this is set to 1 which produces fine grid required for accurate simulations. To increase simulation speed this value can be increased, but for more accurate simulation *'grid.fac'* should be

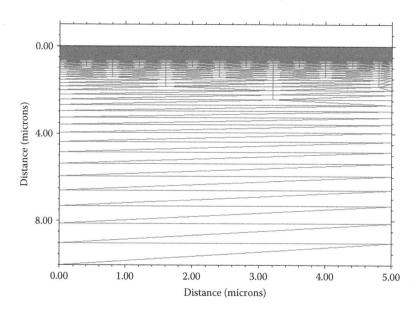

FIGURE 8.1
Initial mesh used for entire device fabrication. Plotted by PLOT.2D GRID statement.

reduced as required. The *dx.min, dx.max, dy.surf, dy.activ,* and *dy.bot* parameters on the mesh statement are multiplied by grid.fac.

Placement of the grid line in the *x* direction is controlled by the parameters '*dx.min,' dx.max'* in the '*mesh'* statement. The depth of the surface region in the vertical grid is controlled by the '*ly.surf'* parameter. The grid spacing between horizontal grid lines in the *y* direction in the surface region is controlled by '*dy.surf'* parameter. This spacing is used between *y* = 0 and *y* = '*dy.surf',* and the spacing is multiplied by '*grid.fac'* when it is used. The depth of the bottom of the active region is controlled by '*ly.activ' parameter,* and the grid spacing between horizontal lines at the bottom of the active region in the *y* direction of the active region is controlled by '*dy.activ'* parameter. The grid spacing varies geometrically between *dy.surf* at *ly.surf* and *dy.activ* at *ly.activ.* This spacing is multiplied by *grid.fac* when it is used. The depth of the bottom of the structure in the default vertical grid is controlled by the parameter '*ly.bot',* and the grid spacing *y* direction at the bottom of the structure is controlled by the '*dy.bot'* parameter. Spacing will be multiplied by '*grid.fac'* when it is used.

```
mesh grid.fac=1.0 dx.min=0.002 dx.max=0.1 ly.surf=0.06
dy.surf=0.001 +
ly.activ=0.5 dy.activ=0.02 ly.bot=10 dy.bot=1
MASK IN.FILE=t.tl1 PRINT GRID="Field,Poly,Contact"
```

8.4 Start Material Initialization

Initial material of length 5 μm of <100> Si wafer with initial dose of boron $1e^{15}$ cm^{-3} has been taken to create the device on the initial material. Usually <100> silicon material is used due to the fact that at the time of fabrication processing, <100> silicon wafer produces the lowest charges at the oxide-silicon interface [5–7]. There is a strong dependence on the built-in charge on the orientation of the silicon crystal. In the case of MOSFET, surface charge is directly related to the sign and magnitude of the threshold voltage. In case of <100> silicon material, the value of the built-in surface charge is lowest [3,8]. It also gives higher mobility in the fabricated device. The '*initialize'* statement will set up the initial structure including background doping level, crystal orientation, and resistivity of the wafer for a simulation. A structure must be initialized after meshing is done and before any processing steps.

```
initialize ratio=1.4 <100> rot.sub=0 boron=1e+15 width=5.0
```

8.5 Defining the Initial Mesh

```
SELECT TITLE="Initial Mesh"
PLOT.2D GRID C.GRID=8
```

It is clear from Figure 8.1 that meshing density is very high on the top portion of the wafer as the device structure will be grown there, so carrier flow and any other changes due to the terminal voltages will take place there.

8.6 N-Buried Layer

An N-buried layer (NBL) is implanted on this wafer which allows source voltage to be raised above the substrate voltage and to avoid leakage current toward the base which could be avoided by silicon-on-insulator (SOI) type devices. SOI structures become very unstable in high voltages as a reverse-biased drain-bulk p-n junction generates a huge amount of heat that cannot be dissipated with an insulator, and self heating becomes a serious concern in terms of device reliability issues [9–10]. Therefore, antimony of dose $1e^{15}$ cm^{-3} (which equals 1×10^{15} cm^{-3}) has been implanted followed by drive in voltage to place the NBL layer at the proper position on the wafer. This N-layer in P-type wafer will create a p-n junction that will stop high bottom leakage current flow due to high supply voltage at the drain end. As both initial wafer and buried layer are doped by boron and antimony of dose $1e^{15}$ cm^{-3}, respectively, a p-n junction of equal depletion depth in both sides will be formed. For the actual structure an epitaxial layer of 14 μm is grown on this wafer where device parameter optimization is possible due to this epitaxial layer.

8.7 Oxidation and Growth of the Initial Oxide

The step required for creating an oxide layer on a semiconductor is called *oxidation*. For oxidation of silicon, oxygen is made to react with silicon at 800 to 1200°C. For wet oxidation the presence of H_2O in the reaction is essential. Though the dry oxidation rate is slow, the quality of oxide grown in this procedure is very good. Usually, dry oxidation is done in the presence of inert gas that acts as a carrier to control oxygen. Inert gas will also ensure that any other gas cannot take part in this reaction. In TSUPREM-4 the diffusion statement is used for this purpose. For fabrication of desired device structure, several steps may be used for the oxidation. The first line of the statement of the program below signifies that the initial temperature of the furnace is 800°C which will rise and reach the final temperature 1000°C for 20 minutes. Similar steps will be followed by changing conditions.

This layer is basically grown on the wafer for masking purposes which is required for the dopant implementation for NBL layer formation. As there is no mask assigned on it before the diffusion step, the entire SiO$_2$ layer will be formed on top of the whole wafer.

```
DIFFUSION TEMPERAT=800 T.FINAL=1000 TIME=20 F.O2=0.5 F.N2=9.5
DIFFUSION TEMPERAT=1000 TIME=65 F.O2=0.5 F.N2=9.5
DIFFUSION TEMPERAT=1000 TIME=5 F.O2=9.5
DIFFUSION TEMPERAT=1000 TIME=190 F.O2=5.975 F.H2=10.4 F.HCL=0.475
DIFFUSION TEMPERAT=1000 TIME=1 F.O2=5.5 F.N2=5
DIFFUSION TEMPERAT=1000 TIME=10 F.N2=10
DIFFUSION TEMPERAT=1000 T.FINAL=800 TIME=50 F.N2=10
print layers
```

8.8 Wafer Masking for Buried Layer Implantation

As the middle portion of wafer material has been chosen for the dopant implantation for NBL layer formation, so the rest of the wafer top needs to be covered by masking material, as a negative photoresist is being used here on the wafer of thickness 1 μm. The mask used here from the mask file (t.tl1) is nbl. It will be developed and be a selective part of the photoresist, and oxide will be etched out simultaneously due to the nature of the photoresist material and the etchant.

```
DEPOSIT PHOTORESIST NEGATIVE THICKNESS=1
EXPOSE MASK=nbl
DEVELOP
etch oxide
```

The above sets of commands are used to display the device structure at any fabrication step as in Figure 8.2. These sets of commands can be used after every step of fabrication to have a look at the device structure formed at that point of time. *Plot.2D* will plot the characteristics, boundaries, junctions, and depletion edges of the two-dimensional simulated structure. The title of the paragraph will be printed along with the simulated output as mentioned here: "*Deposition of negative photoresist.*" Different colors have been assigned for different materials to display at the output. Different doping contour is being plotted by assigning different colors by *FOREACH* command. This procedure has been repeated for different dopants such as boron, phosphor,

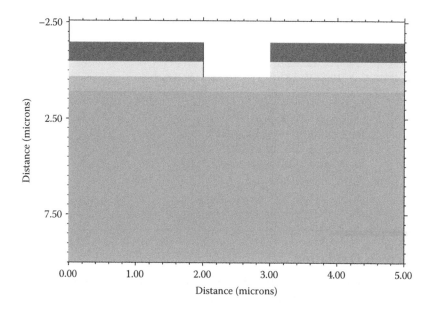

FIGURE 8.2 (See color insert)
Deposition of negative photoresist of thickness 1 μ*m* on grown oxide; a portion (2 to 3 μ*m*) of the photoresist is being etched out by using mask NBL.

arsenic, and antimony. These dopants are commonly used for any semiconductor fabrication procedure and also used for this device fabrication. After every step of fabrication, a created structure has been generated and is now being shown in the figure.

```
SELECT Z=LOG10(BORON) TITLE=" Deposition of negative photoresist "
PLOT.2D
COLOR      SILICON  COLOR=7
COLOR      OXIDE    COLOR=5
COLOR      NITRIDE COLOR=3
COLOR      PHOTORESIST    COLOR=2
COLOR      polysili COLOR=1
COLOR      aluminum COLOR=3
FOREACH X (14 TO 21 STEP 1)
COLOR MIN.V=X MAX.V=(X + 1) COLOR=(X - 1)
END
SELECT     Z=LOG10(phosphor)
FOREACH X (14 TO 21 STEP 1)
COLOR MIN.V=X MAX.V=(X + 1) COLOR=(X - 3)
END

SELECT     Z=LOG10(arsenic)
FOREACH X (19 TO 21 STEP 1)
```

```
COLOR MIN.V=X MAX.V=(X + 1) COLOR=(X - 5)
END
SELECT    Z=LOG10(antimony)
FOREACH X (14 TO 21 STEP 1)
COLOR MIN.V=X MAX.V=(X + 1) COLOR=(X - 7)
END
COLOR     OXIDE      COLOR=10
COLOR     NITRIDE    COLOR=3
COLOR     PHOTORESIST   COLOR=2
COLOR     polysili  COLOR=1
COLOR     aluminum  COLOR=3
```

8.9 Screen Oxidation

A layer of thin oxide is required to form on the wafer at the time of dopant implantation. This layer of oxide is called *screen oxide* and it protects the wafer when dopant bombardment takes place during the ion implantation procedure. Again, few oxidation steps are required in various controlled conditions.

```
DIFFUSION TEMPERAT=800 T.FINAL=900 TIME=10 F.O2=0.5 F.N2=9.5
DIFFUSION TEMPERAT=900 TIME=15 F.O2=0.5 F.N2=9.5
DIFFUSION TEMPERAT=900 TIME=5 F.O2=9.0
DIFFUSION TEMPERAT=900 TIME=5 F.O2=9.5
DIFFUSION TEMPERAT=900 TIME=28 F.O2=9.0 F.HCL=0.19
DIFFUSION TEMPERAT=900 TIME=5 F.O2=9.0
DIFFUSION TEMPERAT=900 TIME=30 F.N2=10.0
DIFFUSION TEMPERAT=900 T.FINAL=800 TIME=37.5 F.N2=10
print layers
```

8.10 Buried Layer Implantation

Due to nbl mask, which is defined as 2000 to 3000 in mask file, which effectively will be 2 to 3 μm as the rule defined in the mask file, there will be an opening of 2 to 3 μm on the wafer. The statement *implant antimony* will implant through it. As few other conditions like dopant angle of dopant implantation, dose and energy of implanted dopant by which it will be implanted into the wafer, and the tilt at which it will be implanted need to be mentioned as stated by the line here. It defines tilt as 7° and dopant dose is 1.0 e^{15} cm^{-3}, and energy of the implanted ion is 100 KeV. After implantation of the layer of screen oxide and the rest of the oxide layer on which photoresist was placed, what was used before this step is being removed by the statement *'etch oxide all'*. The photoresist was placed on oxide, so etching of oxide will automatically remove photoresist.

```
implant antimony pearson tilt=7 dose=1.0e15 energy=100
etch oxide all
```

8.11 Buried Layer Drive-In

As shown in Figure 8.3, a buried layer is implanted very close to the surface of the wafer in the region where it is bombarded. When it is needed to be placed deep inside the wafer, a drive-in voltage is required to drive these dopants through the wafer toward the positive Y axis, as shown in Figure 8.4.

Dopant drive-in operation is performed by high temperature. Due to thermal agitation, the dopants will move in the downward direction of the wafer. So again diffusion statement is required, and at the end of the process a layer

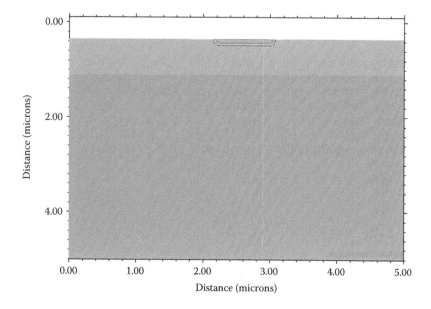

FIGURE 8.3 (See color insert)
Implantation of antimony of pearson tilt = 7, dose = 1.0 e^{15}, and energy = 100, which will be used as NBL.

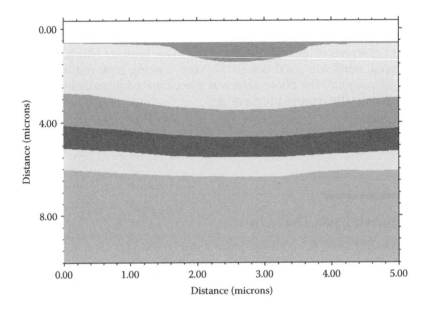

FIGURE 8.4 (See color insert)
Placement of antimony dopant in the wafer after the application of drive-in voltage on it.

of SiO$_2$ will be formed automatically by nature, which is needed to be etched out by the *etch oxide all* statement.

```
DIFFUSION TEMPERAT=800 T.FINAL=1200 TIME=100 F.O2=0.5 F.N2=9.5
DIFFUSION TEMPERAT=1200 TIME=600 F.O2=0.5 F.N2=9.5
DIFFUSION TEMPERAT=1200 T.FINAL=1000 TIME=67 F.O2=9
DIFFUSION TEMPERAT=1000 TIME=20 F.O2=10
DIFFUSION TEMPERAT=1000 TIME=67 F.O2=5.5 F.H2=10.4
DIFFUSION TEMPERAT=1000 TIME=1 F.O2=5.5 F.N2=5.0
DIFFUSION TEMPERAT=1000 T.FINAL=800 TIME=67 F.N2=10
etch oxide all
print layers
```

8.12 P-Type Epitaxial Growth

Drive-in voltage will place the NBL layer in proper place in the wafer. This N-layer in a P-type wafer will create a p-n junction that will stop high bottom leakage current flow due to high supply voltage at the drain end. To ensure this, both the initial wafer and buried layer are doped by boron and

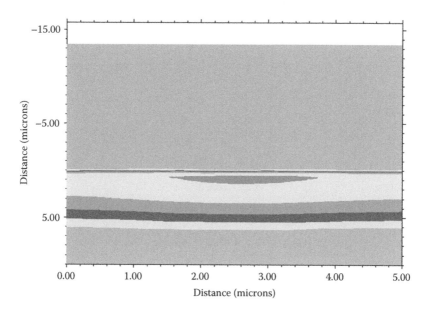

FIGURE 8.5 (See color insert)
Placement of NBL and epitaxial growth on the initial wafer.

antimony of dose 1e^{15} cm^{-3}, respectively. A p-n junction of the same deple-tion depth on both sides will be formed. For the actual structure an epitaxial layer of 14 μm is grown on this wafer, as device parameter optimization is possible in this epitaxial layer. It is always required to grow an epitaxial layer on the initial wafer, as shown in Figure 8.5, to avoid crystal defects in the initial wafer and parameter optimization is convenient in this epitaxial layer.

The statement epitaxy will create an epitaxial layer of 14 μm which will grow on this initial wafer [9–12], as stated below.

```
EPITAXY TIME=14 TEMPERAT=1150 THICKNES=14 dx =.001 ydy=0.0
SPACES=100 +
RESISTIV BORON=45
print layers
```

8.13 Pad Oxide Formation

During the masking process, as shown in Figure 8.6, the masking materials, nitride, and photoresist are not deposited directly on the wafer, as strain would be created on the wafer when the photoresist hardens due to ultravio-let (UV) light. This strain on the silicon material may change the property of the silicon material. So whenever this layer for masking is deposited, it is

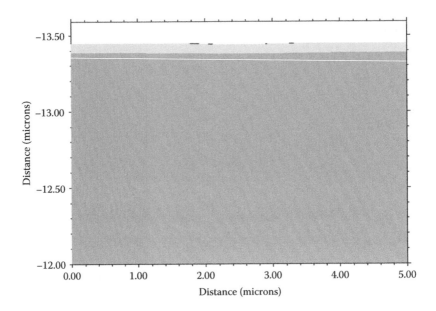

FIGURE 8.6 (See color insert)
Pad oxide on the wafer is shown; here the *y*-axis has been chosen up to –12 μm.

always essential to grow a layer of oxide on the wafer to avoid strain. Hence, a layer of 500 A⁰ SiO₂ has been grown here using the following statements.

```
DIFFUSION TEMPERAT=800 T.FINAL=900 TIME=10 F.O2=9.0
DIFFUSION TEMPERAT=900 TIME=15 F.O2=9.0
DIFFUSION TEMPERAT=900 TIME=18 F.O2=5.5 F.H2=10.4
DIFFUSION TEMPERAT=900 TIME=1 F.O2=5.5 F.N2=5.0
DIFFUSION TEMPERAT=900 TIME=10 F.N2=10
DIFFUSION TEMPERAT=900 T.FINAL=800 TIME=25 F.N2=10
print layers
```

8.14 Gate Under Channel Doping

Gate under substrate region is doped in this portion by boron of dose 2.0 e¹¹cm⁻³. The mask chosen has been named *'gateunderdoping'*. Threshold voltage modification can be done by this doping. The statements needed to perform this doping procedure are given below. With the dose 2.0 e¹¹cm⁻³, threshold voltage achieved is 0.65 V.

```
DEPOSIT PHOTORESIST NEGATIVE THICKNESS=1
EXPOSE MASK=gateunderdoping
```

```
DEVELOP
etch nitride
etch oxide thickness=0.02
implant boron pearson tilt=7 dose=2.0e11 energy=100
etch nitride all
```

8.15 Gate Oxide Formation

For a MOS structure, gate oxide has to be formed. As gate length has been chosen 1.6 to 4.1 μm or effectively 2.5 μm, so the rest of the wafer has to be covered by mask, as shown in Figure 8.7. A mask has been assigned a length of 1600 to 4100 named gateoxet in mask file. As photoresist used here in most of the cases is the negative type, part of the length mentioned in the mask file will be dissolved when it is developed. Now the oxidation steps have been executed to form the gate oxide at the end of the oxidation steps. The entire photoresist and nitride will be removed from the remaining part of the wafer.

```
DIFFUSION TEMPERAT=800 T.FINAL=900 TIME=10 F.O2=0.25 F.N2=10
DIFFUSION TEMPERAT=900 TIME=5 F.O2=0.25 F.N2=10
DIFFUSION TEMPERAT=900 TIME=3 F.O2=9.5
DIFFUSION TEMPERAT=900 TIME=47.5 F.O2=9.5 F.HCL=0.19
```

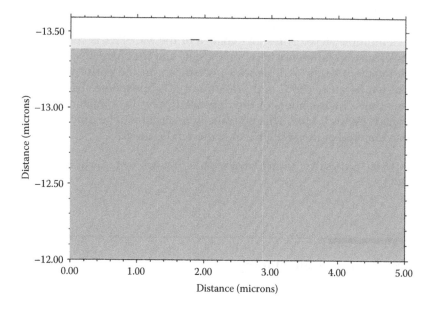

FIGURE 8.7 (See color insert)
Structure of the wafer after formation of the gate oxide.

```
DIFFUSION TEMPERAT=900 TIME=3 F.O2=9.5
DIFFUSION TEMPERAT=900 TIME=15 F.N2=10.0
DIFFUSION TEMPERAT=900 T.FINAL=800 TIME=37.5 F.N2=10
print layers
etch Photoresist
etch nitride all
```

8.16 Gate-Poly Deposition

MOS gate material may be polysilicon or metals [13]. As a gate metal, molybdenum or aluminum can be used. Polysilicon is polycrystalline silicon, a material consisting of small silicon crystals. Though polysilicon gate has severe disadvantages, such as low conductivity which can cause occurrence of delay in circuits as well as unwanted variation of threshold voltage of the MOSFET due to polysilicon depletion effect, polysilicon has several advantages over the metal gate. Polysilicon behaves like a perfect conductor once a poly-layer is doped properly, and it will reduce the delay in channel formation [14–17]. Typically doping concentrations are of the order of 10^{20} atoms cm^{-3}. The main reason for use of the polysilicon gate is that fabrication processes require very high temperature annealing after the initial doping to passivate the radiation damage caused to the silicon crystal structure by the ion implantation [18–19]. Metal gate would melt under such conditions, whereas polysilicon will not. Polysilicon needs a single-step process of etching, whereas a metal gate requires multiple steps. Threshold voltage of the MOSFET is corrected with the work function difference between the gate and the channel.

The statement below will deposit polysilicon on the entire wafer in the ambient temperature 625°C and pressure of 1.0 atmosphere. This statement also mentions the thickness of the deposited polysilicon material which is mentioned here as 0.4 μ*m*. This polysilicon material will be used as polysilicon gate material of the MOSFET, as shown in Figure 8.8.

```
deposition polysili temperature=625 pressure=1.0 thickness=0.4
concentr
```

8.17 Polysilicon Gate Doping

Polysilicon doping is done by the following expression which is essential to increase the conductivity of the gate material.

```
DIFFUSION TEMPERAT=950 TIME=20 INERT
```

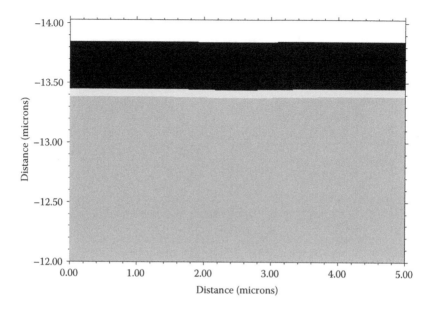

FIGURE 8.8 (See color insert)
Polysilicon material deposition on the grown gate oxide material.

8.18 Gate-Poly Mask

The polysilicon material is deposited on the entire wafer as shown in Figure 8.9. It is necessary to remove the unnecessary extra portion of polysilicon and oxide from the wafer. Here polysilicon is deposited on the total length of the gate oxide. So the same mask *gateoxet* has been called again, and the photoresist used here is a positive type so that from length 1600 to 4100 of the photoresist will remain on it and will use it as a mask for that portion. The remaining part of the photoresist will be dissolved when it is developed. This part nitride and polysilicon will be etched out by *etch nitride* and *etch polysili* statements. After the selective polysilicon etching, the remaining nitride will be removed using the *etch nitride all* statement. Figure 8.10 shows 2.5 µ*m* polysilicon gate formation on the gate oxide.

```
DEPOSITION NITRIDE THICKNES=0.10 CONCENTR
DEPOSIT PHOTORESIST POSITIVE THICKNESS=1
EXPOSE MASK=gateoxet
DEVELOP
etch nitride
etch polysili
etch nitride all
```

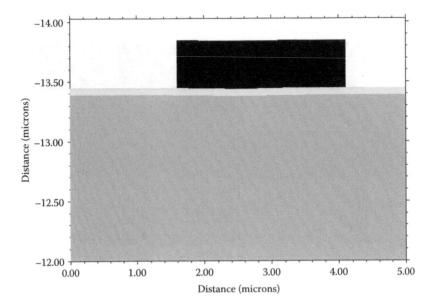

FIGURE 8.9 (See color insert)
Polysilicon gate formation on the gate oxide after selective polysilicon material by etching from the wafer; polysilicon gate length is 2.5 μ*m*.

8.19 Creation of n+ Source and Drain Regions

The next step of fabrication is to create source and drain regions for the n-MOSFET, as shown in Figure 8.10. For n-type doping, arsenic dopant is used. The mask name used to mask the rest of the wafer is Nplus. Two lengths assigned in the Nplus mask are of lengths 900 to 1500 and 4200 to 4900. To cover the rest of the wafer top, it must be masked. So a negative photoresist has been deposited and mask name Nplus has been called, followed by the mask develop stage. From 0.9 to 1.5 μ*m* and 4.2 to 4.9 μ*m* there will not be any photoresist as the mask deposited was negative-type photoresist. Now the nitride and oxide materials from these portions will also be removed to implant arsenic in these regions for source/drain formation in the above-mentioned regions. Here arsenic of dose 6.0 e^{15} cm^{-3} with energy 100 keV has been doped in equal and opposite tilt 7° and –7° to get a source drain shape. After completion, source drain doping, photoresist, and nitride are removed from the entire wafer which was used as a mask at the time of source/drain doping.

```
DEPOSITION NITRIDE THICKNES=0.15 CONCENTR
DEPOSIT PHOTORESIST NEGATIVE THICKNESS=1
EXPOSE MASK=Nplus
DEVELOP
etch nitride
```

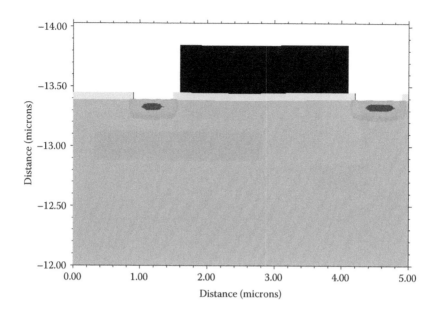

FIGURE 8.10 (See color insert)
Device structure after source and drain formation.

```
etch oxide
implant arsenic pearson tilt=7 dose=6.0e15 energy=100
implant arsenic pearson tilt=-7 dose=6.0e15 energy=100
etch PHOTORESIST
etch nitride all
```

8.20 Creation of p+ Region

The next step is to create a p region that will remain connected to the substrate material and finally work as a bulk material. This p region mask length is chosen as 100 to 700, which means 0.1 to 0.7 μm, and the mask is named pplus. Figure 8.11 shows the formation of the p region that will be used as bulk material.

```
DEPOSITION NITRIDE THICKNES=0.15 CONCENTR
DEPOSIT PHOTORESIST NEGATIVE THICKNESS=1
EXPOSE MASK=pplus
DEVELOP
etch nitride
etch oxide
IMPLANT BORON PEARSON RP.EFF DOSE=1.0e15 ENERGY=30
```

Photoresist and nitride from the remaining part will be removed by *etch photoresist* and *etch nitride all* statements. Figure 8.12 shows the structure achieved

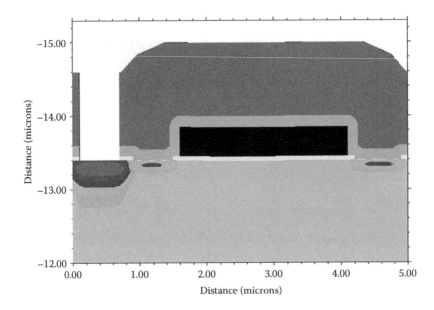

FIGURE 8.11 (See color insert)
Formation of the p region in the wafer of length 0.1 to 0.7 μm.

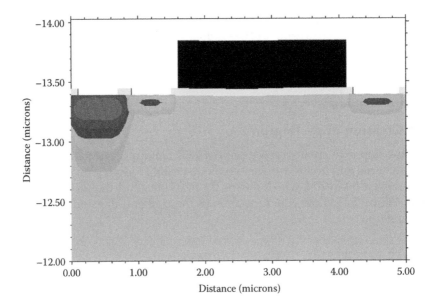

FIGURE 8.12 (See color insert)
The structure achieved by executions of etch photoresist and etch nitride all statements.

by executions of *etch photoresist* and *etch nitride all* statements. In this structure it is evident that bulk, source, and drain regions are formed by doping the wafer. Gate oxide and polysilicon deposition on the gate oxide are also being formed. The next step is to create metal contacts for the different regions for its terminals to connect the device with the outer world. To perform this next step, borophosphosilicate glass (BPSG) deposition and anneal are required. The diffusion statement causes annealing to occur. If the anneal occurs in an oxidizing ambient, then silicon oxidation will occur on the exposed silicon material surface. It is common to specify multiple anneal steps in sequence in order to accurately model a specific furnace process. Semiconductor material needs annealing after every ion implantation step. It will repair the damages caused in the lattice during ion bombardment by the collisions with doping ions. It also allows doping impurities to diffuse further into the bulk.

```
etch PHOTORESIST
etch nitride all
```

8.21 Borophosphosilicate Glass (BPSG) Deposition

Borophosphosilicate glass (BPSG), shown in Figure 8.13, is important in the fabrication of silicon-based lightweight devices and integrated circuits [20]. It

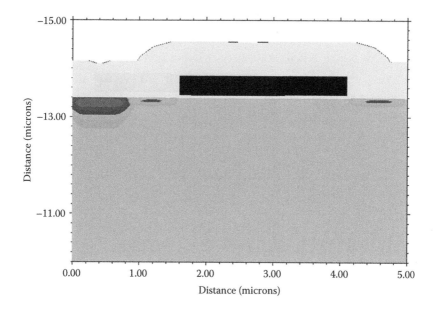

FIGURE 8.13 (See color insert)
Borophosphosilicate glass deposition before the first layer of metal contact to the device.

consists of the final silica glass films. BPSG can be fabricated by several methods like CVD (chemical vapor deposition), sol-gel, and FHD (flame hydrolysis deposition). Usually the CVD procedure is used to form BPSG films [3, 21, 23]. BPSG provides void free fill of 0.2 to 0.8 μ*m* wide spaces between succeeding higher metals or conducting layers. BPSG basically works as an insulating layer for inter-metal layers.

```
deposition oxide thickness=0.7 concentr
```

8.22 BPSG Anneal

Deposited borophosphosilicate glass needs annealing, as shown in Figure 8.14, which will be performed by the following steps:

```
DIFFUSION TEMPERAT=800 TIME=20 F.N2=10.0
DIFFUSION TEMPERAT=800 TIME=15 F.O2=9.5
```

The BPSG layer deposited on it is not smooth due to uneven device structure. It needs chemical-mechanical polishing (CMP) [3, 24, 25]. A nitride layer of thickness 0.15 μ*m* is deposited on it to determine the minimum *y* coordinate and

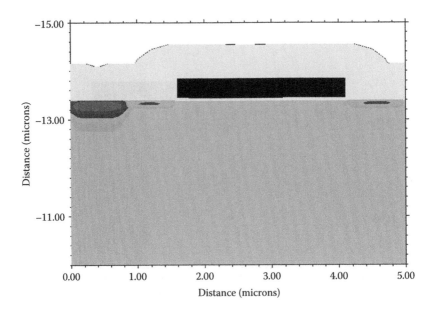

FIGURE 8.14 (See color insert)
Structure of the device after application of annealing step for BPSG.

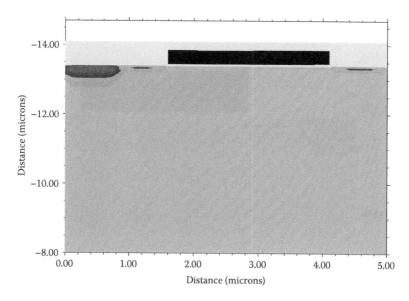

FIGURE 8.15 (See color insert)
Polishing of the device top surface by removal of excess oxide.

is named CMP1. During polishing, the oxide above that minimum y coordinate is etched out to make the surface smooth. Thus four points are defined using the minimum y coordinate for two points, and the uneven portion is removed by etch operations. After the removal of this oxide a smooth surface can be achieved, and this can be seen from Figure 8.15. Statements below will perform this CMP task.

```
DEPOSITION NITRIDE THICKNES=0.15 CONCENTR
extract nitride/oxide y.extract minimum name=CMP1
etch nitride all
ETCH OXIDE START X=0.0 Y=-14.6
ETCH CONTINUE X=5 Y=-14.6
ETCH CONTINUE X=5.0 Y=@CMP1
ETCH DONE X=0.0 Y=@CMP1
```

8.23 Contact Mask Formation

For metal contacts of different terminals like bulk, source, gate, and drain, metallization is required. Aluminum is used most of the time for metal-lization in integrated circuits, because aluminum and its alloys have low

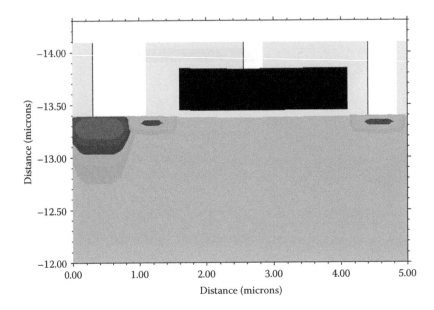

FIGURE 8.16 (See color insert)
Selective etching has been performed to deposit aluminum through it for the first layer of metal contacts.

resistivity (2.7 μΩ for aluminum). Aluminum adheres well to silicon dioxide, though use of aluminum in shallow junctions may create problems like spiking and electromigration. A mask named *contact* has been assigned in the mask file having three lengths assigned to it: one to create source-bulk contact, one for gate, and one for drain contact. Three lengths assigned to the mask contacts are 300 to 1100, 2550 to 2850, and 4400 to 4850, or effectively 0.3 to 1.1 μm, 2.55 to 2.85, and 4.4 to 4.85 μm. The first mask is chosen in such a way that both bulk and source region contact formation are possible, as for an n-MOSFET bulk and source normally remain in the same potential most of the time. The second region from 2.5 to 2.85 μm is chosen for gate contact, and the last region from 4.4 to 4.85 μm is for drain contact. Figure 8.16 shows the structure where selective etching has been performed, after which aluminum is deposited through those regions for the contacts.

```
DEPOSIT PHOTORESIST NEGATIVE THICKNESS=1
EXPOSE MASK=contact
DEVELOP
etch oxide
etch PHOTORESIST
```

8.24 First Layer of Metal (metal-1) Deposition

The statement below will deposit aluminum of thickness 0.3 μm on the entire material.

```
deposition aluminum thickness=0.3 concentr
```

8.25 Metal-1 Mask

As the metal will be deposited on the entire wafer, the undesired part of the metal must be removed from the entire material. So another mask is assigned in the mask file and called here. The mask name is *metal1*, which has three regions assigned in it. The regions are 150 to 1200, 2400 to 3000, and 4300 to 4950, meaning 0.15 to 1.2 μm, 2.4 to 3.0 μm, and 4.3 to 4.95 μm. Only on these regions will the aluminum remain on the wafer, as the contact material and rest of the aluminum will be removed by an etch statement. Figure 8.17 shows the final structure formed after deposition on metal.

```
DEPOSIT PHOTORESIST POSITIVE THICKNESS=1
EXPOSE MASK=metal1
DEVELOP
```

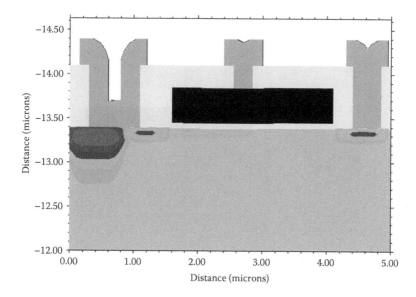

FIGURE 8.17 (See color insert)
First layer of metal deposition through the opening of BPSG for the direct contacts from the device.

```
etch aluminum
etch PHOTORESIST
```

8.26 Inter-Metal Dielectric (IMD) Deposition

For final outer world connection to the device another layer of metallization is required [3]. Another layer of oxide which is called inter-metal dielectric (IMD) is deposited here by the statement below of thickness 0.5 *µm* which is shown in Figure 8.18. From the figure it is obvious that the top surface requires chemical-mechanical polishing again.

```
deposition oxide thickness=0.5 concentr
```

To polish the top surface, a layer of nitride of thickness 0.15 µm is deposited on it to determine minimum y coordinates at which nitride and oxide meet. Until that point, oxide is etched to get a smooth oxide on the top. Figure 8.19 shows the structure in which the top surface is smoothened by CMP operation.

```
DEPOSITION NITRIDE THICKNES=0.15 CONCENTR
extract nitride/oxide y.extract minimum name=CMP2
etch nitride all
```

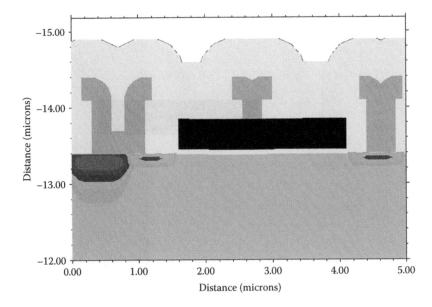

FIGURE 8.18 (See color insert)
Deposition of IMD after formation of the first layer of metal contacts.

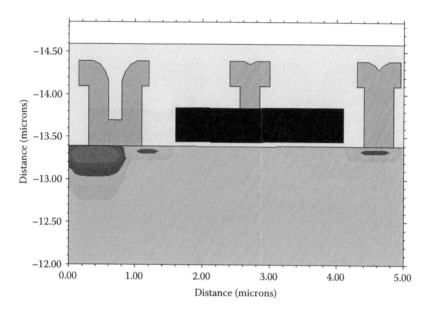

FIGURE 8.19 (See color insert)
Structure achieved by smoothening the top surface by the polishing operation again.

```
ETCH OXIDE START X=0.0 Y=-15.2
ETCH CONTINUE X=5 Y=-15.2
ETCH CONTINUE X=5.0 Y=@CMP2
ETCH DONE X=0.0 Y=@CMP2
```

8.27 Second Layer of Metal (metal-2) Mask

For the final layer of metallization, another mask is required and through its opening another layer of metal is deposited on it. The mask assigned here is named *metal-2* and three regions assigned here are 200 to 1100, 2000 to 3100, and 4200 to 4900 to connect this layer to the previous layer of metal. Figure 8.20 shows the opening regions through which the next layer of metal will be deposited.

```
DEPOSIT PHOTORESIST NEGATIVE THICKNESS=1
EXPOSE MASK=metal2
DEVELOP
etch oxide
etch PHOTORESIST
```

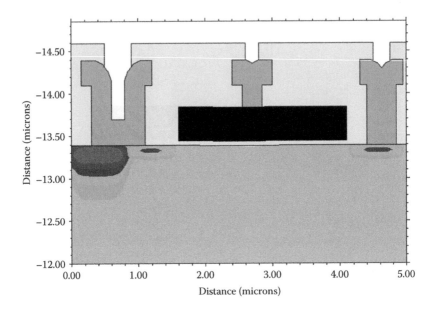

FIGURE 8.20 (See color insert)
Opening region has been created by etching through which the second layer of metal will be deposited.

8.28 Second Layer of Metal (metal-2) Deposition

Now the aluminum of thickness 0.2 μ*m* will be deposited on it, as shown in Figure 8.21 for the next layer of metallization.

```
deposition aluminum thickness=0.2 concentr
etch PHOTORESIST
```

8.29 Metal-2 Final Mask

Figure 8.22 shows a layer of aluminum is deposited on the top, in which a part of metal is essential and the remaining part has to be removed from the top portion. The following statements will remove the undesired metal from it, lengths assigned in the mask file named *metal3* are 200 to 1100, 2200 to 3100, and 4200 to 4900, or effectively 0.2 to 1.1 μ*m*, 2.2 to 3.1 μ*m*, and 4.2 to 4.9 μ*m*. Figure 8.23 shows the final device structure.

```
DEPOSIT PHOTORESIST positive THICKNESS=1
EXPOSE MASK=metal3
```

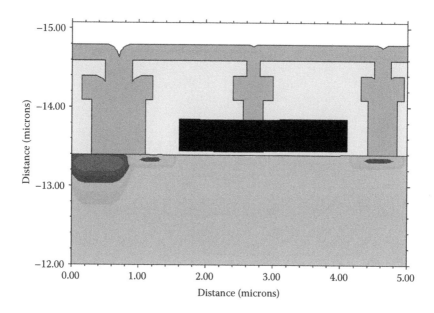

FIGURE 8.21 (See color insert)
Second layer of metal is being deposited and connected to the first layer of the metal through the openings.

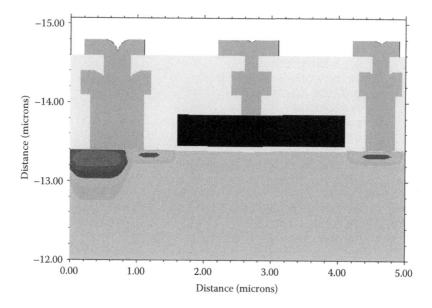

FIGURE 8.22 (See color insert)
Final device structure, y coordinate is taken up to -12.0 μm.

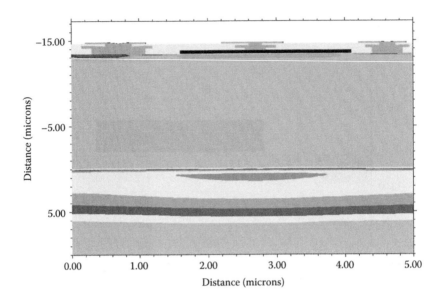

FIGURE 8.23 (See color insert)
Final simulated structure of the 5 μm MOSFET.

```
DEVELOP
etch aluminum
etch PHOTORESIST

SELECT Z=LOG10(BORON)
PLOT.2D    GRID C.GRID=3
COLOR      SILICON COLOR=7
COLOR      OXIDE COLOR=5
COLOR      NITRIDE COLOR=3
COLOR      PHOTORESIST COLOR=2
COLOR      polysili COLOR=1
COLOR      aluminum COLOR=3
FOREACH X (14 TO 21 STEP 1)
COLOR MIN.V=X MAX.V=(X + 1) COLOR=(X - 1)
END

SELECT     Z=LOG10(phosphor)
FOREACH X (14 TO 21 STEP 1)
COLOR MIN.V=X MAX.V=(X + 1) COLOR=(X - 3)
END

SELECT     Z=LOG10(arsenic)
FOREACH X (19 TO 21 STEP 1)
COLOR MIN.V=X MAX.V=(X + 1) COLOR=(X - 5)
END
SELECT     Z=LOG10(antimony)
```

```
FOREACH X (14 TO 21 STEP 1)
COLOR MIN.V=X MAX.V=(X + 1) COLOR=(X - 7)
END
COLOR        OXIDE COLOR=10
COLOR        NITRIDE COLOR=3
COLOR        PHOTORESIST COLOR=2
COLOR        polysili COLOR=1
COLOR        aluminum COLOR=3

savefile medici out.file=LDNBL.str
coordinat
```

8.30 MOSFET.inp

The complete program is given below, and a final full view of the structure is shown in Figure 8.23.

```
$nitial mesh

mesh grid.fac=1.0 dx.min=0.002 dx.max=0.1 ly.surf=0.06
dy.surf=0.001 +
 ly.activ=0.5 dy.activ=0.02 ly.bot=10 dy.bot=1

MASK IN.FILE=t.tl1 PRINT GRID="Field,Poly,Contact"

$start material
initialize ratio=1.4 <100> rot.sub=0 boron=1e+15 width=5.0

$ Plot initial mesh
SELECT TITLE="Initial Mesh"
PLOT.2D GRID C.GRID=8

$ Grow the initial ox... E2010
DIFFUSION TEMPERAT=800 T.FINAL=1000 TIME=20 F.O2=0.5 F.N2=9.5
DIFFUSION TEMPERAT=1000 TIME=65 F.O2=0.5 F.N2=9.5
DIFFUSION TEMPERAT=1000 TIME=5 F.O2=9.5
DIFFUSION TEMPERAT=1000 TIME=190 F.O2=5.975 F.H2=10.4 F.HCL=0.475
DIFFUSION TEMPERAT=1000 TIME=1 F.O2=5.5 F.N2=5
DIFFUSION TEMPERAT=1000 TIME=10 F.N2=10
DIFFUSION TEMPERAT=1000 T.FINAL=800 TIME=50 F.N2=10
print layers

$ NBL mask
DEPOSIT PHOTORESIST NEGATIVE THICKNESS=1
EXPOSE MASK=nbl
```

```
DEVELOP

etch oxide

$ Screen oxidation... E8020
DIFFUSION TEMPERAT=800 T.FINAL=900 TIME=10 F.O2=0.5 F.N2=9.5
DIFFUSION TEMPERAT=900 TIME=15 F.O2=0.5 F.N2=9.5
DIFFUSION TEMPERAT=900 TIME=5 F.O2=9.0
DIFFUSION TEMPERAT=900 TIME=5 F.O2=9.5
DIFFUSION TEMPERAT=900 TIME=28 F.O2=9.0 F.HCL=0.19
DIFFUSION TEMPERAT=900 TIME=5 F.O2=9.0
DIFFUSION TEMPERAT=900 TIME=30 F.N2=10.0
DIFFUSION TEMPERAT=900 T.FINAL=800 TIME=37.5 F.N2=10
print layers

$ Buried layer implant
implant antimony pearson tilt=7 dose=1.0e15 energy=100
etch oxide all

$ Buried layer drive-in... E0381
DIFFUSION TEMPERAT=800 T.FINAL=1200 TIME=100 F.O2=0.5 F.N2=9.5
DIFFUSION TEMPERAT=1200 TIME=600 F.O2=0.5 F.N2=9.5
DIFFUSION TEMPERAT=1200 T.FINAL=1000 TIME=67 F.O2=9
DIFFUSION TEMPERAT=1000 TIME=20 F.O2=10
DIFFUSION TEMPERAT=1000 TIME=67 F.O2=5.5 F.H2=10.4
DIFFUSION TEMPERAT=1000 TIME=1 F.O2=5.5 F.N2=5.0
DIFFUSION TEMPERAT=1000 T.FINAL=800 TIME=67 F.N2=10
etch oxide all
print layers

$ Epi growth, P-type, 30-60 ohm-cm, 14 um

EPITAXY TIME=14 TEMPERAT=1150 THICKNES=14 dx =.001 ydy=0.0
SPACES=100 +
 RESISTIV BORON=45

$ Pad oxide, tox=500A
DIFFUSION TEMPERAT=800 T.FINAL=900 TIME=10 F.O2=9.0
DIFFUSION TEMPERAT=900 TIME=15 F.O2=9.0
DIFFUSION TEMPERAT=900 TIME=18 F.O2=5.5 F.H2=10.4
DIFFUSION TEMPERAT=900 TIME=1 F.O2=5.5 F.N2=5.0
DIFFUSION TEMPERAT=900 TIME=10 F.N2=10
DIFFUSION TEMPERAT=900 T.FINAL=800 TIME=25 F.N2=10
print layers

$ N-tub mask

DEPOSIT PHOTORESIST NEGATIVE THICKNESS=1
EXPOSE MASK=gateunderdoping
```

```
DEVELOP
etch nitride
etch oxide thickness=0.02
implant boron pearson tilt=7 dose=2.0e11 energy=100
etch nitride all

$ Gate oxide-2 200A
DEPOSIT PHOTORESIST NEGATIVE THICKNESS=1
EXPOSE MASK=gateoxet
DEVELOP

DIFFUSION TEMPERAT=800 T.FINAL=900 TIME=10 F.O2=0.25 F.N2=10
DIFFUSION TEMPERAT=900 TIME=5 F.O2=0.25 F.N2=10
DIFFUSION TEMPERAT=900 TIME=3 F.O2=9.5
DIFFUSION TEMPERAT=900 TIME=47.5 F.O2=9.5 F.HCL=0.19
DIFFUSION TEMPERAT=900 TIME=3 F.O2=9.5
DIFFUSION TEMPERAT=900 TIME=15 F.N2=10.0
DIFFUSION TEMPERAT=900 T.FINAL=800 TIME=37.5 F.N2=10
print layers
etch Photoresist
etch nitride all

$ Gate-poly deposition - 4000A

deposition polysili temperature=625 pressure=1.0 thickness=0.4
concentr

$ Poly doping
DIFFUSION TEMPERAT=950 TIME=20 INERT
$ Gate-poly mask
DEPOSITION NITRIDE THICKNES=0.10 CONCENTR
DEPOSIT PHOTORESIST POSITIVE THICKNESS=1
EXPOSE MASK=gateoxet
DEVELOP
etch nitride
etch polysili
etch nitride all

$ N+ mask and implant
DEPOSITION NITRIDE THICKNES=0.15 CONCENTR
DEPOSIT PHOTORESIST NEGATIVE THICKNESS=1
EXPOSE MASK=Nplus
DEVELOP
etch nitride
etch oxide
implant arsenic pearson tilt=7 dose=6.0e15 energy=100
implant arsenic pearson tilt=-7 dose=6.0e15 energy=100

etch PHOTORESIST

etch nitride all
```

```
$ P+ mask and implant
DEPOSITION NITRIDE THICKNES=0.15 CONCENTR
DEPOSIT PHOTORESIST NEGATIVE THICKNESS=1
EXPOSE MASK=pplus
DEVELOP
etch nitride
etch oxide
IMPLANT BORON PEARSON RP.EFF DOSE=1.0e15 ENERGY=30
etch PHOTORESIST

etch nitride all

$ BPSG deposition
deposition oxide thickness=0.7 concentr

$ BPSG anneal
DIFFUSION TEMPERAT=800 TIME=20 F.N2=10.0
DIFFUSION TEMPERAT=800 TIME=15 F.O2=9.5

DEPOSITION NITRIDE THICKNES=0.15 CONCENTR

extract nitride/oxide y.extract minimum name=CMP1

etch nitride all

ETCH OXIDE START X=0.0 Y=-14.6
ETCH CONTINUE X=5 Y=-14.6
ETCH CONTINUE X=5.0 Y=@CMP1
ETCH DONE X=0.0 Y=@CMP1
$ Contact mask

DEPOSIT PHOTORESIST NEGATIVE THICKNESS=1
EXPOSE MASK=contact
DEVELOP
etch oxide

etch PHOTORESIST

$ Metal-1 Deposition
deposition aluminum thickness=0.3 concentr

$ Metal-1 mask
DEPOSIT PHOTORESIST positive THICKNESS=1
EXPOSE MASK=metal1
DEVELOP

etch aluminum
etch PHOTORESIST

$ IMD dep
deposition oxide thickness=0.5 concentr

DEPOSITION NITRIDE THICKNES=0.15 CONCENTR

extract nitride/oxide y.extract minimum name=CMP2
```

```
etch nitride all

ETCH OXIDE START X=0.0 Y=-15.2
ETCH CONTINUE X=5 Y=-15.2
ETCH CONTINUE X=5.0 Y=@CMP2
ETCH DONE X=0.0 Y=@CMP2


$ metal2 mask

DEPOSIT PHOTORESIST NEGATIVE THICKNESS=1
EXPOSE MASK=metal2
DEVELOP
etch oxide

etch PHOTORESIST

$ Metal-2 Deposition
deposition aluminum thickness=0.2 concentr

etch PHOTORESIST

$ Metal-2 final mask
DEPOSIT PHOTORESIST positive THICKNESS=1
EXPOSE MASK=metal3
DEVELOP

etch aluminum
etch PHOTORESIST


SELECT Z=LOG10(BORON)
PLOT.2D     GRID C.GRID=3
COLOR       SILICON COLOR=7
COLOR       OXIDE COLOR=5
COLOR       NITRIDE COLOR=3
COLOR       PHOTORESIST COLOR=2
COLOR       polysili COLOR=1
COLOR       aluminum COLOR=3
FOREACH X (14 TO 21 STEP 1)
COLOR MIN.V=X MAX.V=(X + 1) COLOR=(X - 1)
END

SELECT      Z=LOG10(phosphor)
FOREACH X (14 TO 21 STEP 1)
COLOR MIN.V=X MAX.V=(X + 1) COLOR=(X - 3)
END

SELECT      Z=LOG10(arsenic)
FOREACH X (19 TO 21 STEP 1)
COLOR MIN.V=X MAX.V=(X + 1) COLOR=(X - 5)
END
SELECT      Z=LOG10(antimony)
FOREACH X (14 TO 21 STEP 1)
COLOR MIN.V=X MAX.V=(X + 1) COLOR=(X - 7)
```

```
END
COLOR       OXIDE COLOR=10
COLOR       NITRIDE COLOR=3
COLOR       PHOTORESIST COLOR=2
COLOR       polysili COLOR=1
COLOR       aluminum COLOR=3


savefile medici out.file=LDNBL.str
```

8.31 Mask File Named t.tl1

The mask file named *t.tl1* is given here because it was used during the fabrication of the device. This mask file should be kept in the same folder where the files are kept, especially the actual device fabrication file MOSFET.inp. The first line identifies the file format that contains the character 'TL1' followed by a space and a four-digit binary number. The version of Taurus-Layout that created the file is represented by the binary number. The current version of the Taurus-Layout specifies values from 0000 to 0100. Nine masks are being used in this file named *gateoxet, nbl, Nplus, pplus, contact, metal1, metal2, metal3,* and *gateunderdoping,* and that is why nine is mentioned in the fourth line of this mask file. This information must be provided before the mask names and their dimensions. Total length of the wafer must also be provided, as it is mentioned here 0 to 5000 or 5 μm by the rule assigned here.

```
TL1 0100

1e3
  0 5000
9
gateoxet 1
  1600 4100
nbl 1
  2000 3000
Nplus 2
  900 1500
  4200 4900
pplus 1
  100 700
contact 3
  300 1100
  2550 2850
  4400 4850
metal1 3
  150 1200
```

```
 2400 3000
 4300 4950
metal2 3
  500 900
 2600 2800
 4500 4750
metal3 3
  200 1100
 2200 3100
 4200 4900
gateunderdoping 1
 1600 3700
```

8.32 What Is Medici

Taurus Medici is a 2D device simulator that can model the electrical, thermal, and optical characteristics of any semiconductor device like MOSFETs, bipolar junction transistors (BJTs), heterojunction bipolar transistors (HBTs), power devices, insulated-gate bipolar transistors (IGBTs), high electron mobility transistors (HEMTs), charge-coupled devices (CCDs), and photodetectors [2]. It also can be used for design and optimization of a device to meet performance goals without having to manufacture the actual device, thereby reducing the need for costly experiments. Once the device structure is fabricated using TSUPREM-4, it is saved in the (LDNBL.str) file where detailed description about it is saved in the file named *LDNBL* whose extension is *str*, or structure. Here the Medici simulator has been used to analyze the fabricated MOSFET electrical characteristics. To do that an interface is necessary between TSUPREM-4 and Medici.

8.33 Execution of Command

Execution of Medici is initiated with the following command, medici or medici <filename>. The Medici file name given here is BVNBL.inp, so at the time of execution of this Medici script file, the command would be medici BVNBL.inp.

8.34 Interfacing between TSUPREM-4 and Medici

A meaningful simulated device fabrication is possible if it is based on a real fabrication procedure. It is also essential to analyze the fabricated device performance, especially electrical characteristics, for realistic device structure. For this purpose it is necessary to create an interface between TSUPREM-4 and Medici. The output structure (having extension .str) file is saved during TSUPREM-4 simulation in such a way that it can be used in the Medici script file. It is done by the expression *'savefile medici out.file = LDNBL.str'* of the last line of the TSUPREM-4 script. This file *'LDNBL.str'* is basically created to incorporate the device structure in Medici for its output characteristics analysis. The same file is called to read the simulation meshing information by the following statement.

```
MESH IN.FILE=LDNBL.str TSUPREM4 ELEC.BOT POLY.ELEC Y.MAX=10
```

8.35 Rename Electrodes from TSUPREM-4 to Standard Names

Three metal contacts of the simulated device will be named as 1, 2, and 3 by default at the end of the TSUPREM-4 simulation. After the simulation the program numbered the left-most metal contact (Source and Bulk in this case) as 1. In accordance with this numbering scheme, Table 8.1 represents the electrode names and their coordinate positions.

The minimum and maximum x and y positions of different electrodes will be automatically created at the end of the TSUPREM-4 simulation.

The electrode name 1 will be renamed *Source* by the expression *'RENAME ELECTR OLDNAME = 1 NEWNAME = Source'* in Medici. Similarly, electrodes 2 and 3 will be renamed as Gate and Drain. After renaming the electrodes to the actual standard device name, the mesh file has been saved with the new electrode name by the expression *'SAVE MESH OUT.FILE = BVNBL'* in the ext line. The following commands will plot the device structure in the screen with proper labeling as mentioned in the expressions.

```
RENAME ELECTR OLDNAME=1 NEWNAME=Source
RENAME ELECTR OLDNAME=2 NEWNAME=Gate
RENAME ELECTR OLDNAME=3 NEWNAME=Drain

SAVE MESH OUT.FILE=BVNBL
PLOT.2D GRID FILL TITLE="Structure from TSUPREM-4"
PLOT.1D DOPING LOG X.START=0 X.END=0 Y.START=0 Y.END=2
+    POINTS BOT=1E14 TOP=1E21 TITLE="S/D Profile"
PLOT.1D DOPING LOG X.START=1.8 X.END=1.8 Y.START=0 Y.END=2
```

TABLE 8.1

Electrode Names and Their Coordinate Positions

Electrode Name	Number of Nodes	X-min (microns)	X-max (microns)	Y-min (microns)	Y-max (microns)
1	50	0.1500	1.1500	−14.5918	−13.3914
2	21	2.2000	3.1000	−14.5918	−13.7969
3	32	4.2000	4.9500	−14.5918	−13.3868

```
+    POINTS BOT=1E14 TOP=1E19 TITLE="Channel Profile"
PLOT.2D BOUND FILL L.ELEC=-1 TITLE="Impurity Contours"
CONTOUR DOPING LOG MIN=14 MAX=20 DEL=1 COLOR=2
CONTOUR DOPING LOG MIN=-20 MAX=-14 DEL=1 COLOR=1 LINE=2
```

8.36 Major Physical Models

For accurate simulations, a number of physical models are incorporated into the program. Depending on the type of device structure and the device type models like recombination, mobility, band-gap narrowing, band-to-band tunneling, photogeneration, impact ionization, and lifetime can be incorporated.

- Medici also includes semiconductor statistics like Boltzmann and Fermi-Dirac statistics including the incomplete ionization of impurities.
- Different recombination models like SRH, Auger, direct, surface recombinations, and concentration-dependent lifetimes can be incorporated.
- Carrier mobility and scattering are very important phenomena in the mechanism of electrical transport of the device. Mobility models can be divided into two major categories:
  - Low and high field mobility
  - Surface scattering and electron hole scattering

Low field mobility models are

1. Constant mobility that can be specified with the MUN0 and MUP0 parameters for electron and hole.
2. Concentration dependent mobility can be incorporated with the CONMOB parameter.
3. Either of the two analytic mobility models can be incorporated with the ANALYTIC or ARORA parameters.
4. Carrier-carrier scattering mobility can be incorporated with the CCSMOB parameter.
5. Philips unified mobility can be incorporated with the PHUMOB parameter.

High field mobility models are

1. Field-dependent mobility that can be incorporated with the FLDMOB parameter.

2. Caughey-Thomas mobility can be incorporated with the FLDMOB = 1 parameter.

3. Gallium arsenide–like mobility can be incorporated with the FLDMOB = 2 parameter.

4. Hewlett-Packard mobility can be incorporated with the HPMOB parameter.

Surface scattering mobility models are

1. Surface mobility model can be incorporated with the SRFMOB parameter.

2. Enhanced surface mobility model can be incorporated with the SRFMOB2 parameter.

3. Perpendicular field dependent mobility model can be incorporated with the PRPMOB parameter.

4. Lombardi surface mobility model can be incorporated with the LSMMOB parameter.

5. A number of MOS inversion layer models are available through the parameters UNIMOB, LSMMOB, GMCMOB, SHIRAMOB, and TFLDMOB.

One or more models to be included in a Medici simulation can be specified in *models* statement. For this device simulation models such as lsmmob, fldmob, auger, bgn, btbt, fermi, incomplete, energy.l and high.dop have been incorporated by the '*models lsmmob fldmob auger bgn btbt fermi incomplete energy.l high. dop*' statement below.

```
models lsmmob fldmob auger bgn btbt fermi incomplete energy.l
high.dop
```

8.37 Initial Guess/Convergence and Solution Methods

Depending on the device structure and the range of its operation, one solution method may not be optimal in all cases. Several possibilities can arise for different cases, like at zero bias a Poisson alone is sufficient. For MOSFET, as the device is a unipolar type, only one carrier needs to be solved for its I-V characteristics, though in bipolar and MOSFET breakdown simulations, both carriers are needed. For small geometry devices where the electric field changes rapidly, carrier energy balance may be added to see hot-carrier effect. Solving the lattice heat equation is essential when the device heating effect is important.

The equation that needs to be solved is specified on the *SYMBOLIC* or *SYMB* statement.

8.38 Nonlinear System Solutions and Current-Voltage Analysis

For nonlinear system, the solutions are performed by two widely used iteration methods. The methods are de-coupled solutions (Gummel's method) and coupled solutions (Newton's method). Newton's method with Gaussian elimination of the Jacobian is by far the most stable method of solution. For low current solutions, Gummel's method offers an attractive alternative to inverting the full Jacobian.

Either approach involves solving several large linear systems of equations. The total number of equations in each system is on the order of one to four times the number of grid points, depending on the number of device equations being solved for.

Several ideas are common to all methods of solving the equations. They are convergence rate, error norms, convergence criteria, error norms selection, linear solution options, and initial guess.

The nonlinear iteration converges usually at a linear rate or at a quadratic rate. At a linear rate, the error decreases by about the same factor at each iteration.

The convergence is rapid in the quadratic method, as the error is approximately squared at each iteration. Hence Gummel's method, which is linear in most cases, is less accurate than Newton's method, which is quadratic.

Medici uses six types of initial guesses named initial, previous, local, project, p.local, and post-regrid initial.

For tracing the I-V curves, bias step and number of steps are specified on the SOLVE statement by *VSTEP* and *NSTEP* for a corresponding electrode.

```
SYMB GUMMEL CARR=1 ELECTRON
METHOD ICCG DAMPED itlimit=40 stack=10 cont.stk
SOLVE initial V(Source)=0.0 V(Gate)=0.0 V(Drain)=0.005
SYMB NEWTON CARR=1 ELECTRON
METHOD AUTONR N.DAMP N.DVLIM=0.5

SOLVE PREVIOUS V(Source)=0.0 V(Gate)=2 V(Drain)=0.0
LOG OUT.FILE=BVNBLlog
SOLVE V(Drain)=0.0 ELEC=Drain VSTEP=0.2 NSTEP=50
```

8.39 Post-Processing and Parameter Extraction

After the fabrication of the device, fabricated and simulated device results are necessary, which is possible by the post-processing of the data by the following commands:

Print: The command print will print specific quantities at points within a defined area.

Plot.1D: Plot.1D will plot specific quantity along a line segment through the device.

Plot.2D: Plot.2D command will plot characteristics, boundaries, junctions, and depletion edges.

Contour: It will plot the contours of a physical quantity on a 2D area.

E.line: It will plot potential gradient paths and calculate the ionization integrals.

Label: This command will plot character strings, symbols, and lines as part of a 1D or 2D plot.

```
PLOT.1D X.AXIS=V(Drain) Y.AXIS=I(Drain)
+ TITLE="Ids vs. Vgs" COLOR=2 POINTS OUTFILE=Id_Vd.DAT
LABEL LABEL="Vds=2 V" COLOR=2

EXTRACT MOS.PARA DRAIN=Drain GATE=Gate IN.FILE=BVNBLlog
I.Drain=9e-10
```

8.40 Drain Current versus Drain Voltage Simulation

The complete Medici simulation program, explained step by step, is given below. By executing this program in Medici, drain current has been plotted with respect to drain voltage, shown in Figure 8.24. The gate voltage is fixed at 5 V. From the graph it is evident that initially current is increasing with the increase of drain voltage. Then the current reaches the saturation value, and then it increases slowly with the drain voltage due to channel length modulation.

```
COMMENT MEDICI Input File

MESH IN.FILE=LDNBL.str TSUPREM4 ELEC.BOT POLY.ELEC Y.MAX=10

RENAME ELECTR OLDNAME=1 NEWNAME=Source
RENAME ELECTR OLDNAME=2 NEWNAME=Gate
RENAME ELECTR OLDNAME=3 NEWNAME=Drain

SAVE MESH OUT.FILE=BVNBL

PLOT.2D GRID FILL TITLE="Structure from TSUPREM-4"
PLOT.1D DOPING LOG X.START=0 X.END=0 Y.START=0 Y.END=2
+      POINTS BOT=1E14 TOP=1E21 TITLE="S/D Profile"
PLOT.1D DOPING LOG X.START=1.8 X.END=1.8 Y.START=0 Y.END=2
+      POINTS BOT=1E14 TOP=1E19 TITLE="Channel Profile"
PLOT.2D BOUND FILL L.ELEC=-1 TITLE="Impurity Contours"
```

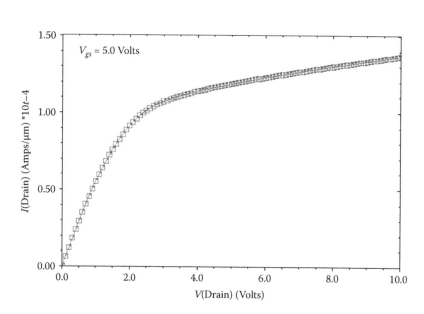

FIGURE 8.24

I_d versus V_{ds} at gate voltage = 5V, device length is 5 μm.

```
CONTOUR DOPING LOG MIN=14 MAX=20 DEL=1 COLOR=2
CONTOUR DOPING LOG MIN=-20 MAX=-14 DEL=1 COLOR=1 LINE=2

models lsmmob fldmob auger bgn btbt fermi incomplete
energy.l high.dop

SYMB GUMMEL CARR=1 ELECTRON
METHOD ICCG DAMPED itlimit=40 stack=10 cont.stk
SOLVE initial V(Source)=0.0 V(Gate)=0.0005 V(Drain)=0.0

SYMB NEWTON CARR=1 ELECTRON
METHOD AUTONR N.DAMP N.DVLIM=0.5
SOLVE PREVIOUS V(Source)=0.0 V(Gate)=5.0 V(Drain)=0.0
LOG OUT.FILE=BVNBLlog
SOLVE V(Drain)=0.0 ELEC=Drain VSTEP=0.1 NSTEP=100

PLOT.1D X.AXIS=V(Gate) Y.AXIS=I(Drain) Y.LOGARITH

PLOT.1D X.AXIS=V(Drain) Y.AXIS=I(Drain)

+ TITLE="Ids vs. Vds" COLOR=2 POINTS OUTFILE=Id_Vd_vg_5V.DAT

LABEL LABEL="Vgs=5.0 Volts" COLOR=2

EXTRACT MOS.PARA DRAIN=Drain GATE=Gate IN.FILE=BVNBLlog
I.Drain=9e-10
```

8.41 Drain Current versus Gate Voltage Simulation

In the Medici program, drain current versus drain voltage simulation has been performed. The drain voltage is fixed at 0.1 V. Drain current in logarithm scale versus gate voltage and drain current in versus gate voltage has been plotted in Figures 8.25 and 8.26, respectively. Threshold voltage is extracted from this curve, which is equal to 0.65 V. On resistance (r_{on}) can be calculated from these data. Different r_{on} can be achieved for different gate voltage. It can be seen that on resistance is decreasing with V_{gs} increase, as higher V_{gs} increase I and $r_{on} = V_{ds}/I$, so automatically r_{on} will decrease at higher V_{gs}, as V_{ds} is fixed at 0.1 V. At the time of on resistance (r_{on}) calculations, V_{ds} should be fixed at 0.1 V.

```
COMMENT MEDICI Input File

MESH IN.FILE=LDNBL.str TSUPREM4 ELEC.BOT POLY.ELEC Y.MAX=10

RENAME ELECTR OLDNAME=1 NEWNAME=Source
RENAME ELECTR OLDNAME=2 NEWNAME=Gate
RENAME ELECTR OLDNAME=3 NEWNAME=Drain

SAVE MESH OUT.FILE=BVNBL
PLOT.2D GRID FILL TITLE="Structure from TSUPREM-4"
PLOT.1D DOPING LOG X.START=0 X.END=0 Y.START=0 Y.END=2
```

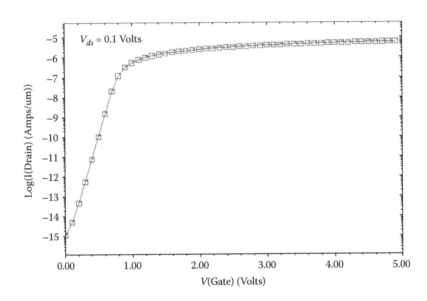

FIGURE 8.25
log (I_d) versus V_{gs} at drain voltage = 0.1 V, device length is 5 μm.

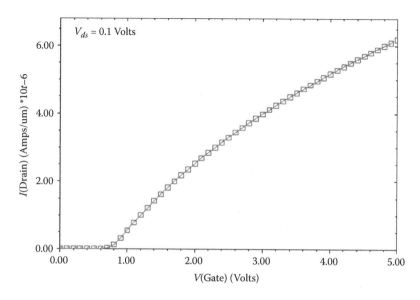

FIGURE 8.26
I_d versus V_{gs} at drain voltage = 0.1 V, device length is 5 μm.

```
+       POINTS BOT=1E14 TOP=1E21 TITLE="S/D Profile"
PLOT.1D DOPING LOG X.START=1.8 X.END=1.8 Y.START=0 Y.END=2
+       POINTS BOT=1E14 TOP=1E19 TITLE="Channel Profile"
PLOT.2D BOUND FILL L.ELEC=-1 TITLE="Impurity Contours"
CONTOUR DOPING LOG MIN=14 MAX=20 DEL=1 COLOR=2
CONTOUR DOPING LOG MIN=-20 MAX=-14 DEL=1 COLOR=1 LINE=2


models lsmmob fldmob auger bgn btbt fermi incomplete energy.l
high.dop

SYMB GUMMEL CARR=1 ELECTRON
METHOD ICCG DAMPED itlimit=40 stack=10 cont.stk

SOLVE initial V(Source)=0.0 V(Drain)=0.0005 V(Gate)=0.0

SYMB NEWTON CARR=1 ELECTRON
METHOD AUTONR N.DAMP N.DVLIM=0.5

SOLVE PREVIOUS V(Source)=0.0 V(Drain)=0.1 V(Gate)=0.0

LOG OUT.FILE=BVNBLlog

SOLVE V(Gate)=0.0 ELEC=Gate VSTEP=0.1 NSTEP=50

COMMENT Plot results
```

```
PLOT.1D X.AXIS=V(Gate) Y.AXIS=I(Drain) Y.LOGARITH
+   TITLE="Ids vs. Vgs" COLOR=2 POINTS OUTFILE=Id_Vg_vd_pt1V.DAT
LABEL LABEL="Vds=0.1 Volts" COLOR=2

PLOT.1D X.AXIS=V(Gate) Y.AXIS=I(Drain)
+   TITLE="Ids vs. Vgs" COLOR=2 POINTS OUTFILE=Id_Vg_vd_pt1V.DAT
LABEL LABEL="Vds=0.1 Volts" COLOR=2

EXTRACT MOS.PARA DRAIN=Drain GATE=Gate IN.FILE=BVNBLlog
I.Drain=9e-10
```

8.42 Conclusion

Device fabrication technology is a complex process that involves developing process-dependent patterns at each step using different masks. For
this it is required to define the mask lengths that require accurate calculations of junction depths and pattern areas that vary with process steps.
For scaled devices, the temperature, time, and ion implantation dose needs
to be predefined by accurate estimation to obtain desired specification with
minimum variation. Complete fabrication procedure needs many oxidation
steps and annealing steps for eliminating the lattice defects arising because
of ion bombardment at a different stage of fabrication, which tends to induce
device parameter and specification variation. Usually, a thin layer of protec-
tive oxide, also known as padding oxide, is grown on the wafer surface for
protection before the ion implantation steps. While fabricating a device, all of
the process dependent variations need to be accounted for with extreme care,
or acquired results will deviate from the desired results. Thus a simulation
of the entire fabrication process helps us optimize the mask lengths, temperature, implantation dose, etc., before proceeding toward the actual process,
thereby helping reduce production cost and time.

The threshold voltage of the device presented here is 0.65 V, which can be further modified by varying the gate oxide thickness and under-the-gate substrate
doping. Higher meshing densities in appropriate regions are considered for
more accurate simulation results. Meshing is chosen in such a way that meshing
density is higher near the surface of the wafer, as most of the phenomena occur
near the surface and boundary regions. The operation of the device fabricated
by TSUPREM-4 can be analyzed in a TCAD Medici device simulator by incorporating a different physical model and appropriate biasing conditions in the
simulator program of the device. Medici simulations are very fast, widely used,
and well accepted in industry. Before commencing analysis of a device, the
TCAD Medici simulator must be calibrated with standard experimental data.

References

1. Taurus TSUPREM-4 User Guide, Version D-2010.03, March 2010.
2. Taurus Medici User Guide, Version F-2011.09, September 2011.
3. Gary S. May and Simon M. Sze, *Fundamentals of Semiconductor Fabrication*, Wiley, New York.
4. Samar Saha, MOSFET test structures for two-dimensional device simulation, *Solid-State Electronics*, 38(1), 69–73 (1995).
5. E.H. Nicollian and J.R. Brews, *MOS Physics and Technology*, Wiley, New York, 1982.
6. J.D. Meindl et al., Silicon epitaxy and oxidation, in F. Van de Wiele, W.L. Engl, and P.O. Jesper, Eds., *Process and Device Modeling Integrated Circuits Design*, Noorhoff, Leyden, 1977.
7. B.E. Deal, Standardization terminology for oxide charge associated with thermally oxidized silicon, *IEEE Trans. Electron Devices*, ED-27, 606 (1980).
8. S.K. Gandhi, *VLSI Fabrication Principles*, Wiley, New York, 1983.
9. Kalyan Koley, Binit Syamal, Atanu Kundu, N. Mohankumar, and C.K. Sarkar, Subthreshold analog/RF performance of underlap DG FETs with symmetric and asymmetric source/drain extensions, *Microelectronics Reliability*, 52(11), 2572–2578 (2012).
10. Atanu Kundu, Binit Syamal, Kalyan Koley, N. Mohankumar, and C.K. Sarkar, RF parameter extraction of bulk FinFET: A non quasi static approach, *IEEE International Conference on Electron Devices and Solid-State Circuits* (EDSSC'10) in Hong Kong, China, December 15–17 (2010).
11. C.W. Pearce, Crystal growth and wafer preparation and epitaxy, in S.M. Sze, Ed., *VLSI Technology*, McGraw-Hill, New York, 1983.
12. W.F. Beadle, J.C.C. Tsai, and R.D. Plumber, Eds., *Quick Reference Manual for Engineers*, Wiley, New York, 1985.
13. Sung-Mo Kang and Yusuf Leblebici, *CMOS Digital Integrated Circuits: Analysis and Design*, 3rd ed., McGraw-Hill, New York, 2003.
14. J.C. Bean, The growth of noble silicon material, *Physics Today*, 39(10), 36 (1986).
15. T. Yamamoto et al., An advanced 2.5 nm oxidized nitride gate dielectric for highly reliable 0.25 μm MOSFETs, *Symp. VLSI Technol. Dig. Tech. Pap.*, p. 45 (1997).
16. H.N. Yu et al., 1 μm MOSFET VLSI technology. Part I—An overview, *IEEE Trans. Electron Devices*, ED-26, 318 (1979).
17. D. Pramanik and A.N. Saxena, VLSI metallization using aluminum and its alloy, *Solid State Tech.*, 26(1), 127 (1983); 26(3), 131 (1983).
18. K.A. Pickar, Ion implantation in silicon, in R. Wolfe, Ed., *Applied Solid State Science*, vol. 5, Academic Press, New York, 1975.
19. W.G. Oldham, The fabrication of microelectronic circuit, in *Microelectronics*, 237(3), pp. 111–114. Freeman, San Francisco, 1977.
20. Behzad Razavi, *Design of Analog CMOS Integrated Circuits*, McGraw-Hill, New York, 2002.
21. M.C. King, Principles of optical lithography, in N.G. Einspruch, Ed., *VLSI Electronics*, Vol. 1, pp. 73–81, Academic, New York, 1981.

22. J.H. Bruning, A tutorial on optical lithography, in D.A. Doane, et al., Eds., *Semiconductor Technology*, p. 119, Electrochemical Society, Penningstone, 1982.
23. W.L. Brown, T. Venkatesan, and A. Wagner, Ion beam lithography, *Solid State Technol.*, 24, 8, 60 (1981).
24. J.P. Joly, Metallic contamination of silicon wafers, *Microelectron. Eng.*, 40, 285 (1998).
25. J.C. Irvin, Evaluation of diffused layers in silicon, *Bell Syst. Tech. J.*, 41, 2 (1962).

Index